# DUSTY AND SELF-GRAVITATIONAL PLASMAS IN SPACE

# ASTROPHYSICS AND SPACE SCIENCE LIBRARY

VOLUME 193

# DUSTY AND SELF-GRAVITATIONAL PLASMAS IN SPACE

by

PAVEL BLIOKH

VICTOR SINITSIN

and

VICTORIA YAROSHENKO

*Institute of Radio Astronomy,*
*National Academy of Sciences of the Ukraine,*
*Kharkov, Ukraine*

KLUWER ACADEMIC PUBLISHERS

DORDRECHT / BOSTON / LONDON

Library of Congress Cataloging-in-Publication Data

Bliokh, P. V. (Pavel Viktorovich)
Dusty and self-gravitational plasmas in space / by Pavel Bliokh, Victor Sinitsin, and Victoria Yaroshenko.
p. cm. -- (Astrophysics and space science library ; v. 204)
1. Cosmic dust. 2. Plasma astrophysics. I. Sinitsin, Victor. II. Yaroshenko, Victoria. III. Title. IV. Series.
QB791.B55 1995
523.01--dc20 95-34350

ISBN 0-7923-3022-6 (hb : alk. paper)

Published by Kluwer Academic Publishers,
P.O. Box 17, 3300 AA Dordrecht, The Netherlands.

Kluwer Academic Publishers incorporates
the publishing programmes of
D. Reidel, Martinus Nijhoff, Dr W. Junk and MTP Press.

Sold and distributed in the U.S.A. and Canada
by Kluwer Academic Publishers,
101 Philip Drive, Norwell, MA 02061, U.S.A.

In all other countries, sold and distributed
by Kluwer Academic Publishers Group,
P.O. Box 322, 3300 AH Dordrecht, The Netherlands.

*Printed on acid-free paper*



Printed in the Netherlands

# Contents

# Preface

The diverse and often surprising new facts about planetary rings and comet environments that were reported by the interplanetary missions of late 1970s – 1980s stimulated investigations of the so-called dusty plasma. The number of scientific papers on the subject that have been published since is quite impressive. Recently, a few surveys and special journal issues have appeared. Time has come to integrate some of the knowledge in a book. Apparently, this is the first monograph on dusty and self-gravitational plasmas. While the circle of pertinent problems is rather clearly defined, not all of them are equally represented here. The authors have concentrated on cooperative phenomena (i.e. waves and instabilities) in the dusty plasma and the effects of self-gravitation. At the same time, in an attempt to present the vast material consistently, we have included such topics as electrostatics of the dusty plasma and gravitoelectrodynamics of individual charged particles. Also mentioned are astrophysical implications, mostly concerning planetary rings.

We hope that the book shall be of interest and value both to specialists and those (astro)physicists who have just discovered this area of plasma physics.

We are thankful to many scientists actively working in the field of dusty plasma physics who have generously let us become acquainted with their results, sometimes prior to publication of their own papers: U. de Angelis, N. D'Angelo, O. Havnes, A. Mendis, M. Rosenberg, P. Shukla, F. Verheest, and E. Wollman.

We appreciate the efforts of Galina Golubnitchaya who prepared the English manuscript and patiently introduced the many corrections and revisions.

P. Bliokh, V. Sinitsin, and V. Yaroshenko
Institute of Radio Astronomy
Kharkov, Ukraine.

...All are of the dust, and all turn to dust again.
*Ecclesiastes, Ch. 3, 20.*

Dust to the dust! but the pure spirit shall flow.
*Percy Bysshe Shelly*

# Introduction

The terms *plasma* and *cosmic dust* were introduced in the scientific vocabulary at approximately the same time, namely late 1920s – early 1930s. Tonks and Langmuir (1929) applied the former to a mixture of charged particles, and Trampler (1930) and Stebbins *et al.* (1934) reported on the presence of dust in dark "holes" of the Milky Way.

Neither the researchers investigating laboratory plasmas nor the astrophysicists studying cosmic dust could predict at the time that the objects of their study would show so many interdependent properties as to give rise to the concept of a *dusty plasma* and a new branch of plasma physics.

The plasma is generally defined as a gas with a considerable content of electrically charged particles which are electrons and ions. Since the major part of matter in the universe is in the gaseous state, plasmas are equally widespread in space, making up, for example, the intergalactic or interstellar media, the solar corona or planetary ionospheres. However, the gas in space also contains macroscopic components like fine grains and larger material fragments. These are acted upon by the ambient plasma and various kinds of radiation. As a result, the macroscopic grains become electrically charged and start interacting with one another, as well as with ions and electrons via long-range electric fields. This is how the special medium called the dusty plasma is formed (apparently the name was first suggested by H. Alfvén). Among the astrophysical objects certainly containing dusty plasmas are gas-and-dust clouds, comet tails, planetary rings, etc.

The interaction of macroscopic particles in a dusty plasma does not occur via electric fields alone but also via gravitation which is an equally long-range force. Which of the forces prevail is determined by the particle mass and charge, i.e. ultimately by the size of particles of a given kind. While submicron and micron-size grains interact mainly through electric forces, gravitational attraction may become noticeable and even dominant for particles of a few hundred microns. A very special case is that of equally important gravitational and electric forces, or a self-gravitational plasma.

The motion of individual charged grains through external fields (gravitational, magnetic or electric) has been recently studied quite intensely, mainly in connection with the discoveries of the *Pioneer*, *Voyager*, *Giotto* and *Vega* space missions. The work has even been marked by the appearance of a new term, namely gravitoelectrodynamics. Meanwhile, the forces acting on a particle in a dispersed medium of sufficient density include, along with external fields, all the fields produced by the ensemble of charged particles. The cooperative processes possible in the dusty plasma-like medium have not been given as much attention. They are the main subject of this book.

Chapter 1 is dedicated to the effects responsible for electric charging of macroscopic particles in space, specifically collisions with electrons and ions, secondary electron emission and photoelectronic emission. The electric potential acquired by a particle depends on its radius, chemical composition, and intensity of the incident corpuscular flows (e.g., the solar wind in the case of the interplanetary medium). The typical values are units to a few tens of volts, however in the presence of energetic electrons ($\mathcal{E} \sim 1 \div 10\,\mathrm{keV}$) the grain potential may reach hundreds or thousands volts. With potentials as high as this, electrostatic forces may influence the grain geometry. The charged dust grains in space appear to always be nonspherical. When calculating the potential distribution in such grains, one can predict mechanical stresses comparable with the strength limit of their material ($10^4$ to $10^6\,\mathrm{dyne/cm^2}$), which may result in ruptures. Along with the electrostatic fragmentation, the opposite process goes on in the dust clouds, namely coagulation of particles into larger grains. It develops differently for electrically neutral and charged grains.

Chapter 2 describes the dynamics of individual charged grains in the vicinity of a magnetized planet. The particle motion is governed by external gravitation and electromagnetic forces. The analysis of grain trajectories about the planet is additionally complicated by rotation of the planet around its axis. A rotating magnetized sphere generates an electric field known as the unipolar induction field which takes part in controlling the motion of charged particles. Strictly speaking, gravitation forces are also altered in the vicinity of a massive body if the body is rotating but the effect is weak for planets of the solar system.

The laws governing the motion of corpuscles about planets are quite well known for the cases of heavy fragments or, on the contrary, of extremely lightweight charged particles. In the first case, these are Kepler's laws, while the trajectories of ions and electrons are determined by external magnetic and electric fields. We are discussing a broad spectrum of particle masses, including cases when the electromagnetic and gravitational forces can prove commensurate. The resulting trajectories may be highly intricate, carrying the impact both of Keplerian orbital motion and Larmorian gyration. The deviations from Kepler's laws of motion that are detected for the particles of Saturn rings have allowed determining the sign of their electric charge (which has proven to be negative) and the charge-to-mass ratio.

The joint action of gravitation and magnetic fields might be the reason for some features of the distribution of micrometeor dust about Jupiter (the so called magnetogravitational traps) discovered by the *Pioneer* and *Voyager* space probes. Interesting gravitoelectromagnetic effects have also been predicted for the terrestrial magnetosphere. As follows from theoretical calculations, submicron-size dust grains ($a \simeq 0.1\,\mu$) should be swept out of the magnetosphere over a short time, while larger grains ($a \geq 1\,\mu$) make up a long-living ring in the vicinity of the synchronous orbit.

External fields can also affect the particle geometry. The grains polarized in the unipolar induction field coalesce to form radially extended chains. The presence of such formations in the upper Saturnian atmosphere has been confirmed in polarized light observations.

Along with external gravitation and electromagnetic fields, small-size grains are acted upon by the solar radiation. The spatial distribution of dust in the interplanetary space is the joint result of the Poynting-Robertson and Lorentz effect. Within the ecliptic plane, the number density of the dust grains decreases with the heliocentric distance $r$ as $r^{-1/3}$, which law has been confirmed by zodiacal light observations of the HELIOS space probe. Theoretical estimates of the radiative pressure on a charged particle in a plasma show it to be much higher than the pressure corresponding to light scattering by free particles. Gravitoelectrodynamic effects have also changed the classic parameter of celestial mechanics like the Roche limit. The critical orbit radius of a charged particle near a magnetized planet becomes dependent on its electric parameters. Another characteristic new effect is the appearance of so-called magneto-gravitational and gyro-orbital resonances.

In Chapter 3, the study of individual trajectories and particle dynamics is replaced by the analysis of cooperative phenomena in dusty and self-gravitational plasmas. The conditions are formulated under which an ensemble of charged particles with a broad spectrum of masses will support specifically plasmatic, i.e. cooperative effects. Special attention is given to the situations (e.g. in Saturn rings) when the cooperative motion of heavier particles is predominantly controlled by gravitation, while the behavior of light-weight components is governed by electrostatic forces. Dispersion relations identifying free oscillation modes are derived and analyzed, and the Debye length concept is extended to the case of self-gravitational plasmas. The dielectric constant tensor is derived. A question of considerable physical interest is that of energy losses of a probe particle moving through the self-gravitational plasma. In such a medium, electric disturbances can be excited even by a neutral particle, which is impossible for the conventional ion-electron plasma. Meanwhile, the energy losses of a charged massive particle associated with wave excitation can either increase (as compared with the conventional plasma) or drop down to zero for specific values of the charge-to-mass ratio, depending on the charge sign. Among possible effects of the collective phenomena discussed, we have considered electromagnetic wave scattering in gas-and-dust clouds of the interstellar medium. By observing the scattered radiation, it might be possible to estimate the electric charge of interstellar dust grains, which has not been done until now.

Low frequency waves in the dusty plasma are subject to attenuation of a special kind, nonexistent in the "pure" plasma. The three specific attenuation mechanisms discussed are the nonuniform electric field around massive charged grains, variations of the grain charge with time, and continuous mass spectrum (this is an analog of the Landau damping).

Nonlinear waves can exist in the self-gravitational dusty plasma in the form of electromagnetic or gravitational solitons. Among the external fields acting upon space plasmas, magnetic fields stand out. The behavior of dusty and self-gravitational plasmas placed in a d.c. magnetic field is given special consideration. The dielectric constant tensor for the self-gravitational magnetoplasma is derived and the eigenwave spectrum analyzed for the simple mass spectrum of the dust involving only two grain species, i.e. heavy- and light-weight. Even this oversimplified model shows quite a number of unusual properties, like an Alfvén wave with a frequency dependent phase velocity, a new low frequency mode and substantially new conditions for the appearance of hybrid resonances.

Along with randomly moving particles, space plasmas often involve particle streams with ordered velocities. The regular speeds of particles characterized by different masses and charges are different, which may give rise to various instabilities. They are analyzed in Chapter 4. Unstable states accompanied by generation of noise or individual waves can arise in the case of purely gravitational as well as purely electrostatic interaction between the streams. In the general case of a self-gravitational plasma, the instability growth rates can be higher or lower as compared with the "pure" modes.

The growth rate can even change its sign to become attenuation owing to the combined interaction, i.e. an unstable stream-controlled state can turn stable and vice versa. Note another distinction of dust grains from electrons and ions. Their electric charge can be changed by external agents, thus bringing about parametric excitation of waves in the plasma. Parametrically excited density waves in a planetary ring partially shadowed by the planet itself should be observable owing to specific optical effects (brightness variations along the ring azimuth). Chapter 5 is devoted to the astrophysical applications of the theory developed, mainly to electric oscillations and waves in the rings of giant planets (primarily Saturn). Saturn's rings are a complex system of interacting corpuscular streams. The structure of planetary rings is determined by gravitation forces about the planet and collisions between individual particles. This concept is capable of providing an explanation to the long lived hierarchical system of rings and ringlets enclosed in one another. Meanwhile, no generally accepted theory has been suggested for the short-living, radially extended formations ("spokes") discovered with the aid of *Voyager* space probes. We are considering in this book all the hypotheses regarding the "spokes" known from the literature, including the wave theory of the present authors. Apparently, of undisputable importance is the role of electromagnetic particle—particle and particle—planetary magnetic field interactions. The wave theory provides a natural explanation for the quasiperiodic structure of spokes and their grouping in wave packets. The latter appear as envelope solitons of the electrostatic waves, with account of nonlinear effects.

The wave processes occurring in narrow rings are different from similar effects in the main rings of Saturn. These differences are also discussed in Chapter 5.

## References

Stebbins, J., Huffer, C.H. and Whitford, A.E. 1934, *Publ. Washburn. Obs.* **15**, (V).

Tonks, L. and Langmuir, I. 1929, *Phys. Rev.* **33**, 195.

Trampler, R.J. 1930, *Lick Obs. Bull.* **14**, 154.

## Reviews

de Angelis, U. The Physics of Dusty Plasmas. 1992 *Physica Scripta* **45**, 465.

Goertz, C.K. Dusty Plasmas in the Solar System. 1989, *Rev. of Geophys.* **27**, 271.

Mendis, D.A. and Rosenberg, M. Cosmic Dusty Plasmas. 1994, *Ann. Rev. Astron. Astrophys.* **32**, (*in press*).

Northrop, T.G. Dusty Plasmas. 1992, *Physica Scripta* **45**, 475.

Tsytovich, V.N., Morfill, G.E., Bingham, R., and de Angelis, U. Dusty Plasmas (Capri Workshop, May 1989). 1990. *Comments. Plasma Phys. Controlled Fusion.* **13**, 153.

## Special Issues of Journals

Goree, J.A. (Ed.) 1994, *IEEE Trans. on Plasma Science* **22**, No 2- Special Issue on Charged Dust in Plasmas.

Shukla, P.K., de Angelis, U. and Stenflo, L. (Eds.) 1992, *Physica Scripta* **RS19**, Dusty Plasmas — Proceedings of the Special Session, Spring College on Plasma Physics, Trieste, Italy, June 1991.

# 1 Electrostatics of Dusty Plasmas

The physical properties of a plasma are greatly dependent on the state of the macroscopic particles it may contain, e.g. dust grains. Placed in the plasma environment, the grains are subject to collisions with electrons and ions, thus acquiring a nonzero electric charge. The charge may also result from other physical effects, such as photoelectron emission, secondary electron or field emission, etc. The resultant charge carried by a dust grain may vary both in magnitude and sign, depending on parameters of the ambient plasma and spectral density of the ultraviolet and other short wave radiation. Generally, the charge of an isolated grain and that of a grain in a dusty cloud should be different. Moreover, the electric potential of a grain may prove a multivalued function of the ambient plasma parameters, suggesting more than one stationary value of the potential under the same external conditions. Which of the potentials and hence charge values is actually realized depends on the particle and plasma prehistory.

Closely related to the problem of grain charge are such effects as electrostatic disruption and its reverse process, i.e. grain coagulation responsible for the formation with time of a specific mass spectrum of the macroscopic particles. Of certain interest is the destiny of the fragments resulting from electrostatic disruption of a grain, although strictly this is a dynamical problem not belonging to electrostatics. Nonetheless, it will be discussed in this Chapter as well.

## 1.1 Isolated Grain in a Plasma

An "isolated" particle in a plasma is an idealization. Actually, in all cases where the term is used it is simply assumed that the number density of dust grains is very low compared with that of electrons and ions. The low density $n_d$ of grains can be characterized quantitatively by the inequality $d \gg \lambda_D$, where $d \sim n_d^{-1/3}$ is the average separation between the dust particles and $\lambda_D = (T/n_0 e^2)^{1/2}$ the Debye length of the ambient plasma (for simplicity, we have assumed the electron and the ion temperature, as well as the unperturbed densities to be equal, $T_e = T_i = T$ and $n_{0,e} = n_{0,i} = n_0$).

In some cases it proves necessary to distinguish between dust in a plasma ($d \gg \lambda_D$) and a "dusty plasma" characterized by $d \ll \lambda_D$ (see Table 1). The latter inequality shows that the number of dust grains contained in the Debye sphere is high, $(4\pi/3)\lambda_D^3 n_d \gg 1$, which is exactly the criterion for an ionized gas to be considered as an ideal plasma.

Table 1: $d/\lambda_D$ for a few cosmic environments (after Mendis and Rosenberg, 1994).

| | $n_{o,e}$[cm$^{-3}$] | $T$[°K] | $n_d$[cm$^{-3}$] | $d/\lambda_D$ | $a[\mu]$ |
|---|---|---|---|---|---|
| Interstellar clouds | $10^{-3}$ | 10 | $10^{-7}$ | 0.3 | $0.01-10$ |
| Noctilucent clouds | $10^3$ | 150 | 10 | 0.2 | $\sim 1$ |
| Saturn's E-ring | 10 | $10^5-10^6$ | $10^{-7}-10^{-8}$ | $\leq 1$ | $\sim 1$ |
| Halley's Comet: | | | | | |
| inside ionopause | $10^3-10^4$ | $< 10^3$ | $10^{-3}$ | $\leq 1$ | $.1 \sim 10$ |
| outside ionopause | $10^2-10^3$ | $\sim 10^4$ | $10^{-9}-10^{-7}$ | $\geq 10$ | $.01 \sim 10$ |
| Saturn's spokes | $0.1-10^2$ | $2 \times 10^4$ | 1 | $\leq 0.01$ | $\leq 1$ |
| Saturn's F-ring | $10-10^2$ | $10^5-10^6$ | $< 30$ | $\leq 10^{-3}$ | $\leq 1$ |

### 1.1.1 Grain Charge in a Maxwellian Plasma

We have already mentioned that dust grains immersed in a plasma acquire an electric charge $Q$. The charging process can be described with the continuity equation

$$\frac{dQ}{dt} = \sum_{\beta} J_\beta, \tag{1}$$

where $J_\beta$ are the electric currents of different origin flowing to and from the grain ($J_\beta$ is a positive current if it transports a positive sign charge to the grain). The physical effects responsible for the appearance of $J_\beta$ are many. First, they are the electron, $J_e$, and the ion, $J_i$, currents owing to interception by the grain of the respective particle species from the ambient plasma. For a spherical grain of radius $a$, these can be estimated as

$$J_{e,i} \simeq 4\pi a^2 n_{e,i} Q_{e,i} v_{T,e,i}, \tag{2}$$

where $Q_{e,i}$ are the electron and ion charge magnitudes, respectively (in what follows, we assume singly charged positive ions only, i.e. $Q_e = -e = -1.6 \times 10^{-19} C$ and $Q_i = e$); $v_{T,e,i} = (T_{e,i}/m_{e,i})^{1/2}$ are the respective thermal velocities of the microparticles, and $n_{e,i}$ their number densities around the dust grain. As a matter of fact, Equation (2) should also involve the so-called sticking coefficients, $S_e$ and $S_i$, generally different in magnitude for electrons and ions, but they have been all set equal to 1 here. Besides, the equation does not allow for the effects produced by the dust grain motion through the plasma at a velocity $\mathbf{w}$. They are indeed insignificant if $w \ll v_{T,i}$. However, if $w$ greatly exceeds the ion thermal velocity, then it should be written in place of $v_{T,i}$ in Equation (2). As a result, the ion current $J_i$ would increase, and hence the grain charge $Q$ would either increase in magnitude ($Q > 0$) or decrease ($Q < 0$) as compared with the grain at rest (Whipple, 1985; Alpert *et al.*, 1965). The effects may be of opposite sense in the case of moderate grain velocity $w$, $w/v_{T,i} \leq 2$, owing to lower cross-section areas for the grain-ion collisions (Northrop *et al.*, 1989).

The currents $J_\beta$ figuring in Equation (1) may not only be produced by external flows of charged particles but arise from electrons emitted by the grain itself. For most of the objects that we intend to consider below, thermal emission may be ignored since it is only effective when the body temperature is sufficiently high. The remaining currents to be considered are that of photoelectrons, $J_{ph}$, the current of secondaries, $J_s$ and the flux of field-emitted electrons, $J_f$. All of these currents carry some negative charge away from the grain, hence it either becomes positively charged or reduces its negative charge. We will discuss the importance of $J_{ph}$, $J_s$, and $J_f$ below, concentrating now on the electron and ion currents from the plasma. First, we formulate a limitation on the grain size that we suppose to be satisfied, namely that the grains are small compared with the Debye length in the plasma, i.e. $a \ll \lambda_D$, which is true in most realistic cases.

To understand the basic effects, consider a charge neutral dust grain in a plasma. The net current to the grain owing to collisions with the plasma particles is $\sum J_\beta = J_i - J_e \simeq 4\pi a^2 n_0 e(v_{T,i} - v_{T,e})$. In contrast to Equation (1), we do not distinguish between $n_e$ and $n_i$, assuming these densities equal to $n_0$ in virtue of the plasma quasineutrality. With this simplification, the right-hand side of Equation (1) is independent of $t$, which implies a linear law of growth with time for the negative charge on the grain (as of the initial time moment $t = 0$), $Q(t) = -4\pi a^2 t e n_0(v_{T,e} - v_{T,i})$. In fact, the rate of this growth will soon decrease as the electric field produced by the negatively charged grain will tend to reduce the number of electrons around the grain ($n_e < n_0$), while increasing the number of ions ($n_i > n_0$). The difference of currents, $J_i - J_e$ will decrease accordingly, reducing the rate of variation of $Q(t)$. The result will be an equilibrium state with $dQ/dt = 0$ and $J_i = J_e$. It is not difficult to estimate the time required for the initially neutral grain to reach an equilibrium charge value $Q_0$, as well as the magnitude of that charge. Since at the early stage (i.e. $J_{e,i} = J_{0,e,i}$) we have $J_{0,e}/J_{0,i} = v_{T,e}/v_{T,i} = (m_i/m_e)^{1/2} \gg 1$ (indeed, the ratio is $J_e/J_i \simeq 43$ even for protons, the lightest of all ions), the time derivative of $Q(t)$ can be written as $(dQ/dt))_{t=0} = -4\pi a^2 e n_0 v_{T,e}$. The time $t_0$ for $Q(t)$ to reach the steady state level $Q_0$ (assuming a constant rate of charge growth) then is $t_0 \simeq Q_0/(4\pi a^2 e n_0 v_{T,e})$. On the other hand, the electron current to the grain will cease as soon as the electrostatic potential on the grain surface becomes $\psi_0 \sim -T/e$. Recalling that the capacitance of an isolated quasispherical grain is $C_0 = a$ (in CGS units) and $Q_0 = a\psi_0$, we obtain

$$t_0 \sim T(4\pi a e^2 n_0 v_{T,e})^{-1} \simeq (\lambda_D/v_{T,e})(\lambda_D/a) \simeq \omega_{p,e}^{-1}(\lambda_D/a), \qquad (3)$$

where $\omega_{p,e} = (4\pi e^2 n_0/m_e)^{1/2}$ is the electron plasma, or Langmuir frequency. As a matter of fact, this is an easily predictable result as $\omega_{p,e}^{-1}$ is the characteristic time of charge relaxation in a collisionless plasma, while the dimensionless parameter $\lambda_D/a$ characterizes the grain ability to accumulate the charge (the smaller is its radius $a$, the rarer are absorptions of the plasma electrons). Equation (3) suggests that *larger-size* particles are *less* inertial in the processes controlling variations of the grain charge in the plasma. Since we have not taken into account the reduction in the electron flux resulting from the increased negative charge $Q(t)$, Equation (3) should be rather regarded as a lower estimate of $t_0$.

The approximate magnitudes of the equilibrium potential, $\psi_0$, and charge, $Q_0$, that have been found are only rough estimates requiring correction. In a more rigorous approach let us consider velocity distribution functions $f_{e,i}(\mathbf{r}, \mathbf{v})$ of the electrons and ions moving through the field $\psi(\mathbf{r})$ of a charged grain. (In what follows, the zero subscript denoting equilibrium values of $\psi$ and $Q$ will be omitted almost everywhere, while the variable potential or charge will be given with the appropriate argument, e.g. $\psi(r)$ or $\psi(t)$.) The radial current densities $j_{e,i}$ are given by

$$j_{e,i} = Q_{e,i} \int_D v_r f_{e,i}(\mathbf{r}, \mathbf{v})\, d^3\mathbf{v}, \qquad (4)$$

where $v_r$ is the radial velocity component. The equilibrium state of the plasma far from the grain is characterized by the Maxwell-Boltzmann distributions

$$f_{e,i}(\mathbf{r}, \mathbf{v}) = f_{m,e,i}(v) \exp[-Q_{e,i}\psi(\mathbf{r})/T_{e,i}], \qquad (5)$$

with

$$f_{m,e,i} = n_{0,e,i}\left(\frac{m_{e,i}}{2\pi T_{e,i}}\right)^{3/2}\exp[-m_{e,i}v^2/2T_{e,i}], \tag{6}$$

While seeming symmetric with respect to $e$ and $i$, Equations (4-6) actually lack the symmetry. This will become apparent when we have determined the domain of integration $D$ in Equation (4) accounting for the nonradial velocity components.

If the dust grain is charged negatively $Q < 0$, then positive ions move through the field of an attraction center allowing for orbital trajectories that do not cross the grain surface. Such ions do not contribute to the charging current, and hence should be excluded from the integral of Equation (4). The domain of integration should be limited to such velocities $v_r$ and $v_\theta$ ($v_\theta$ being the azimuthal component) at which the total energy of the plasma particle is non-negative,

$$E_{tot} = \frac{1}{2}m_{e,i}(v_r^2 + v_\theta^2) + Q_{e,i}\psi(r) \geq 0. \tag{7}$$

The potential energy of ions in the field of a negatively charged grain is negative, $Q_i\psi(r) < 0$, hence Equation (7) sets limitations on the allowed values of $v_r$ and $v_\theta$. Under the same conditions, the potential energy of electrons is positive, $e\psi(r) > 0$, therefore $E_{tot} > 0$ and the domain of integration is the infinite velocity space. As a result, the electron and ion current densities calculated from Equation (4) are (Northrop, 1992)

$$J_e = -4\pi a^2 e n_0 (T_e/2\pi m_e)^{1/2}\exp(-e\psi/T_e) \tag{8}$$

$$J_i = 4\pi a^2 e n_0 (T_i/2\pi m_i)^{1/2}(1 - e\psi/T_i). \tag{9}$$

Equation (9), known as the orbit limited current, again is an approximation. With $e\psi/T_{e,i} \ll 1$ the symmetry of Equations (8) and (9) is restored.

Combining Equation (8) and (9) with the balance of currents condition, $J_e = J_i$, we obtain the improved equation for the equilibrium potential $\psi$,

$$1 - e\psi/T_i = (m_i T_e/m_e T_i)^{1/2}\exp(e\psi/T_e). \tag{10}$$

In the case of an electron-proton plasma with equal temperatures of electrons and ions (i.e. $T_e = T_i = T$ and $(m_i/m_e)^{1/2} \simeq 43$) the solution of Equation (10) is

$$\psi \simeq -2.51T/e, \tag{11}$$

which is in agreement with the above given estimate $\psi \sim -T/e$.

Equations (8) and (9) permit describing $Q(t)$ as a function of time rather than simply estimating the characteristic charging time $t_0$. The function is specified by the differential Equation (1) combined with Equations (2), (8) and (9), where the potential $\psi$ is replaced by $\psi(t) = Q(t)/a$, *viz.*

$$\frac{dQ}{dt} = J_{0,e}[(1 - Q/q_0)(m_i/m_e)^{-1/2} - \exp(Q/q_0)], \qquad (12)$$

with $J_{0,e} = 4\pi a^2 e n_0 v_{T,e}$ and $q_0 = aT/e$. Integration of Equation (12) gives the $t$ *vs* $Q$ dependence which can be written, in terms of dimensionless variables $t^* = t/t_0$ and $Q^* = Q/q_0$, as

$$t^* = \int_0^{Q^*} \frac{dx}{(1-x)(m_e/m_i)^{1/2} - \exp(x)}.$$

This is plotted in Figure 1. To determine the scaling factors $t_0$ and $q_0$, one needs to know the grain size and the plasma parameters. E.g., for the above considered hydrogen plasma $\left((m_i/m_e)^{1/2} \simeq 43\right)$ with $T = 10\,\mathrm{eV}$ and $n_0 = 100\,\mathrm{cm}^{-3}$ (i.e. $\omega_p \simeq 5.64 \times 10^5\,\mathrm{s}^{-1}$ and $\lambda_D \simeq 2.3 \times 10^2\,\mathrm{cm}$) the values are $t_0 \simeq 4\,\mathrm{s}$ and $q_0 = 6.2 \times 10^3\,\mathrm{e}$.

### 1.1.2 Non-Maxwellian Velocity Distributions in Plasmas

To describe various plasma components, we have used in the preceding Sections the notion of temperature, thus assuming Maxwellian velocity distributions with respective parameters $T_e$ and $T_i$ for the plasma electrons and ions. This is generally done by most writers treating electric charging of dust grains in a plasma. Meanwhile, observations suggest that the fraction of grains characterized by some velocity $v$ does not decrease at greater $v$s as quickly as it follows from Equation (6) (see, e.g. (Summers and Thorne, 1991)). Therefore, non-Maxwellian distributions are also considered, in particular power law functions with characteristic high energy tails. Some of these show a fair agreement with the observations, like the generalized Lorentzian (or $k$-distribution) (Christon *et al.*, 1988),

$$f_k(E) = n_0 \left(\frac{m}{2\pi k E_0}\right)^{3/2} \frac{\Gamma(k+1)}{\Gamma(k-1/2)} \left(1 + \frac{E}{kE_0}\right)^{-(k+1)}, \qquad (13)$$

where $E$ is the particle energy; $k$ the spectral index and $\Gamma$ is the gamma-function. Obviously enough, Equation (13) goes over to (5) at $k \to \infty$, provided that $E = mv^2/2$; $E_0 = T_{e,i}$ and $m = m_{e,i}$. For finite values of

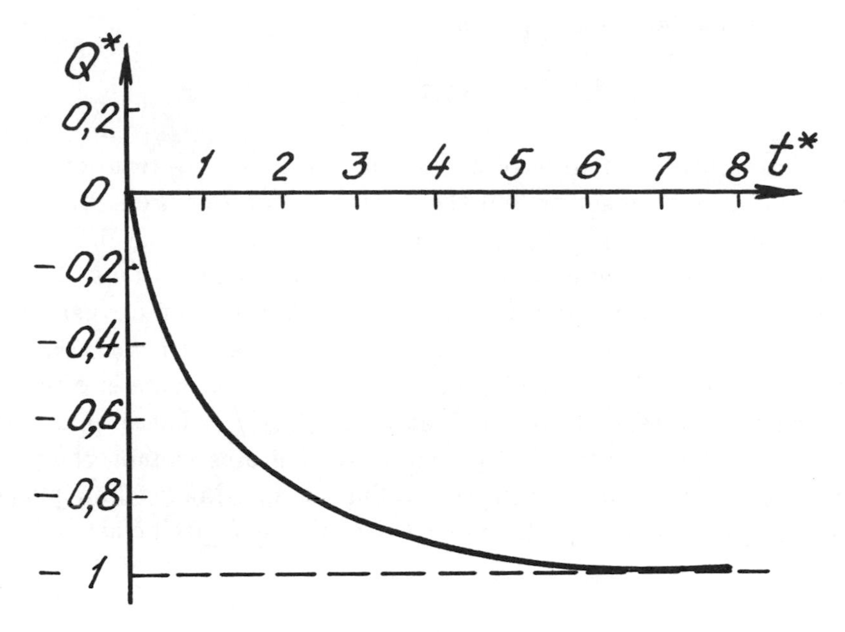


Figure 1: Grain Charging in a Hydrogen Plasma ($m_i/m_e = 1840$). The Dimensionless Variables are $Q^* = Q/q_0$ and $t^* = t/t_0$.

$k > 3/2$, the relation between $E_0$ and $T$ takes the form $E_0 = T(2k-3)/2k$, where $k$ generally may assume different values for electrons and ions.

The presence of a high-energy "tail" in the distribution function may result in an increased magnitude of the grain's negative potential $\psi$. In the case of the k–distribution, Equation (10) for the equilibrium potential which can be re-written as

$$\left(\frac{m_i}{m_e}\right)^{1/2} = R_T^{1/2}\left(1 - \frac{e\psi}{T_i}\right)\exp\left(-R_T\frac{e\psi}{T_i}\right), \quad R_T = \frac{T_i}{T_e} \tag{14}$$

is replaced by

$$\left(\frac{m_i}{m_e}\right)^{1/2} = R_T^{1/2}(k-1)\left(\frac{2}{2k-3}\right)^k\left(\frac{2k-3}{2k-2} + \frac{e\psi}{T_i}\right)\left(\frac{2k-3}{2} + R_T\frac{e\psi_0}{T_i}\right)^{k-1}, \tag{15}$$

where $k$ has been assumed $k_e = k_i = k$. For the particular case $k = 2$, a very simple solution to Equation (15) can be indicated (Rosenberg and Mendis, 1992; Mendis and Rosenberg, 1992), namely

$$\psi = -\frac{T_i}{2e}\left[\left(\frac{m_i}{m_e}\right)^{1/4} - 1\right], \tag{16}$$

which implies for the electron-proton plasma

$$\psi \simeq -2.75T_i/e. \tag{17}$$

Indeed, the negative potential of a grain placed in an electron-ion plasma is greater in absolute magnitude if the electrons and ions are k–distributed (compare Equation (17) and Equation (11)). This increase in the grain potential $|\psi|$ is even better pronounced in plasmas with heavy ions. Thus, for instance, the equilibrium grain potential $\psi$ in a Maxwellian oxygen plasma is $\psi \simeq -3.6T_i/e$, while in the case of a k–distribution with $k_e = k_i = 2$ it becomes $\psi \simeq -6.1T_i/e$. Assuming $k_e = k_i = 5$ (which value is often used to describe space plasmas), we arrive at $\psi \simeq -4.5T_i/e$. These quantitative distinctions of the potential and charge of dust grains cannot change the essence of the physical effects in Maxwellian or non-Maxwellian plasmas, however they may be of importance for the analysis of specific astrophysical objects.

### 1.1.3 Photoelectrons and Particle Charging in Space

A material particle in space does not only collide with plasma electrons and ions but is practically always influenced by electromagnetic radiation. Within the solar system, the major source of radiation acting over an extremely wide frequency range is the Sun itself.

The grain material affected by electromagnetic radiation may release electrons. This photoemission is a quantum-mechanical effect in which the electron acquires extra energy and momentum as a result of absorbing the photon. However, a free electron can never absorb a photon in view of the limitations set by the energy and momentum conservation laws. Therefore, photoemission is possible only for the electrons bound to their environment. For an electron in an atom or a molecule this binding is characterized by the ionization energy, while in condensed matter like the grain material the characteristic value is the work function. The energy conservation law in photoemission takes the form of the Einstein relation

$$E = \hbar\omega - \phi,$$

where $E$ is the kinetic energy of the emitted electron, $\hbar\omega$ the photon energy and $\phi$ the work function. Apparently, photoemission can occur if the radiation frequency is high enough, i.e. the photon is sufficiently energetic,

$\hbar\omega > \phi$, which condition sets the low-frequency or red boundary to photoemission. The magnitude of the work function is determined by the chemical composition of dust grains and normally equals a few electron-volts (e.g., in the case of silicon $\phi = 4.8\,\mathrm{eV}$). With account of numerical values of $\hbar = 1.05 \times 10^{-34}\mathrm{J \cdot s}$ and $1\,\mathrm{eV} = 1.6 \times 10^{-19}\mathrm{J}$, the limiting frequency and wavelength of the radiation capable of producing photoemission can be written as $\omega\,[\mathrm{s}^{-1}] > 1.5 \times 10^{15}\phi\,[\mathrm{eV}]$ and $\lambda\,[\mu] \leq 1.3 \times 10^{-4}/\phi\,[\mathrm{eV}]$, which figures correspond to the ultraviolet range. (For silicon the values are $\omega \geq 7.7 \times 10^{15}\,\mathrm{s}^{-1}$ and $\lambda \leq 250\,\mu$). This knowledge of the "active" electromagnetic spectrum allows estimating the energy of the electrons released, hence their velocity and density of the corresponding current. The photocurrent carries electrons away from the particle, thus making it more positive. The Coulombian field of the appearing charge tends to decelerate the emitted electrons, and had there been no alternative current sources, grain charging would stop with the surface potential reaching $\phi \sim E/e$. In the presence of other current producing mechanisms, an equilibrium is determined, as before, by the current balance condition $J_e + J_{ph} = J_i$. The resultant equilibrium potential may be either positive or negative, depending on the relative strength of individual currents.

Despite the apparent simplicity of the scheme described, estimating the grain charge numerically may prove rather difficult, owing to the uncertain knowledge of the various physical parameters. This is particularly true for dust particles found in deep space, i.e. in the interstellar medium. The problem is considerably more definite for the dust of the solar system where parameters of the electromagnetic radiation and physical properties of the plasma are quite well known. Speaking of the interplanetary medium, we mean parameters of the solar wind carrying flows of electrons, protons and $\alpha$-particles with respective number densities $n_e$, $n_p$ and $n_\alpha$. However, numerical estimates cannot be obtained even in this case without simplifying assumptions. Consider, for example, the equilibrium electric potential of dust grains of different chemical compositions (specifically, carbon and silicon) and sizes flying through the interplanetary plasma at different distances from the Sun. Phases of the solar activity are also taken into account. The basic equation for the grain charge owing to photoemission can be written as

$$\frac{dQ}{dt} = \pi a^2 \int_{\omega_{\min}}^{\omega_{\max}} d\omega \int_{E_{\min}}^{E_{\max}} dE\, q_{abs}(a, \omega, \mu) F(\omega) Y(\omega) f(E, \omega). \tag{18}$$

This equation corresponds to the simple scenario of grain charging that has been outlined above, however with allowance for a broad spectrum $F(\omega)$ of elecromagnetic frequencies rather than monochromatic radiation producing photoelectrons. The frequency dependent factor $Y(\omega)$ in the integrand is the electron yield efficiency. Strictly speaking, the yield depends not on the photon frequency alone but on the chemical composition and even size of the dust particle. These dependences are also taken into account through yet another parameter, the absorption factor $q_{abs}$ involving the radiation frequency $\omega$, particle radius $a$ and refractive index $\mu$ of the material as arguments. The value is defined as the ratio of the absorption cross-section to the geometrical cross-section area of the dust particle. Quite often, dust grains are considered as absolutely black absorbing bodies, i.e. $q_{abs} = 1$. The factor may be different from 1 at such frequencies where the radiation wavelength $\lambda$ is comparable with $a$, and hence diffraction effects are essential.

The photoelectrons released from a dust grain are characterized by a frequency-dependent energy distribution function $f(E,\omega)$ which is a weight factor in the integral over energies (i.e. electron velocities) in Equation (18). The integration limits are set as follows: $E_{max}(\omega) = \hbar\omega - \phi$, whereas $E_{\min} = 0$ for $\psi < 0$ or $E_{\min} = e\psi$ for $\psi \geq 0$. As for the explicit form of $f(E,\omega)$, several different representations are known in the literature; however a good fit to measured results (for different materials exposed to solar radiation) is provided by the Maxwellian function $f(E,\omega) = (E/E_c^2)\exp(-E/E_c)$ with $E_c \simeq 1.5\,\text{eV}$. This function was employed by Mukai (1981) to perform calculations after Equation (18).

Analysis of the spectral function $F(\omega)$ and the electron yield $Y(\omega)$ shows that the photons of greatest importance for the photoelectron emission are those with energies $6.7 \leq \hbar\omega \leq 9.2\,\text{eV}$ (i.e. $10^{16} \leq \omega \leq 1.4 \times 10^{17}\,\text{s}^{-1}$, or $0.2 \geq \lambda \geq 0.14\,\mu$). Since diffraction effects are weak at $ka \gg 1$ (where $k = 2\pi/\lambda$), $q_{abs}$ may be considered a constant value slightly less than 1 for sufficiently large grains $a > 1\,\mu$. As the grain size decreases, $q_{abs}$ increases reaching a maximum near $a \sim 0.03\,\mu$ (where $ka \sim 1$) and drops off quite rapidly with a further decrease of $a$.

To evaluate changes in the grain charge owing to collisions with electrons, protons and $\alpha$-particles in the solar wind, a slightly different approach is needed compared with Subsection 1.1.1 where the grain was assumed to be at rest with respect to the ambient plasma. For grains moving along Keplerian orbits through the interplanetary medium, the following inequality may be adopted, $v_{T,i} \ll w \ll v_{T,e}$, where $w$ is the solar wind velocity. Therefore, the electron flux from the plasma may be assumed isotropic, while the ion current is unidirectional. In contrast to Subsection 1.1.1 where the sticking coefficients (i.e. numerical factors $S_{e,i}$ in the right-hand part of Equation (2)) were all assumed equal to 1, sticking should be considered in some detail here. While in Subsection 1.1.1, we discussed low temperature plasmas with $T_e$ and $T_i$ about a few electron-volts, the electron temperature in the solar wind may reach 100 eV and more (distance from the Sun about 1 a.u., maximum solar activity). At these energies the secondary electron emission is essential, which effect may be described as a reduction of the sticking coefficient $S_e$ in the right-hand part of Equation (2). Contrary to this, the ion factor $S_i$ becomes somewhat higher.

As a result of calculations in the scheme described, Mukai (1981) obtained estimates of the equilibrium potential of graphite and silicon grains at different separations from the Sun and for different activity phases (Figure 2). In the vicinity of the Earth's orbit, $D \sim 1$ a.u., particles of both kinds acquired a positive potential of a few electron-volts, with $\psi_{sil} > \psi_{graph}$. To extrapolate the results to greater distances, parameters of the solar wind were assumed to vary with $D$ as follows: $n_e$, $n_p$, and $F(\omega) \sim D^{-2}$; $T_e \sim D^{-1.22}$, while $w$ was independent of $D$. Then the contribution of photoemission to the resultant potential of the dust grain should decrease at greater distances $D$. Accordingly, the potential $\psi$ should gradually change its sign from positive to negative, where the negative magnitude may be evaluated as in Subsection 1.1.1. The simple extrapolation formulas summarizing the $\psi = \psi(D)$ dependences calculated for large grains ($a \geq 1\,\mu$) are

$$\psi\,[\text{volts}] = -1.6 \log D\,[\text{a.u.}] + 7.7\,(\text{silicon})$$

and

$$\psi\,[\text{volts}] = -1.3 \log D\,[\text{a.u.}] + 4.8\,(\text{graphite}).$$

Closer to the Sun (where the separations are measured in solar radii, $R_{\odot}$, rather than astronomical units), still another grain charging mechanism becomes active, namely thermoelectron emission. The fraction of thermally emitted electrons becomes comparable with that of photoelectrons at $D \sim 4R_{\odot}$ to $5R_{\odot}$.

If the separation is further reduced, the importance of thermoemission greatly increases; however, the high temperature in the region causes sublimation of the grain material, and hence the dust grains end their existence as such.

To quickly assess the role of photoelectrons, Equation (18) may be replaced by simpler, semi-empirical formulas, like those representing the pho-

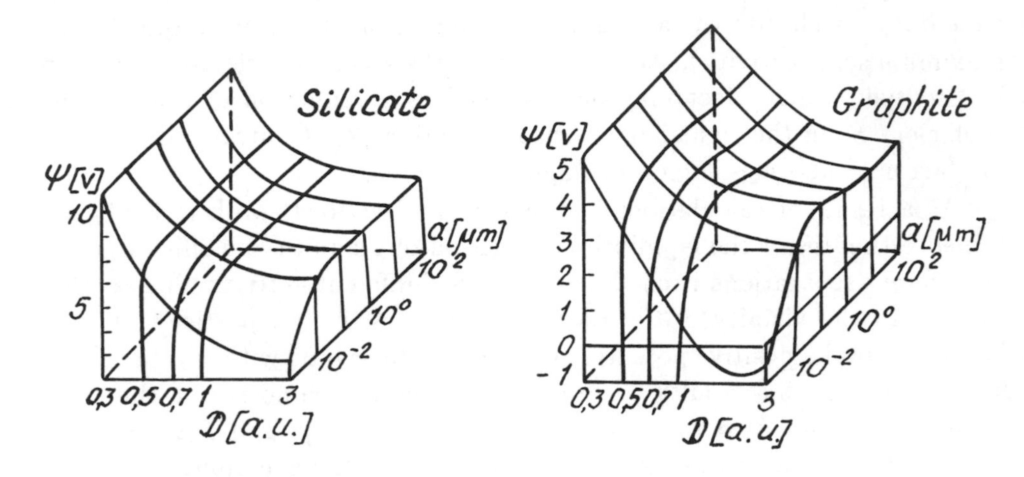


Figure 2: The Electrostatic Potential $\psi$ of a Grain of Radius $a$ at a Heliocentric Distance $D$, Medium Phase of the Solar Wind Activity (after Mukai, 1981).

toelectron current $J_{ph}$ due to solar radiation (Goertz, 1989), *viz.*

$$\begin{aligned} J_{ph} &\simeq \pi a^2 K && \text{for } \psi \leq 0 \\ J_{ph} &\simeq \pi a^2 K \exp(-e\psi/\chi_0) && \text{for } \psi > 0. \end{aligned} \tag{19}$$

Here $K = \eta(2.5 \times 10^{-14})D^{-2}$ is the flux of photoelectrons from $1\,\mu^2$ per 1 s at distance $D$ [a.u.] from the Sun, and the photoemission efficiency $\eta$ is $\eta \simeq 1$ for metals and $\eta \simeq 0.1$ for dielectric materials, with $\chi_0 \simeq 1\,\text{eV}$.

### 1.1.4 Secondary Electron Emission. Ambiguity in Charge Sign and Magnitude

Secondary electrons can be released if the surface of a solid or liquid is bombarded by (primary) electrons. In our case we speak of the emission from the surface of dust particles. The importance of the secondary emission effect increases with the plasma temperature, such that the current of secondaries $J_s$ should be included in the right-hand part of Equation (1) governing the equilibrium grain charge when $T$ reaches a few electron-volts. The current $J_s$

carries electrons away from the grain, reducing the magnitude of its negative potential. Moreover, at temperatures about a few tens electron-volts $J_s$ can become greater than the current of plasma electrons on the grain (each primary electron releases more than one secondary). As a result, the grain charge may change its sign. Then the positively charged particle will tend to return the secondary electrons back, establishing a dynamic equilibrium when the potential of some magnitude $\psi > 0$ nullifies the net current $I = J_i + J_s - J_e$.

The secondary emission is characterized by the yield parameter $\delta$ which is defined as the current ratio $\delta = J_s/J_e$, with $J_s$ involving both "true" secondaries released from the grain material and some primary electrons that are reflected from the grain surface. The yield parameter depends on the energy $E$ of primary electrons, reaching a maximum at $E = E_m$. The functional dependence often quoted in the literature (see, for instance, Meyer-Vernet (1982)) is

$$\delta(E) = 7.4\delta_m E/E_m \exp[-2(E/E_m)^{1/2}], \tag{20}$$

with $\delta_m$ and $E_m$ being characteristic parameters of the grain material. Typical magnitudes of these are $\delta_m \sim 0.5$ to 30 and $E_m \sim 0.1$ to 2 keV. The energy spectrum of secondary electrons is continuous through the range from 0 to the maximum energy in the primary flux. It is generally approximated by a Maxwellian distribution function with the temperature parameter $T_s = 1$ to 5 eV. To calculate $J_s$, we also need the energy distribution of primary electrons which is assumed to be another Maxwellian function with a temperature $T$. By integrating Equation (20) with account of this distribution, we can obtain $J_s$ as a function of the plasma temperature $T$, containing $\delta_m$, $E_m$ and $T_s$ as parameters, e.g.

$$\begin{aligned} J_s &= 14.8\frac{\delta_m}{E_m}TJ_e & \psi < 0 \\ J_s &= 14.8\frac{\delta_m}{E_m}TJ_e(1 + e\psi/T_s)\exp(-e\psi/T_s) & \psi > 0 \end{aligned} \tag{21}$$

(de Angelis, 1992). It is important to note that the $J_i$, $J_c$ and $J_s$ dependences upon the grain surface potential $\psi$ are different. With high negative values of the potential (i.e. $|\psi| \gg T/e$), the currents intercepted by the grain are predominantly those of positively charged particles, i.e. $J_i \neq 0$, and $J_e$, $J_s \simeq 0$. As the negative potential reduces in magnitude, $J_i$ decreases, while $J_e$ and $J_s$ grow. At $\psi > 0$ the current $J_i \simeq 0$, and only $J_e$ and $J_s$ remain nonzero, where $J_s$ is considered a positive current and $J_e$ a negative one. As $\psi > 0$ is increased, $J_e$ grows continuously, while $J_s$ reaches a maximum at some $\psi \sim T/e$ and then falls off rather quickly (as the secondary electrons become unable to overcome the Coulombian attraction of the positively charged grain). The non-monotonous character of $J_s = J_s(\psi)$ can give rise to qualitative changes in the setting of the equilibrium potential $\psi_0$ determined by the balance of currents

$$J(\psi_0) = J_i(\psi_0) + J_s(\psi_0) - J_e(\psi_0) = 0. \tag{22}$$

With sufficiently high values of the secondary yield parameter $\delta_m$, Equation (22) may allow for more than one (specifically, three) solutions, the number of solutions also being dependent on the plasma temperature. This can be seen in Figure 3 reproduced from Goertz (1989). Along with the currents $J_i$, $J_e$ and $J_s$, it shows the net current $J$ which turns to zero at $\psi_{01} \simeq -1.2\,\mathrm{V}$, $\psi_{02} \simeq 1.5\,\mathrm{V}$ and $\psi_{03} \simeq 4.3\,\mathrm{V}$. The highest and the lowest of these potentials are stable solutions of Equation (22), while the intermediate magnitude $\psi_{02}$ is unstable. To show this, consider small variations of the current, $\delta J$, resulting from variations $\delta Q$ of the grain charge and the associated changes of its surface potential, $\delta\psi = \delta Q/a$. In the vicinity of $\psi = \psi_0$,

$$\delta J = J(\psi_0 + \delta\psi) - J(\psi_0) \simeq \left.\frac{dJ}{d\psi}\right|_{\psi_0} \delta\psi = \frac{1}{a}\left.\frac{dJ}{d\psi}\right|_{\psi_0} \delta Q,$$

which implies $d(\delta J)/dt = \gamma d(\delta Q)/dt = \gamma\delta J$ and $\delta J(t) = (\delta J)_m e^{\gamma t}$, with $\gamma = (1/a)\,(dJ/dt)_{\psi=\psi_0}$ and $\delta J_m$ being the initial current disturbance. Thus, the change in current increases exponentially with $(dJ/d\psi)_{\psi_0} > 0$ (instability) or decreases if $(dJ/d\psi)_{\psi_0} < 0$ (stability). Quite apparently, the instability condition, $dJ/d\psi > 0$ is satisfied not for the point $\psi = \psi_0$ alone but for the entire increasing section of the curve $J = J(\psi)$ with $dJ/d\psi = 0$ at the ends. Which of the stable potentials, $\psi_{01}$ or $\psi_{02}$, is actually acquired by the grain depends on the prehistory of the equilibrium state. If the initial surface potential was lower than $\psi_{01}$, $\psi < \psi_{01}$, then the evolution of the grain charge ends at point $\psi_{01}$. With $\psi > \psi_{02}$, the equilibrium is reached at $\psi_{03}$. This implies that the charges on some of the grains may be of opposite signs.

The equilibrium potential varies with variations in the external parameters (i.e. plasma temperature). With sufficiently high values of the secondary yield parameter $\delta_m$, these variations may show characteristic instabilities similar to those just discussed. Calculated dependences of the

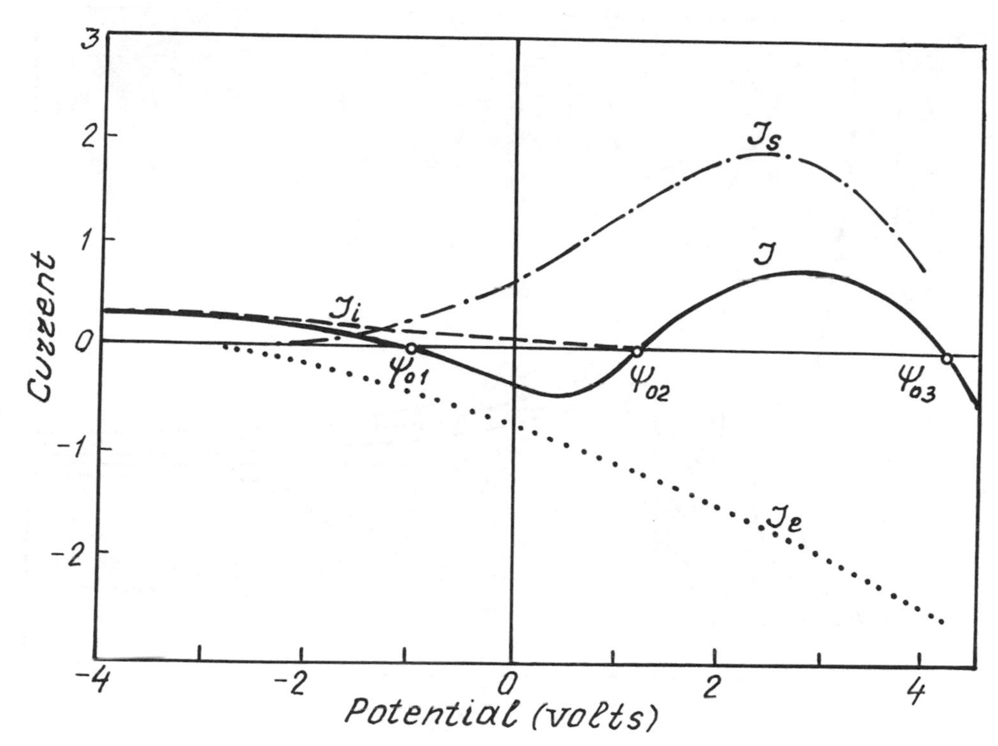


Figure 3: The Charging Current Density due to Plasma Electrons ($J_e$), Plasma Ions ($J_i$) and Secondaries ($J_s$), and the Net Current ($J$) as Functions of the Grain Surface Potential $\psi$ for $\delta_m = 10$, $T = 1\,\mathrm{eV}$ and $a = 1\,\mu$. The Equilibrium Surface Potential is Given by the Zeros of $J$ (after Goertz, 1989).

equilibrium grain surface potential on the plasma temperature are shown in Figure 4 for different values of $\delta_m$ (Goertz, 1989). At $\delta_m \geq 5$ there are three rather than one value of the potential $\psi$ for each temperature $T$. Note the remarkable points where $d\psi/dT = \infty$ (e.g. $T = T_1$ and $T = T_2$ of the curve for $\delta_m = 8$). If the plasma temperature increases from values below $T_2$, then the grain will retain a negative potential up to the point $T = T_2$. (In this region the charging current $J_s$ is low and the grain charging process develops according to the scheme of Subsection 1.1.1). At $T = T_2$ the grain potential changes abruptly from the negative $-0.1\,\mathrm{V}$ to the positive value $+2.4\,\mathrm{V}$, decreasing smoothly after that.

If the plasma temperature varies in the opposite direction, from $T > T_1$ down, then $T_1$ is the point where the potential switches from $+2.3\,\mathrm{V}$ to $-0.2\,\mathrm{V}$. This is how the characteristic hysteresis loop is formed (see the

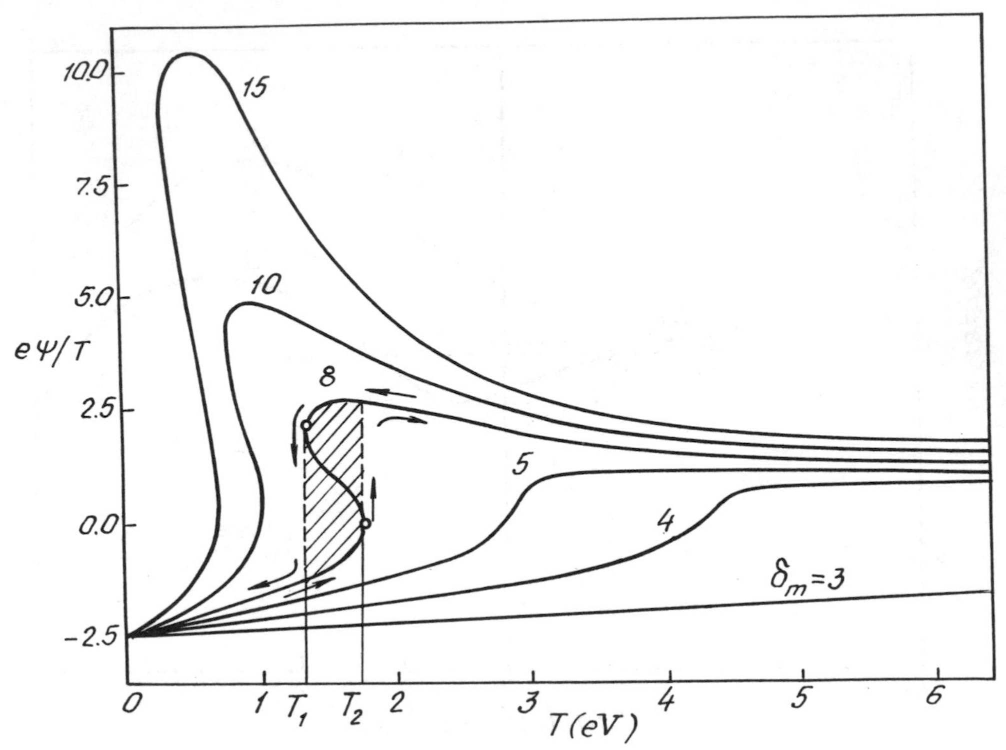


Figure 4: The Equilibrium Surface Potential $\psi$ as a Function of Plasma Temperature $T$ for Different Values of the Secondary Yield Parameter $\delta_m$ (after Goertz, 1989).

hatched area in Figure 4), embracing the negatively inclined portion of the $\psi = \psi(T)$ curve between $T_1$ and $T_2$. The hysteresis-type temperature dependence of the grain potential can give rise to some peculiar effects in a plasma with a decreasing temperature (Goertz, 1989; Horanyi and Goertz, 1990). Before discussing these, let us consider first how quick are the "abrupt" changes of $\psi$ at the critical points $T_1$ and $T_2$. The scale time $t_0$ of the potential variation can be estimated from Equation (3). The value happens to depend on the grain size $a$, specifically the larger is the grain, the quicker its response to plasma parameter variations. For this reason dust grains of different size can acquire potentials of different magnitude and sign in a plasma with a fluctuating temperature. Consider an ensemble of dust grains of sizes $a_1$ and $c_2 > a_1$ flowing through a "hot" plasma layer. The ambient plasma is at a temperature $T_0$ satisfying the inequalities $T_1 < T_0 < T_2$, while the hot layer at $T_x > T_2$ (Figure 5). Initially, the grain potentials corresponded to the *lower* part of the hysteresis loop of Figure 4, i.e. all the grains were at

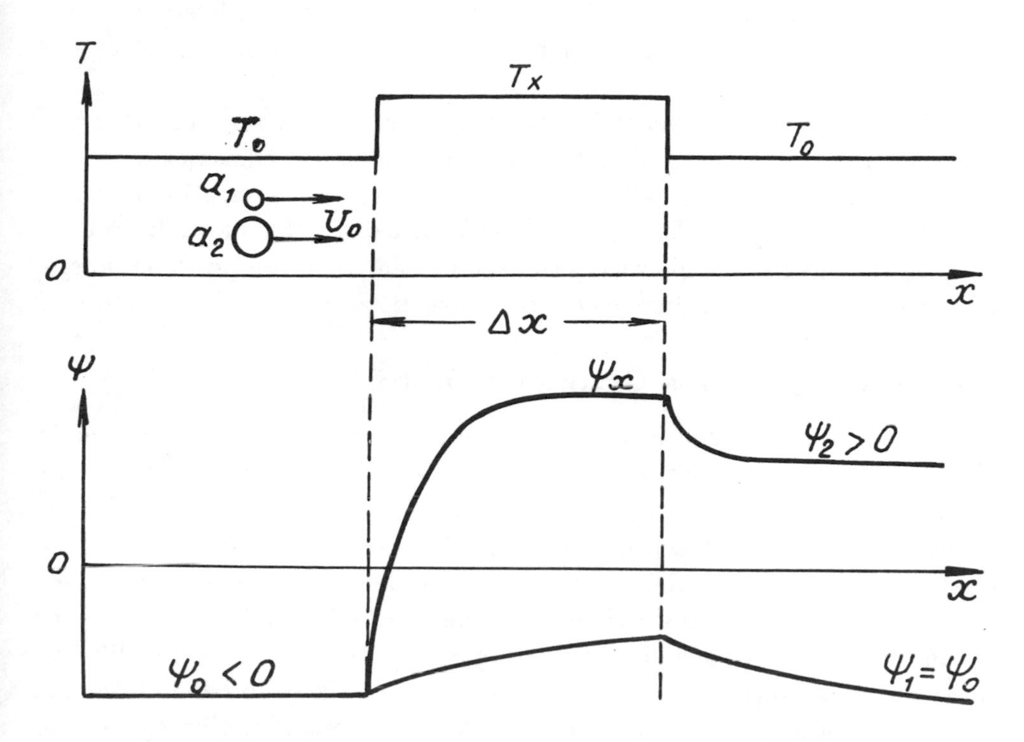


Figure 5: Variation of the Electric Potential of a Grain Passing through a Hot Plasma Layer. The Charges Acquired by Grains of Different Size Differ in Sign ($v_0 t_{0,2} < \Delta x < v_0 t_{0,1}$; $a_1 < a_2$, $t_{0,1} > t_{0,2}$).

the same potential $\psi_0 < 0$. When the grains enter the layer, their potential increases, tending to the new value $\psi_x$ matched with the layer temperature $T_x$. In a "thick" layer $\Delta x > v_0 t_{01}$ (where $v_0$ is the grain velocity and $t_{01}$ the scale time of charge variation for the grains of size $a_1$), the grains will assume the new equilibrium potential before they leave the layer. When they fly into the ambient plasma of temperature $T_0$ again, their potentials will be different from $\psi_0$ as the potential changes back following the *upper* curve of the hysteresis loop. In a thin layer, $\Delta x < v_0 t_{02}$, the flight time is too short for the grains to reach the new equilibrium potential. While in the layer, the grains are at a higher potential than $\psi_0$; however later they return to the initial state. Now it is easy to analyze the case $v_0t_0 > \Delta x > v_0t_{02}$. The smaller grains do not have enough time to reach their new equilibrium, so upon leaving the hotter plasma they restore the initial potential. Meanwhile, the larger grains "switch" to the upper curve of the hysteresis loop $\psi = \psi(T)$, changing not only the magnitude but the sign of their charge as well.

There is yet another reason for which dust grains may acquire charges of different sign. It is associated with the increased emission efficiency when the grain size happens to be smaller than the characteristic penetration depth of the primary electrons. The current of secondaries, $J_s$, then flows from both sides of the electron-bombarded grain which can acquire a positive charge, whereas larger grains will have $Q < 0$ (Chow *et al.*, 1993).

## 1.2 The Charge of a Grain in a Dust Cloud

While discussing isolated dust grains, the ambient plasma potential could be assumed zero at great distances from the grain as it followed the law $\Phi(r) = (\psi a/r)\exp[-(r-a)/\lambda_D]$. The electric potential around a single grain in a plasma is shown in Figure 6. According to Equation (11), the depth of the potential "well" is $\psi \simeq -2.51\,T/e$ and the scale length of variation is determined by the Debye screening effect. The potential difference $\psi - \Phi(\infty) = \psi$ is supported by the grain charge $Q = \psi a$. It is not difficult to assess the changes in the potential distribution likely to result from the appearance of many dust grains in the "Debye sphere" of radius $\lambda_D$ around the first one. Each of the grains would support a potential "well" of its own, of the same depth as an isolated grain. However, the potential $\Phi(r)$ far from the grain surface will not reach the zero level, being influenced by the negative potential of the neighbor grains. Let $\Phi$ be the average potential in the plasma at large distances from a given dust grain. In a sense, it plays the part of $\Phi(\infty)$ of the isolated grain. As is shown in Figure 6, $\psi < \Phi < 0$ and the difference $\psi - \Phi$ is less than the $\psi - \Phi(\infty)$ of the isolated particle. Accordingly, the charge required to support this new potential difference would be lower,

$$Q \simeq a(\psi - \Phi).$$

The equality is approximate because, when placed in a dust cloud, the capacitance of a spherical grain is no longer equal to its radius. A capacitor consisting of two concentric spheres of respective radii $a$ and $R$ $(a < R)$ is characterized by the capacitance $C = (1/a - 1/R)^{-1}$. For an isolated grain in a vacuum, $R = \infty$ and $C = C_0 = a$. In the case of a grain surrounded by a

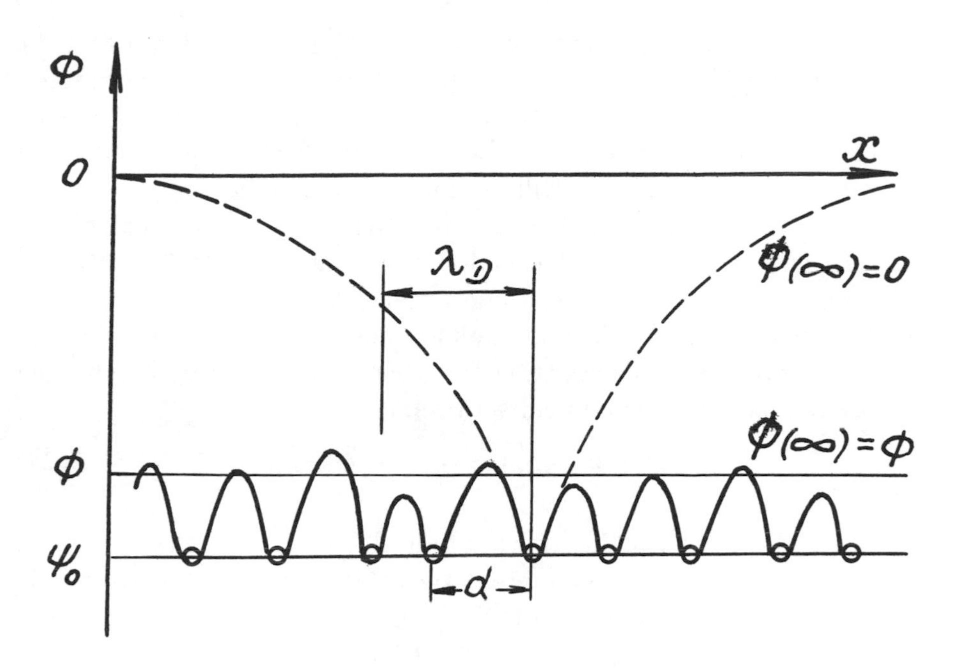


Figure 6: Plasma Potential Distributions in the Vicinity of an Isolated Grain (Dashed Curve) and Near a Grain in a Dusty Environment of High Density ($d \ll \lambda_D$).

"pure" plasma, the voltage drops across an effective layer of thickness $\sim \lambda_D$, so it can be assumed $R \simeq a + \lambda_D$ and $C \simeq C_0(1 + a/\lambda_D)$. Since $a \ll \lambda_D$, the capacitance is $C \simeq C_0$. For a grain in a dust cloud the Debye screening length is replaced by the intergrain separation $d$ (if $d < \lambda_D$). Accordingly, $C \simeq C_0(1 + a/d)$, which yields $C \simeq C_0$ again if $a \ll d$. However $a$ might not be so much less than $d$ in a cloud of high density (Whipple *et al.*, 1985). In what follows, we will always assume $C \simeq C_0 = a$.

### 1.2.1 Grain Charge in an Unbounded Dust Cloud

The effect of reduced grain charge in a dust cloud was first noted by Goertz and Ip (1984). We will follow their pattern of analysis in a somewhat simpler form here, using essentially the inequalities relating the three basic length parameters,

$$a \ll d \ll \lambda_D. \tag{23}$$

Generally speaking, the particle radius $a$, the average separation $d$ and the Debye length should be complemented by yet another length parameter, characterizing the size of the entire dust cloud. However, for a time we shall consider an unbounded cloud. Besides, the effects of photoemission and secondary electron emission will be ignored for simplicity.

While considering the isolated grain, it proved sufficient to analyze the balance of currents condition which allowed evaluating $\psi$ and $Q$. In the case of a dust cloud, it is necessary to determine the "far field" potential $\Phi$, taking into account the net effect of the ensemble of grains. To do this, we shall have to solve Poisson's equation for a medium in which the charged grains play the part of external field sources, i.e.

$$\Delta\Phi + 4\pi e(n_i - n_e) = -4\pi \sum_\beta Q_\beta n_{d,\beta}, \tag{24}$$

Making use of the first inequality in Equation (23) we can treat the grains as "point-size" sources, and hence represent the solution $\Phi(\mathbf{r})$ to Equation (24) in terms of the Green function $(1/4\pi)|\mathbf{r} - \mathbf{r}_\beta|^{-1}\exp(-|\mathbf{r} - \mathbf{r}_\beta|/\lambda_D)$, where $\mathbf{r}_\beta$ is the location of the charge $Q_\beta$. Introducing polar spherical coordinates with their center at the dust grain, we can evaluate the potential $\Phi$ of all the surrounding grains at the zero point as

$$\Phi = \sum_\beta Q_\beta \left.\frac{\exp(-r_\beta/\lambda_D)}{r_\beta}\right|_{r_\beta \neq 0}. \tag{25}$$

The sum can be evaluated quite easily if all the grains and their charges are assumed identical, $Q_\beta = Q$, and the summation is replaced by integration over the spherical sheath $r_{\min} < r < r_{max}$ ($r_{\min} \simeq d$ and $r_{\max} \to \infty$) with the volume charge density $Qn_d$. In other words, the charge of all the grains beyond the nearest neighbors is assumed uniformly distributed in space, hence

$$\begin{aligned}\left.\sum_\beta (1/r_\beta) Q_\beta \exp(-r_\beta/\lambda_D)\right|_{r_\beta \neq 0} &\simeq 4\pi Q n_d \int_d^\infty (1/r)\exp(-r/\lambda_D) r^2\, dr = \\ &= 4\pi Q \exp(-d/\lambda_D)\lambda_D(\lambda_D + d) n_d.\end{aligned}$$

Making use of the second inequality in Equation (23) we arrive at an even simpler formula,

$$\Phi|_{r=0} \simeq 4\pi Q \lambda_D^2 n_d, \tag{26}$$

whence the new value of $Q$ is

$$Q = Q_0/(1+\chi). \tag{27}$$

Here

$$\chi = 4\pi n_d \lambda_D^2 a \tag{28}$$

and $Q_0$ is the charge acquired by an isolated grain of size $a$ in the same plasma (variations of the plasma parameters have been neglected). Since the grain under consideration is not different from its neighbors, the origin of coordinates is not a preferred point either. With an account of Equation (27), the plasma potential is a function of $\chi$, *viz.*

$$\Phi \simeq 4\pi Q_0 \lambda_D^2 n_d/(1+\chi) = Q_0\chi/a(1+\chi). \tag{29}$$

If the grain size and all plasma parameters were fixed, $\Phi$ would be a function of the density ratio $n_d/n_0$ alone. Within this approximation, the potential difference $U = \psi - \Phi$ can be easily estimated, $U = Q/a = Q_0/a(1+\chi)$. It is also a function of $n_d/n_0$ (see Figure 7). The numerical estimates of Goertz and Ip (1984) suggest that some objects in space might be characterized by large values of the parameter $\chi$, such that the reduction of their charge may be significant. E.g., Saturn's ring $F$ of optical depth $\tau_{opt} \simeq \pi n_d a^2 h \simeq 1$ is characterized by $n_d \simeq 30\,\mathrm{cm}^{-3}$ (the physical thickness $h$ has been taken $h = 10\,\mathrm{km}$ and the average radius of dust grains $a = 10^{-4}\,\mathrm{cm}$). The literature data for the ambient plasma are $n_0 \sim 10^2\,\mathrm{cm}^{-3}$ and $T \sim 4\,\mathrm{eV}$, which yields $\lambda_D \sim 150\,\mathrm{cm}$. As can be seen, the grain separation is $d \sim 0.3\,\mathrm{cm} \ll \lambda_D$, whence Equation (28) gives $\chi \simeq 800$. Thus, the grain charge may be reduced by almost three orders of magnitude (indeed, an isolated grain would acquire a charge of nearly 7000 e, whereas its charge in a dust cloud would be only 9 e). In the area of radial structures known as "spokes" ($\tau_{opt} \simeq 0.1$; $h \simeq 30\,\mathrm{km}$; $a = 0.5\,\mu$ and $\lambda_D \simeq 100\,\mathrm{cm}$) the parameter is $\chi = 25$, hence the average grain charge may reduce to about 200 e. Meanwhile, in ring $E$ of optical thickness $\tau_{0pt} \simeq 10^{-6}$ and $h \simeq 100\,\mathrm{km}$ the parameter is $\chi \simeq 3 \times 10^{-6}$ and the charge virtually is not reduced.

The numerical estimates given suggest that the analysis of particle charging processes in a cloud environment has not been fully consistent. The idea of an isolated grain intercepting 7000 electrons from a plasma with the electron number density $n_0 \sim 10^2\,\mathrm{cm}^{-3}$ cannot cause objections. Indeed, the capture of any *finite* number of electrons from an *unbounded* volume of a plasma will have no effect on the electron density. The situation is different if

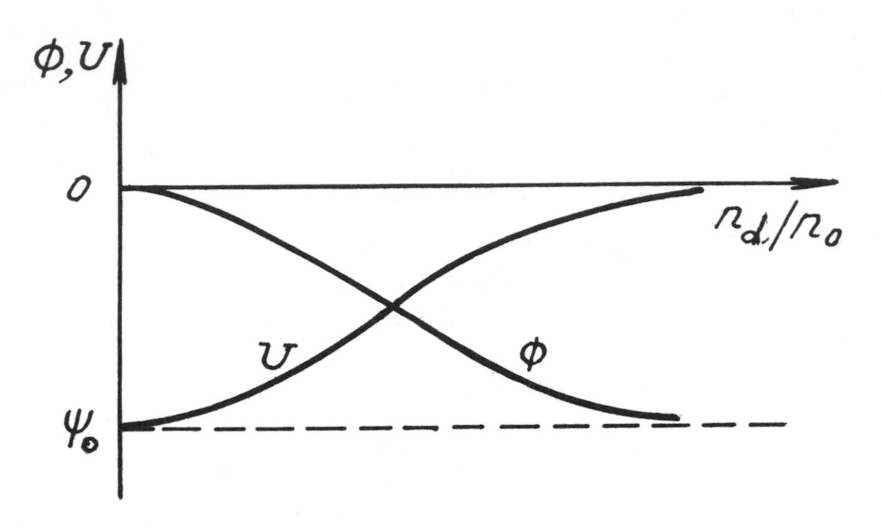


Figure 7: Plasma Potential $\Phi$ and the Relative Dust Grain Potential $U = \psi - \Phi$ as Functions of the Normalized Dust Concentration $n_d/n_0$.

the number of dust grains is infinitely great (as in an unbounded dust cloud). It is then difficult to imagine how the *average* charge of dust grains in 1 cm$^3$ could be greater than the number density of plasma electrons (in the case of Saturn's ring $F$, $Qn_d \sim 270$ electrons/cm$^3$ and $en_0 \sim 100$ electrons/cm$^3$, respectively), bearing in mind that the grain charge $Q$ is produced at the expense of plasma electrons alone.

When electrons are intercepted by dust grains, their number density in the ambient plasma is reduced. Meanwhile, the positive ions intercepted by a grain are neutralized on the surface and go back to the plasma in the form of charge neutral particles, i.e. the number density of plasma ions also decreases. Thus, it seems necessary to consider not only the effect of the ambient plasma on macroscopic particles but the back action of those grains on the plasma too (Whipple, 1981). Still, this formulation of the problem is not fully consistent either, since the *closed* system "plasma + dust" cannot remain in a stationary condition for an infinitely long time (like the plasma itself at that). Indeed, the balance of currents condition $J_e = J_i$ conserving the grain charge ($dQ/dt = 0$) results in a continuous reduction of $n_e$ and $n_i$. Hence, a steady-state regime in the presence of nonzero currents $J_e$ and $J_i$ is not possible without involvement of an ionizing agent (e.g. short-wavelength radiation which is always present in space). Accordingly, the complete set of equations should include, along with Poisson's equation and the balance of currents, equations to describe electron and ion production by the radiation, with allowance for recombination effects, grain charging among them. Having replaced the discrete sum in Equation (25) by an integral over the entire plasma volume, we ignored the fact that the plasma cannot penetrate into a dust grain. In other words, the results relate to a model of permeable grains, in contrast to the impermeable grains model where the electric charge is concentrated on the grain surface. The latter approach could allow lifting some of the limitations set by Equation (23) (Whipple *et al.*, 1985)

The back action of the dust cloud manifests itself through modification of all the grain charging currents, $J_\alpha$. Accordingly, Equations (8) and (9) should be written as

$$\begin{aligned} J_e &= -4\pi a^2 e n_{0,e} \left(T_e/2\pi m_e\right)^{1/2} \exp\left(-eU/T_e\right) \\ J_i &= 4\pi a^2 e n_{0,i} \left(T_i/2\pi m_i\right)^{1/2} \left(1 - eU/T_i\right), \end{aligned} \tag{30}$$

i.e. the grain potential $\psi$ replaced by $U = \psi - \Phi$. Besides, the average electron and ion number densities are not equal as the quasineutrality condition applies to the "plasma + dust" system, and hence $n_{0,i} - n_{0,e} = n_d Z$, where $Z = Q/e$ is the charge number a dust grain. Equations (19) and (21) for the currents of secondary and photoemission should be modified in the same way.

It might be instructive to analyze the grain charge behavior in dust clouds of various densities $n_d$. This can be done by balancing all the charging currents, i.e.

$$J_i + J_s + J_{ph} - J_e = 0.$$

Figure 8 shows the grain charge $Z$ as a function of the dust grain concentration normalized by the total density of plasma particles $n_0 = n_{0,e} + n_{0,i}$ (de Angelis, 1992). In plasmas of relatively high density of electrons and ions the grain charge is always negative and reduces in absolute magnitude as the dust concentration is increased. The charge may become positive in a plasma of low density, owing to the photocurrent and secondary emission. These effects are but weakly dependent on the dust concentration.

### 1.2.2 Semi-Infinite Dust Cloud. Boundary Layer Effects

The model of an infinite space filled with a dusty plasma is not only inconsistent in the sense described but generally too ideal a scheme to represent

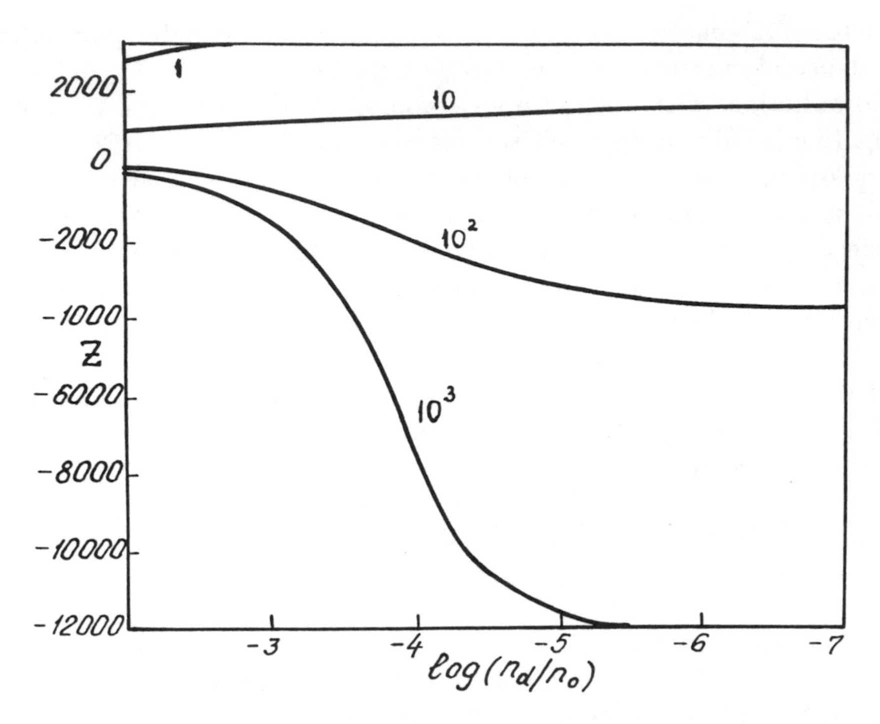


Figure 8: Grain Charge Number $Z$ as a Function of $n_d/n_0$. The Curves Are Labeled by the Values of the Number Density of Plasma Particles, $n_0 = n_{0,e} + n_{0,i}\,[\mathrm{cm}^{-3}]$. The Ion and Electron Temperatures Are $T = T_i = T_e = 20\,\mathrm{eV}$ and the Grain Radius $a = 1\,\mu$ (after de Angelis, 1992).

intrinsically bounded objects like, e.g. planetary rings. Therefore, more realistic models have been suggested (e.g. Havnes *et al.*, 1984, 1987; Goertz *et al.*, 1988; Goertz, 1989), taking into account the effects of boundaries and finite size of the dusty objects.

To begin, consider a semi-infinite plane-parallel dust layer where the number density of grains varies as

$$\begin{aligned} n_d(z) &= n_{0,d}\exp(-z^2/H^2) \quad & z \geq 0 \\ n_d(z) &= n_{0,d} = \mathrm{const} \quad & z \leq 0. \end{aligned} \tag{31}$$

It should permit analyzing the effects associated with the boundary, while probably being applicable to layers of finite thickness $L \gg \lambda_D$ too. If the electrons and ions inside the layer reach an equilibrium state characterized by a temperature $T$, then the particle density is given by the Boltzmann distribution with the potential $\Phi(z)$,

$$\begin{aligned} n_e(z) &= n_0 \exp(e\Phi/T) \\ n_i(z) &= n_0 \exp(-e\Phi/T). \end{aligned} \tag{32}$$

Generally, the ion thermalization owing to ion-ion collisions is a much slower process than thermalization of the electrons; therefore in a layer of finite thickness and negative potential $\Phi < 0$ Equation (32) can be replaced by $n_i(z) = n_0 = \text{const}$ (the ions fly through the layer). The Poisson equation (24) in a plane-stratified medium takes the form

$$\frac{d^2\Phi}{dz^2} = 4\pi\{[n_e(z) - n_i(z)]e - n_d(z)Q(z)\}. \tag{33}$$

The balance of currents equation $J_e = J_i$ allowing for variations of $n_e$ and $n_i$ with $z$ (Equations (31) and (32)) gives the relation between $\Phi$ and the potential on the grain surface $\psi$, which for $n_i(z) = n_0$ is

$$\Phi = \frac{T}{e}\left(\frac{m_i}{m_e}\right)^{1/2} \exp\left(\frac{e\psi}{T}\right) - \frac{T}{e} + \psi. \tag{34}$$

By substituting this into Equation (33) we can arrive at a second-order nonlinear differential equation for the potential $\psi(z)$, to be solved with the boundary conditions

$$\psi = -2.51T/e \quad \text{at } z \to \infty \qquad (\Phi = 0) \tag{35}$$

and

$$\frac{d\psi}{dz} = 0 \qquad \text{at } z \to \infty \qquad \left(\frac{d\Phi}{dz} = 0\right). \tag{36}$$

The first condition follows from Equation (11), since the number density $n_d$ of dust grains is vanishingly low at $z \to \infty$, such that the grains may be considered as isolated.

Numerical results concerning the plasma potential $\Phi(z)$ and the potential difference $U(z) = \psi - \Phi$ were reported by Havnes *et al.* (1984) in a graphical form, one of the graphs being reproduced here (Figure 9). The horizontal axis is distance $z$ in terms of the Debye length $\lambda_D \simeq 150\,\text{cm}$ (this value of $\lambda_D$ corresponds to parameters of the plasma far from the dust cloud, $n_0 = 100\,\text{cm}^{-3}$ and $T = 4\,\text{eV}$). The dust grain density $n_d$ is normalized to $n_{0,d}$; the potential difference $U$ is normalized to its value for an isolated

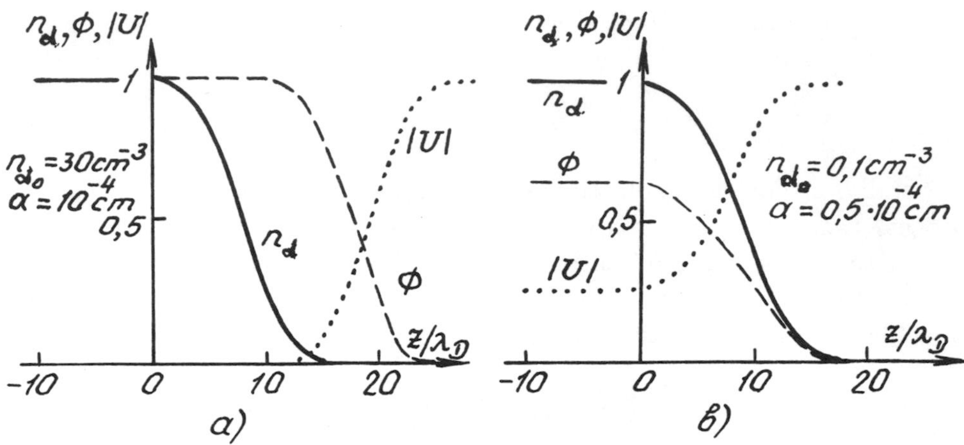


Figure 9: The Cloud Dust Density $n_d$, Plasma Potential $\Phi$ and Dust Particle Potential $U = \psi_0 - \Phi$ as Functions of Position Relative the Cloud Edge ($T = 4\,\text{eV}$, $n_0 = 100\,\text{cm}^{-3}$, $\lambda_D = 148\,\text{cm}$ and $H = 1000\,\text{cm}$) (after Havnes *et al.* (1984)).

grain (i.e. $U_0 \simeq -2.51T/e \simeq -10.04\,\text{V}$), and the plasma potential $\Phi$ is normalized so as to satisfy the complete screening condition $\psi - \Phi = 0$. According to Equation (34), this yields the normalizing factor $\Phi_n = \psi_n = -(T/e)\log(m_i/m_e)^{1/2} \simeq 15.03\,\text{V}$. Different kinds of dusty objects have been considered, namely the relatively dense dust of Saturn's ring $F$ ($n_{0,d} = 30\,\text{cm}^{-3}$); the area of spokes ($n_{0,d} = 0.1\,\text{cm}^{-3}$). The curves of Figure 9a are for the densest dust, while Figure 9b is for the dust of lowest density. The layer boundary is not sharp with $H \gg \lambda_D$. The curves $\Phi = \Phi(z)$ and $U = U(z)$ strongly resemble the $\Phi = \Phi(n_d/n_0)$ and $U = U(n_d/n_0)$ dependences discussed above, which is quite understandable as the density of dust decreases in the $z > 0$ region, while the plasma density remains practically unchanged.

Comparison of Figures 9a and 9b shows the electric field at the cloud boundary to be totally screened in the first case, so that $U = 0$ and $\Phi/\Phi_n = 1$ for $z < 0$. In the second case the field is nonzero inside the layer. The total charge density of the dust expressed in $e$ units, $n_d(z)aU(z)/300\,\text{e}$, ($U(z)$ is in volts) shows a characteristic sharp maximum near the boundary (see Figure 10), whose height and width are determined by the growth rate of $n_d(z)$ at $z < 0$ and the parameter $p = a\,[\mu]n_{0,d}/n_0$. The charge density inside the layer approaches $100\,\text{e/cm}^3$, which value corresponds to the charge density of ions, $n_0e$ (recall again that $n_i(z)$ has been assumed constant,

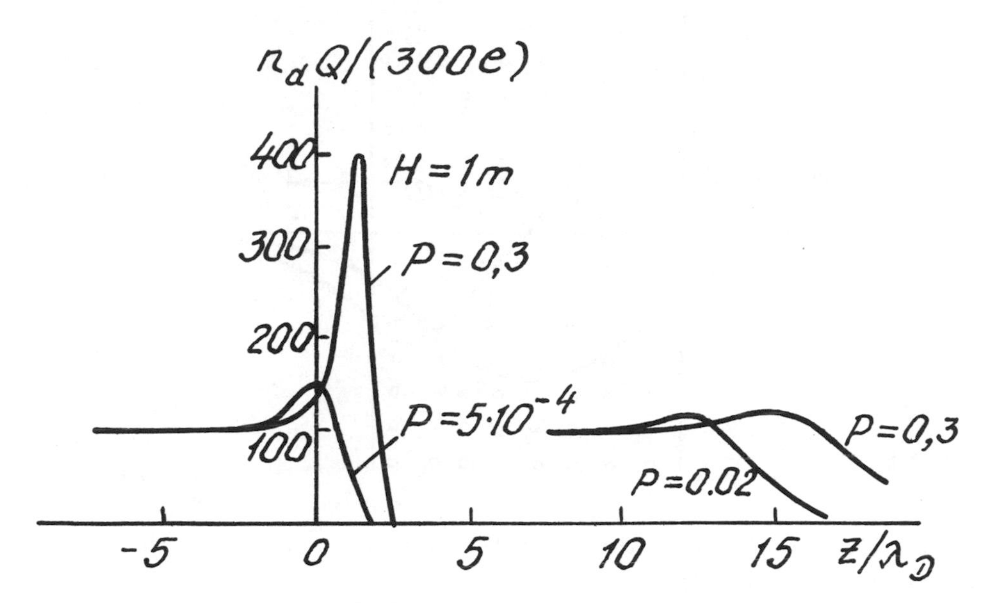


Figure 10: Total Dust Charge Density $n_d(z)Q(z)/(300\,\mathrm{e})$ as a Function of Position Near the Cloud Edge (after Havnes *et al.* (1984)).

$n_i = n_0$). This infers that nearly all of the plasma electrons are absorbed by the dust grains. The quasineutrality of the dusty plasma is violated near the edge where $\Phi(z)$ varies sharply. It can be expected that the ambient plasma — dusty layer transition with $U \to 0$ and $n_{0,d} \to \infty$ should be similar to a surface charge layer at a solid boundary. The quasineutrality is restored at great distances from the edge. If the boundary of the dusty layer is not sharp at all (i.e. $H \gg \lambda_D$), then the deviations from quasineutrality are small throughout, since the plasma potential at every level $z$ is determined by the local dust density $n_d(z)$. Then the right-hand side of Poisson's Equation (33) can be set equal to zero,

$$[\exp(e\Phi/T) - 1]\,e - p(\psi - \Phi) = 0 \tag{37}$$

Combined with Equation (34), this condition expressing the quasineutrality leads to a set to determine $\psi$ and $\Phi$, as well as $Q$ (Havnes *et al.*, 1987; Goertz *et al.*, 1988). With sufficiently large values of the parameter $p$, $p \geq 10^{-2}$, the solution to Equations (34) and (37) is $\psi \simeq \Phi \simeq -(T/e)\log(m_i/m_e)^{1/2}$ that was employed before to normalize $\Phi$.

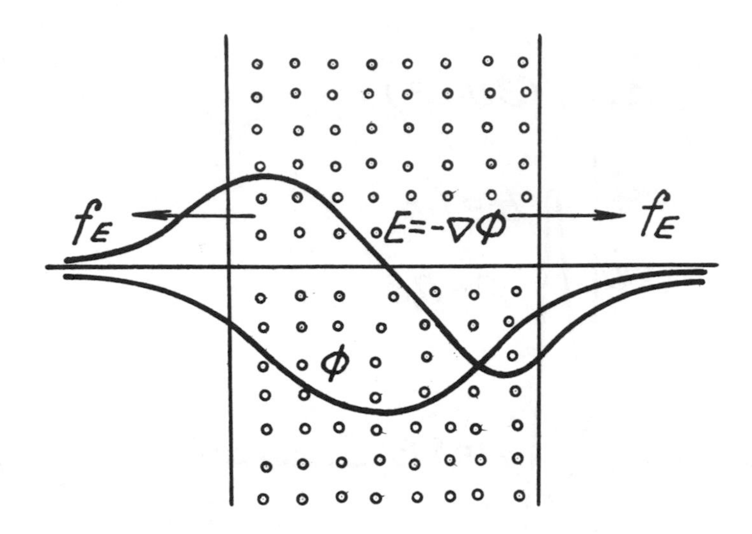


Figure 11: Potential Variations in a Dusty Plasma Layer and the Electric Forces Acting on Electrons.

### 1.2.3 Dust Layer of Finite Thickness

Consider now some specific electrostatic effects in a dusty layer of finite thickness. As was shown earlier, the plasma potential $\Phi$ decreases as the number density of dust grains, $n_d$, is increased. This leads to the appearance of a potential well in the layer and the corresponding electric field $\mathbf{E} = -\nabla\Phi$. The negatively charged dust grains are influenced by the force $\mathbf{f}_E = Q\mathbf{E}$ tending to eject them from the layer (Figure 11). This force arises from the potential gradient, and hence the gradient of $n_d$,

$$\mathbf{f_E} = -Q\nabla\Phi = -Q(n_d)\frac{\partial\Phi}{\partial n_d}\nabla n_d. \tag{38}$$

No such force exists in a uniform dust cloud where $\nabla n_d = 0$. The electrostatic force $\mathbf{f}_E$ is similar in action to the forces of gas kinetic pressure, $\mathbf{f}_p = -T\nabla n_d/n_d$, also tending to eject the dust grains from regions of elevated density.

Dust grains are attracted to one another owing to gravitation, so in the absence of other forces a dust cloud should compress until all the particles coagulated (gravitational collapse). In a gaseous medium heated to some temperature $T$, the gravitational forces can be compensated by the pressure forces $\mathbf{f}_p$. A similar compensating effect can be produced in a cold dust layer by the forces of electrostatic repulsion. By analyzing the forces acting on charged dust grains in the gravitation field of a central body, one can show a planetary ring to necessarily have a finite thickness (Havnes and Morfill, 1984). This problem will be discussed below, in the sections dedicated to the dusty plasmas of planetary rings.

The electric fields appearing in and about the layer certainly do not affect the dust grains alone but also influence the plasma electrons and ions. The electrons arriving in the dusty layer are decelerated by the field, while the ions experience acceleration. The electron component is quickly thermalized, which results in a Boltzmann distribution for $n_e$ with the potential $\Phi$. If the ions are so accelerated that they are not thermalized during their transit through the layer, then the ion fluxes arriving through two boundaries make up two opposite ion beams in the plasma layer, which may result in the generation of various wave modes.

## 1.3 Electrostatic Fragmentation of Charged Grains

### 1.3.1 Electric Disruption of a Spherical Particle. Critical Radius

The like charges carried by elements of the grain surface are mutually repulsed as if the grain were subjected to "negative pressure" on the part of the electric field it has produced. The electrostatic tensile forces can ultimately exceed the strength of the grain material, in which case a large grain may disintegrate into two or more fragments.

The force of interaction (or self-action) between charged bodies can be estimated from variations of the system's total electrostatic energy $W$ during small virtual displacements of its parts. In a virtual displacement of a conductor along the coordinate $\zeta$, the $\zeta$-component of the force is $f_\zeta = -\partial W/\partial\zeta$. To apply this relation to a surface element $\Delta S$, let us assume the element to be displaced outwardly along the surface normal by a distance $\Delta\zeta$. Then the electrostatic energy in the space around the conductor will reduce by $\Delta W = (1/8\pi)E^2\Delta S\Delta\zeta$. The energy density, $E^2/8\pi$, can be expressed in terms of the surface charge density, $\sigma = E/4\pi$, so that the force acting on a unit surface area may be written as

$$f_\zeta = \frac{1}{\Delta S}\frac{\Delta W}{\Delta\zeta} = 2\pi\sigma^2. \tag{39}$$

For the sake of simplicity, we have considered conducting grains only. In fact, any finite value of the material conductivity ultimately will be able to

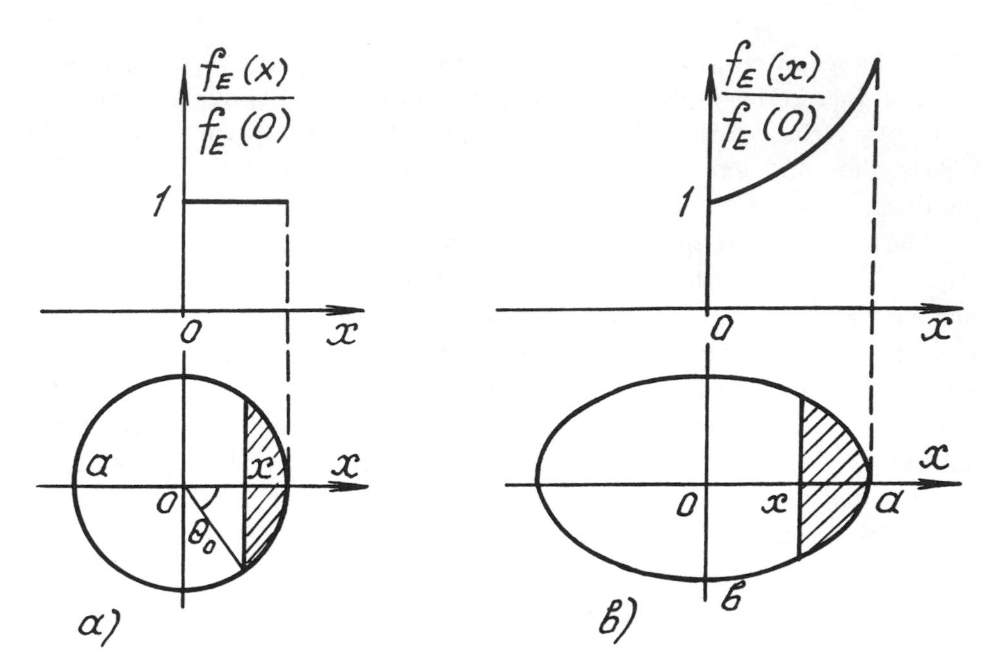

Figure 12: The Electrostatic Tensile Force at Different Cross-Sections of a Sphere (a) and Prolate Spheroid (b).

level the potential on the grain surface, ousting the electric field from its interior (unless fast transient processes are involved).

First, consider the tensile force of a large grain. Imagine a sphere cut by a plane at some distance $x$ from its center (Figure 12). The breakaway force acting on the spherical segment can be evaluated by projecting the radial components $f_r \equiv f_\zeta$ to the $x$-axis (normal to the section plane) and integrating along the surface. The result is

$$F_E = 2\pi^2 a^2 \sigma^2 \sin^2 \theta_0,$$

where $\theta_0$ is the angular half-width of the segment. The tensile force per unit cross-section area, $f_E$, is independent of $x$,

$$f_E = 2\pi\sigma^2 = \frac{Q^2}{8\pi a^4} = \frac{\psi^2}{8\pi a^2}. \tag{40}$$

By comparing this with the tensile strength $f_t$ of the material we obtain the condition determining the disruption threshold,

$$f_t < f_E = \frac{\psi^2}{8\pi a^2}, \tag{41}$$

while the equation $f_t = f_E$ gives the critical grain radius $a_c$ (the grain disintegrates if $a < a_c$). If $a_c$ is measured in microns, the potential in volts and $f_t$ in dyne/cm$^2$, then $a_c = 6.65\psi f_t^{-1/2}$.

### 1.3.2 Non-Spherical Grains

The force $f_E$ being independent of the cross-section position, the fracture can occur at any place. Obviously, this is a result of the high symmetry assumed for the grain. In fact the grains are most probably asymmetric, hence the tensile force should be different at different cross-sections. The electric field in the vicinity of a grain of irregular geometry is enhanced near points of highest curvature, especially near pointed tips if such are present. The surface charge density, $\sigma$, is enhanced in the same areas and, accordingly, so is the tensile force. Thus, tips will be the first to break away and the effect of electrostatic fractures is to remove surface irregularities and reduce the asymmetry. Consider a few examples where the grain geometry is rather regular, while the spherical symmetry is absent. The charge density at the surface of an ellipsoid is given by the equation (Landau and Lifshitz, 1982)

$$\sigma = \frac{Q}{4\pi abc}\left(\frac{x^2}{a^4}+\frac{y^2}{b^4}+\frac{z^2}{c^4}\right)^{-1/2}, \tag{42}$$

where $a$, $b$ and $c$ are semiaxis lengths of the ellipsoid along the $x$, $y$ and $z$ directions, respectively. The charge densities at three extreme points of the ellipsoid along $x$, $y$ and $z$ are

$$\sigma_a\Big|_{\substack{x=a\\y=0\\z=0}} = \frac{Q}{4\pi bc}; \quad \sigma_b\Big|_{\substack{x=0\\y=b\\z=0}} = \frac{Q}{4\pi ac}; \quad \sigma_c\Big|_{\substack{x=0\\y=0\\z=c}} = \frac{Q}{4\pi ab}. \tag{43}$$

Let $a > b > c$, then $\sigma_a > \sigma_b > \sigma_c$, i.e. the charge density is the highest near the sharper end of the ellipsoid and the lowest near the obtuse end. Therefore, the fragments that may appear as a result of disruption will not be as oblong as the prime grain.

A more detailed analysis was performed for the prolate spheroid, $a > b = c$. Öpik (1956) and Mendis (1989) considered the spheroid problem similarly to the above-discussed sphere. By placing the secant plane, perpendicular to a symmetry axis, at different distances $x$ from the center, one can see the tensile force indeed increase near the pointed end. The increase in $f_E$ may prove significant for high axial ratios $a/b \gg 1$ (Figure 12b).

The tensile strength of different materials varies over a wide range from $\simeq 10^4$ dyne/cm$^2$ for loose aggregates (e.g. snow flakes) to $10^8 \div 10^{10}$ dyne/cm$^2$ for silicates or metals (Grün *et al.*, 1984) and the critical radius changes accordingly.

It might seem that the process of disruption, once started at a sufficiently high potential $|\psi|$ when the critical radius $a_c$ is greater than the grain radius, should continue in an avalanche manner as the new smaller fragments would be inevitably unstable. In fact this is not so. In the case of markedly nonspherical particles, everything may end with the breakaway of particularly acute protuberances. If, however, the characteristic radius of curvature $a$ should remain smaller than $a_c$ and the disruption would not stop, then a new stabilizing mechanism would ultimately become active to limit the electric field intensity $E$ and the tensile force.

One such new effect is the field emission which consists of tearing electrons (ions if $\psi > 0$) out of the grain under the action of a high electric field. The critical field strength for field emission is about $10^8$ V/cm. Since the field around a grain is $E = \psi/a$, its magnitude may exceed $E_{fe}$, provided $a$ is sufficiently small. As a result, the negative charge and potential of the grain would sharply decrease in magnitude, halting the electrostatic disruption (Öpik, 1956). Mendis and Rosenberg (1992) discussed some observational results that might be related to electrostatic disruption. As a possible reason for the appearance of the so-called striae in some comet tails, abrupt electrostatic disintegration of prolate grains can be suggested (Secamina and Farrell, 1980; Hill and Mendis, 1980). The particles are disrupted as a result of the sharp increase in their electric charge (occurring over characteristic times about 100 s) taking place in streams of energetic electrons ($E \sim 1 \div 10$ keV). The grain potential may become as high as $|\psi| = 60 \div 300$ V. The tensile strength of grains estimated on the assumption of this mechanism as the only cause for fragmentation is quite realistic, namely $10^4 \div 10^6$ dyne/cm$^2$. Another evidence for the reality of this mechanism was provided by concentration measurements of very small dust grains ($m \simeq 10^{-20} \div 10^{-17}$ g) in the vicinity of Halley's comet that were performed by the Soviet spacecraft *Vega*. The experiments revealed a sharp peak in the spatial distribution of small particles some 180,000 km away from the comet core (i.e. at the boundary of the so-called cometopause). The peak might result from electrostatic disruption of larger grains owing to the abrupt change of the plasma temperature from $T \leq 1$ eV within the cometopause to $T \geq 10$ eV beyond. The jump in $T$ brings about an increase of the negative potential of grains from a few volts to a few tens of volts, preconditioning grain disruption by electrostatic forces.

Similar processes are likely to occur in planetary magnetospheres (Fechtig *et al.*, 1979). The experiment on board HEOS-2 revealed a two to three orders of magnitude increase in the flux intensity of meteoroids ($m = 10^{-15} \div 10^{-12}$ g) at about 10 Earth radii. The surface potential of a macroscopic body crossing the Earth's auroral region (which is filled with a plasma of energetic particles, $T \simeq 1$ to 10 keV) may exceed the critical value. Meanwhile, Mendis and Rosenberg (1992) noted that electrostatic disruption of a body $m \sim 10$ g (or $r \simeq 1.36$ cm), even at $|\psi| \sim 100$ kV, is probable only for materials of very low tensile strength ($F_t \leq 2.4 \times 10^3$ dyne/cm$^2$). Therefore, the increased flux intensity of meteoroids might not result from complete disintegration of some large bodies but rather from disengagement of individual small fragments.

### 1.3.3 Forces Acting on the Detached Particle. Levitation

To describe electrostatic disruption of a grain, it does not suffice to just formulate the necessary conditions but actually requires following the motion of the debris. At first sight, the problem should not present any difficulties. Once the electrostatic forces have exceeded the tensile strength of the material, the likely charged fragments are bound to move apart to greater separations. In fact the situation is not that simple. Under certain conditions the forces of repulsion may be replaced by those of attraction, even with charges of the same sign.

By way of example, consider a small particle detached from a larger parent body. For the sake of simplicity, both particles will be assumed spherical. The effect of the plasma environment will not be considered either. Thus, the system consists of a sphere of radius $a$ and a very small (size $a_0 \ll a$) detached particle of electric charge $q$. If the potential of the larger sphere were zero, then the electric field around the two particles could be evaluated as a sum of two fields. The first would be produced by the detached grain and the second by the image charge $q' = -(a/x)q$ located at point $x' = a^2/x$ inside the larger sphere ($x$ is the coordinate of the smaller particle treated as a point-size grain). A nonzero electric potential $\psi$ at the surface of the larger sphere can be produced by a charge $Q = a\psi$ placed at the center. The potential energy $U_E$ of the fragment's interaction with the central charge and its own image can be written as

$$U_E(x) = \frac{qQ}{x} + \frac{qq'}{x - x'} = \frac{aq\psi}{x} - \frac{aq^2}{x^2 - a^2}, \tag{44}$$

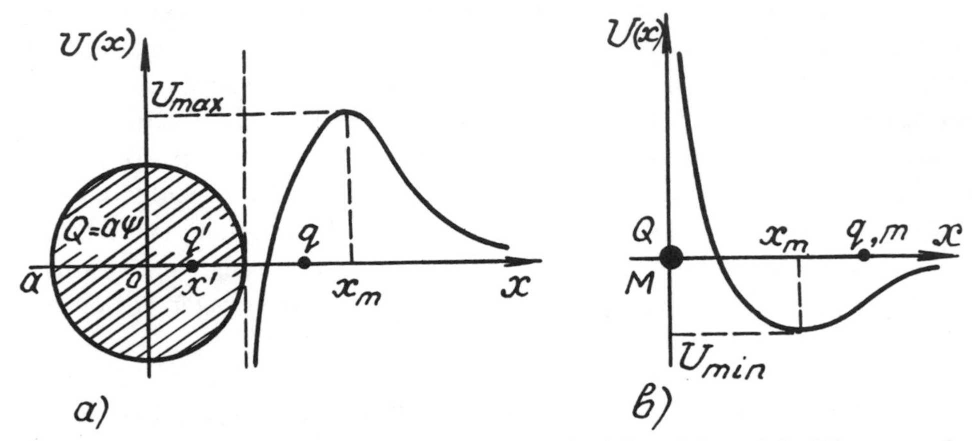


Figure 13: The Potential Energy of a Detached Particle, with Allowance for the Induced Charge (a) and Gravitational Interaction (b).

where the first term corresponds to a force of repulsion, since $q$ and $Q$ are of the same sign, while the second term represents attraction between the charge and its image. The variation of $U_E(x)$ is shown diagrammatically in Figure 13a. Attraction prevails at $a \leq x \leq x_m$ $(f = -\partial U_E/\partial x < 0)$, while repulsion at $x > x_m$, $(f = -\partial U_E/\partial x > 0)$. The zero point $x = x_m$, $(\partial U_E/\partial x)_{x=x_m} = 0$ is a point of unstable equilibrium. The detached particle cannot leave the potential barrier of height $U_{max}$ (a particular case is the work function of electrons in a metal). The infinite increase of the force of attraction at $x = a$ in the Figure is unreal. It has resulted from neglecting the size of the smaller fragment.

If both particles are point-size, then there is no question of the image charge; however, forces of attraction between particles of the like charges exist in this case too. They are produced by the gravitational interaction characterized by the energy $U_G = -GMm/x$, with $G$ being the gravity constant, and $M$ and $m$ the particle masses. The total potential energy, $U_{tot} = U_E + U_G$ may be either positive, or negative, but the system cannot have an equilibrium configuration since both terms in the energy are characterized by the same (Coulombian) range dependence.

The case will be different if the plasma environment is taken into account. While it has no effect on the gravitational interaction, the electric

Coulombian interaction will be replaced by that of the screened potential. Then the total energy of interaction would become

$$U_{tot} = U_E + U_G = \frac{Qq}{x} \exp(-x/\lambda_D) - G\frac{Mm}{x}. \tag{45}$$

Suppose, $Qq > GMm$. Then repulsion forces prevail for $x \ll \lambda_D$. However, attraction dominates at sufficiently great distances. As a result, there is some distance $x_m$ where the total energy reaches its minimum which corresponds to a stable equilibrium (Figure 13b). Thus, allowance for the gravitation interaction of charged particles makes the so-called levitation possible.

In fact, the question of levitation is not as simple, the presence of a minimum of $U_{tot}(x)$ being a necessary but not a sufficient condition. Indeed, the detached fragment first "slides down the hill" of the potential barrier and gains a kinetic energy sufficient for leaving the potential well and going to infinity. In order to stay in the well, the grain must lose much of its initial energy on the way from the point of separation from the parent body to $x_m$. This loss of energy should result from interaction of the charged particle with the plasma it passes through. Besides, the electric charge of the particle may change during its motion. While it stayed on the surface of the larger body, its share was a small fraction of the surface charge $\sigma \simeq Q/4\pi a^2$, namely $q_0 \simeq \pi a_0^2 \sigma = (a_0^2/4a)\psi$, where $a_0$ is the radius of the separated grain. At a distance $z \gg \lambda_D$ the equilibrium potential of the small grain becomes equal to $\psi$, hence its charge increases to $q \simeq a_0\psi \simeq (4a/a_0)q_0 > q_0$, though this requires a certain time (cf. Section 1.1). Consider the motion of a charged dust grain through the plasma layer around a larger body, accounting for variations in $q$ (Nitter and Havnes, 1992). The size of the celestial body with a dusty-plasma layer around is assumed to be much greater than $\lambda_D$ (specifically, it may be the Moon or an asteroid). This allows us to greatly simplify the problem, treating it as one-dimensional. First, let us determine the electric potential $\Phi(x)$ in the plasma layer ($x$ is distance from the body surface) using the Poisson Equation (24) with a zero right-hand side which corresponds to a low number density of dust grains. The electrons, as usual, are supposed to obey the Boltzmann distribution $n_e = n_0 \exp(e\Phi/T_e)$, while the ions are "cold". They enter the plasma layer from without, with the drift velocity $v_{i,dr}$. In a plasma with $\Phi < 0$ the ions are accelerated, hence their number density decreases by virtue of the continuity equation,

$$n_i = n_0 \left(1 - \frac{2e\Phi}{m_i v_{i,dr}^2}\right)^{-1/2}. \tag{46}$$

This equation for $n_i$ results in an estimate of the surface potential $\psi$ that is somewhat different from the above-given Equation (11),

$$\psi = \frac{T_e}{e}(-2.84 + \log M), \tag{47}$$

where $M = v_{i,dr}/(T_e/m_i)^{1/2}$ is Mach's number for the ions incident on the body, relative the ion-acoustic wave. Most of the numerical estimates in the literature relate to the solar wind conditions where $M \simeq 5$.

The equilibrium charge of dust grains in the plasma can be estimated, under the same assumptions about $n_e$ and $n_i$, with allowance for ion acceleration in the electric field of the grain at a potential $U = \psi - \Phi$ relative the plasma. Having found $\Phi(x)$ and $U(x)$, Nitter and Havnes (1992) analyzed the motion of a dust grain of mass $m$ under the action of the force $F = F_E + F_G = a_0U(d\Phi/dx) - mg$. The first term represents the electric force in the field of potential $\Phi$, while the second is the attraction in the gravity field of the larger body that remains constant within close ranges. The two forces are characterized by different dependences on the grain radius, $a_0$, since $q = a_0U$, whereas $m \simeq (4/3)\pi a_0^3\rho$, with $\rho$ being the mass density of the grain material. The $U(x; a_0)$ and $\Phi(x; a_0)$ dependences established can be used to estimate the equilibrium position $x_m$ (for each value of the radius $a$), where $F_E = F_G$ and $F = 0$. As follows from the numerical analysis, the balance of forces $F_E = F_G$ can never occur for large grains (with the plasma parameters and size of the parent body typical for the lunar environment), in view of the dominance of gravitational attraction. The equation $F(x) = 0$ can have real roots in the case of smaller dust grains, specifically two roots, $x_{01}$ and $x_{02} > x_{01}$. The equilibrium is unstable at $x = x_{01}$ but stable at $x = x_{02}$. The stable solution becomes the only one for still smaller grains.

The dynamics of the separated particle is described by an equation of motion with a variable grain charge, in fact by a complex set of nonlinear differential equations for the particle coordinates, electric charge and currents. Numerical analysis of the set showed the existence of oscillatory solutions about the equilibrium point. A remarkable feature of those solutions was the considerable attenuation. E.g., oscillations with a period of $\sim 2000$ s were sizably reduced in amplitude after 3 or 4 periods, while forces of friction were not included in the equation of motion. The attenuation may be regarded as a parametric effect as it is associated with space and time variations of the grain charge.

Finally, it can be shown that the solar wind plasma can support formation of a suspension of "floating" dust grains, consisting of particles smaller than $1\,\mu$ in the case of the Moon or $\leq 80\,\mu$ in the case of an asteroid about 2 km in diameter. The smallest particles are accelerated in the plasma sheath to the extent of leaving the vicinity of the parent body, even if their initial velocity were zero. Therefore, an interval of grain sizes can be indicated, $a_{max} \geq a_0 \geq a_{min}$, capable of supporting a "dusty atmosphere" around large bodies in the solar wind. The size spectrum of particles leaving the parent surface may be much wider.

## 1.4 Coagulation of Neutral and Charged Grains

Coagulation of dust grains during collisions produces larger particles. The process is of considerable importance for the formation of the spectrum of masses in space. The rise of dust grains of different size from a gaseous medium passes through several stages. The first is condensation of liquid droplets or solid particles from a uniform cold gas. This is a slow process resulting in the appearance of grains smaller than a few microns. Later on, atoms and molecules condense on the nuclei that arose before, and the grains increase in size. Meanwhile, the pressure of the neutral gas decreases, reducing the rate of condensation. Further growth of the grains in size is only possible through coagulation of the already-existing particles. It is this stage of the process that will be discussed below. The effects of electrostatic attraction (or repulsion) on coagulation of colliding particles were discussed by Horanyi and Goertz (1990).

### 1.4.1 Neutral Particles: Mass Spectrum Formation

The primary condensation is controlled by the action of a source of constant intensity, $S(m) = \tilde{S}\delta(m - m_0)$, which model suggests all the particles to be of the same mass, $m = m_0$, and the original gas reserve unlimited (otherwise $S(m)$ would be time dependent). In the absence of coagulation processes, the mass-spectrum would have remained delta-like, whereas the total number of particles certainly would increase. In fact particles of greater mass $m > m_0$ are formed continuously, owing to coagulation. A particle of mass $m$ is born through a collision of two smaller particles, $m' < m$ and $(m - m') < m$ (only binary collisions are considered). Let the effective cross-section of this process be $A(m',\ m - m')$. On the other hand, a particle of mass $m$ disappears if it collides with any other particle of mass $m'$, since $m + m' > m$. The corresponding cross-section is $A(m, m')$. Formation of the mass spectrum $n(m, t)$ can be described, with allowance for these processes and the action of the source, by the Smoluchowsky kinetic equation

$$\frac{\partial n(m,t)}{\partial t} = I_{col}\{m,n\} + \tilde{S}\delta(m-m_0), \tag{48}$$

where

$$\begin{aligned} I_{col}\{m,n\} &= \frac{1}{2}\int_0^m A(m', m-m')n(m',t)n(m-m',t)\,dm' - \\ & n(m,t)\int_0^\infty A(m,m')n(m',t)\,dm' \end{aligned} \tag{49}$$

is the collision integral.

The development of particle coagulation is entirely determined by $A(m,m')$. It has been assumed that the relative particle velocities $w$ are independent of their masses and, besides, that the probability of coagulation is equal to unity, once the particles have collided. In the case of electrically neutral grains, the collisional cross-section area is equal to their common geometrical area $\pi(a+a')^2$. It is convenient to normalize the value by $A_0 = \pi a_0^2$, with $a_0 = (3m_0/4\pi\rho)^{1/3}$, where $a_0$ is the radius of primary particles and $\rho$ the mass density of their material:

$$A(m,m') = A_0\left[(m/m_0)^{1/3} + (m'/m_0)^{1/3}\right]^2 w, \tag{50}$$

Numerical results concerning Equation (48) are shown in Figure 14 (taken from Horanyi and Goertz, 1990). The individual curves in the Figure are mass distribution functions taken at different time moments.

The greater is the time lapse after "turn-on" of the source, the greater the rightward shift of the curve. Gradual formation of a power-law distribution with an exponent $\simeq -1.8$ can be clearly seen in the Figure (upper part of the graphs). From the side of greater masses, the power-law distribution is replaced by an exponential one which drops off quite abruptly.

In fact, the theory of coagulation based on the Smoluchowski kinetic equation (see, for instance, Kats and Kontorovich, 1990) permits the exponent of the power-law distribution $n(m) \sim m^\gamma$ to be deduced theoretically, presuming a steady-state self-similar solution for Equation (48). The equation can be written in the form of a continuity relation for the mass density $mn(m)$, *viz.*

$$\frac{\partial(nm)}{\partial t} + \frac{\partial}{\partial m}j(m,n) = m\tilde{S}\delta(m-m_0). \tag{51}$$

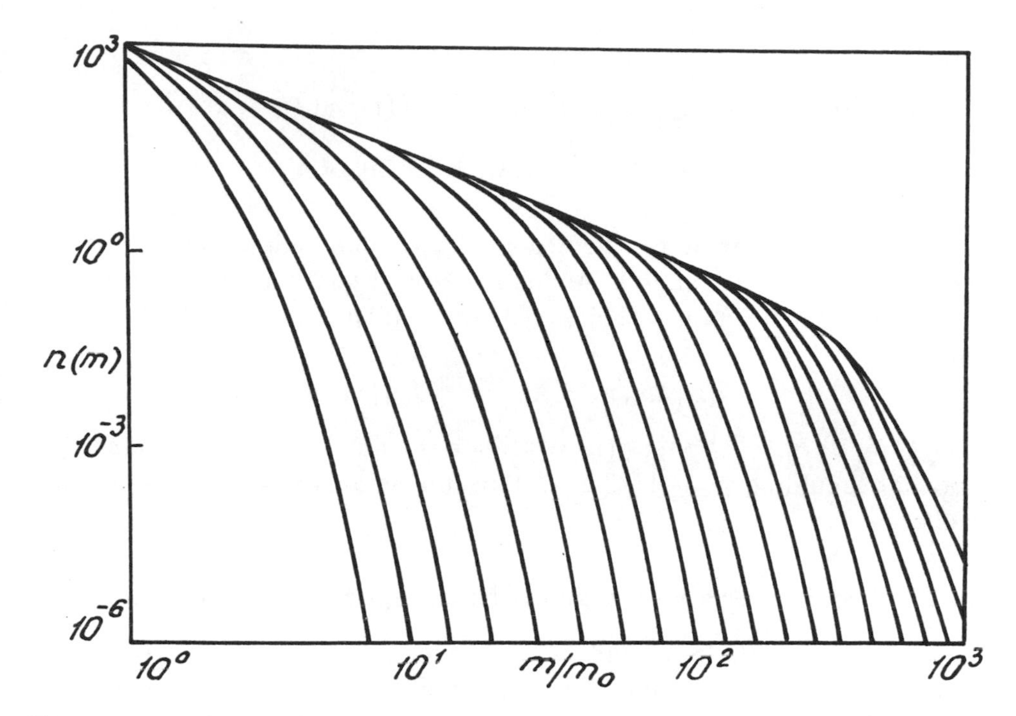


Figure 14: The Evolution of the Mass Distribution of Uncharged Grains (after Horanyi and Goertz, 1990).

The mass density flux $j(m,n)$ is related to the collision integral of Equation (49) as

$$j(m,n) = -\int_0^m m' I_{col}\{m',n\}\, dm', \tag{52}$$

and the steady-state ($\partial(n,m)/\partial t = 0$) solution of Equation (51) corresponds to

$$j(m,n) = \tilde{S} \int_0^m m'\delta(m' - m_0)\, dm' = \begin{cases} \tilde{S} m_0 & \text{for } m > 0 \\ 0 & \text{for } m < m_0 \end{cases}. \tag{53}$$

Thus, the steady-state regime is characterized by a constant flux of mass in the mass-spectrum domain, starting from a minimal value $m_0$ determined by properties of the source. A constant-flux solution of Equation (48) is indeed possible in the class of power law distributions $n(m) \sim m^\gamma$ (Kadomtsev and Kontorovich, 1974). If the mass is changed by a factor $\lambda$, the mass density becomes $n(\lambda m) = \lambda^\gamma n(m)$, while the flux $j$ should remain invariant, $j(\lambda m, n) = j(m,n)$. As for the collision integral, it is transformed according to

$$I_{col}\{\lambda m, n\} = \frac{1}{2}\int_0^{\lambda m} A(m', \lambda m - m')n(m')n(\lambda m - m')\, dm' - \\ -n(\lambda_m)\int_0^{\infty} A(\lambda_m, m')n(m')\, dm'. \quad (54)$$

The integration variable can be denoted $\lambda_{m'}$, which allows making use of the homogeneity of $A(m, m')$. As follows from Equation (50), $A(m, m')$ is a homogeneous function of index 2/3, i.e. $A(\lambda m, \lambda m') = \lambda^{2/3}A(m, m')$, whence

$$I_{col}\{\lambda m, n\} = \lambda^{1+2/3+2\gamma} I_{col}\{m, n\}$$

and $j\{\lambda m, n\} = \lambda^{2+1+2/3+2\gamma} j(m, n)$. Since the flux is a constant value, the exponent should be $\gamma = -11/3 \simeq 1.82$, in full agreement with the numerical result.

### 1.4.2 Effect of Grain Charge on Mass Spectrum Formation

If the colliding particles $m$ and $m'$ carry electric charges $Q$ and $Q'$ of the same sign, then the effect of the Coulombian repulsion is to reduce the cross-section area,

$$A^c(m, m') = A(m, m')\left[1 - QQ'/[(a + a')\mu w^2]\right], \quad (55)$$

where $\mu = mm'/(m + m')$ is the effective mass. The reduction is easily predictable. The rate of particle amalgamation with formation of a power-law mass spectrum is slowed down, with the exponential drop-off shifting toward lower masses.

Particles of opposite charge signs are pulled closer together by the electrostatic force and their collisional cross-section $A(m, m')$ is increased. However, generally the alterations in the process of coagulation are not as straightforward as in the case of like charges.

First, let us recall how dust grains from the same plasma environment can acquire charges of opposite signs. The reason may be either emission of photoelectrons by particles of different chemical composition (see Subsection 1.1.3) or secondary electron emission in the field of a multiple-valued surface potential (see Subsection 1.1.4). Of special interest is the latter possibility associated with temperature variations of the ambient plasma. Under certain conditions the charge sign proves dependent on the particle size (mass). Let $m = m_{cr}$, which is some critical (with respect to changes in sign) mass. Then the coagulation cross section $A(m, m')$ would decrease, as compared with neutral particles, for all $m, m' < m_{cr}$ or $m, m' > m_c r$ but increase if $m$ and $m'$ lay to opposite sides of $m_{cr}$.

The distribution function $n(m, t)$ evolves so as to yield a power-law spectrum as before with $m \leq m_{cr}$. There is a break point and a sharp decrease near $m = m_{cr}$, after which the spectrum for $m > m_{cr}$ flattens over a wide range, up to the exponential drop off from the side of very large masses. It seems interesting to compare the ratios $n^c(m, t)/n(m, t)$ and $n^{c*}(m, t)/n(m, t)$ for great values of $t$ when the spectra have generally been formed ($n$ is the distribution of neutral grains; $n^c$ that of particles with like charges, and $n^{c*}$ the distribution function of particles $m > m_{cr}$ and $m < m_{cr}$ whose charges have opposite signs). This comparison is shown in Figure 15 (after Horanyi and Goertz, 1990). The abrupt change at $m = m_{cr}$ is clearly seen for charges of opposite signs, and sharp variations of concentration for great values of $m$. The number of large grains increases in the case of oppositely charged particles, decreasing for charges of like sign.

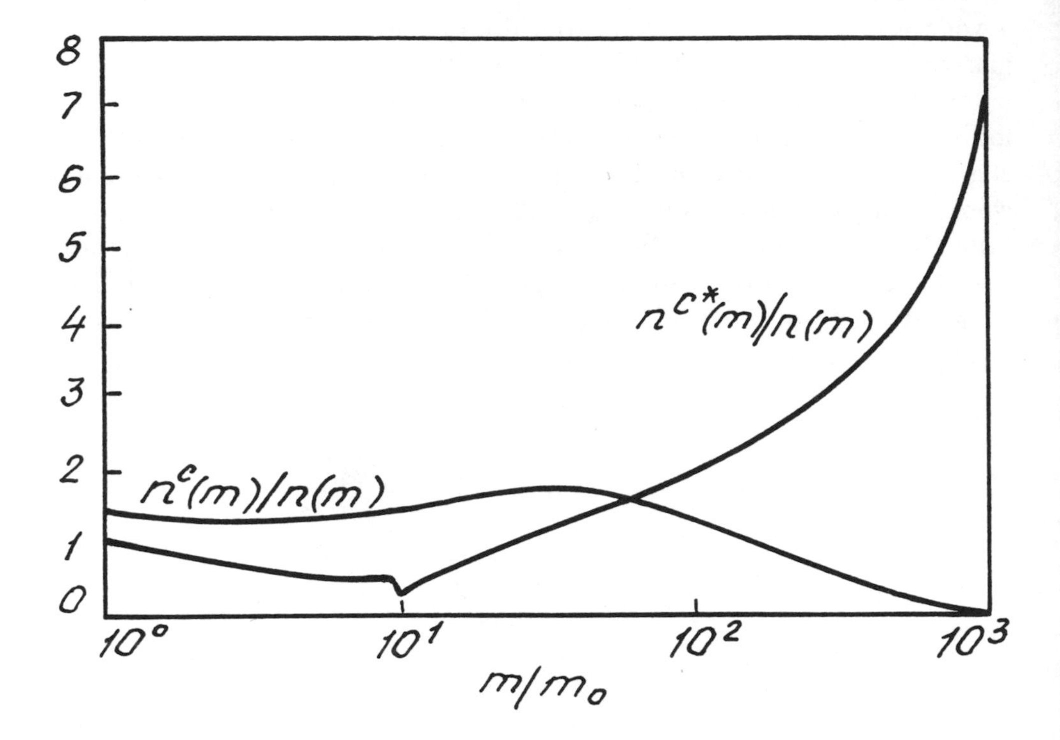


Figure 15: The Ratios of the Constant-Potential ($n^c(m)/n(m)$) and Opposite-Sign Potential ($n^{c*}(m)/n(m)$) Distribution Functions, with $n(m)$ Being the Distribution of Uncharged Particles. (In the Latter Case the Potential Differs in Sign for the Larger and Smaller Particles) (after Horanyi and Goertz, 1990).

## References

Alpert, Y.L., Gurevich, A.V. and Pitaevskii, L.P. 1965. Space Physics with Artificial Satellites (Consultants Bureau: Boulder, CO).

de Angelis, U. 1992 *Physica Scripta* **45**, 465.

Chow, V.W., Mendis, D.A. and Rosenberg, M. 1993, *J. Geophys. Res.* **98**, 19065.

Christon, S.P., Mitchel, D.G., Williams, D.J., Frank, L.A., Huang, C.Y. and Eastman, T.E. 1988, *J. Geophys. Res.* **93**, 2562.

Fechtig, H., Grün, E. and Morfill, G.E. 1979, *Planet. Space Sci.* **27**, 511.

Goertz, C.K. 1989, *Rev. Geophys.* **27**, 271.

Goertz, C.K. and Ip, W.H. 1984, *Geophys. Res. Lett.* **11**, 349.

Goertz, C.K., Shan, L. and Havnes, O. 1988, *Geophys. Res. Lett.* **15**, 84.

Grün, E., Morfill, G.E. and Mendis, D.A. 1984. In R. Greenberg and A. Brahic (Eds.) *Planetary Rings*, (Univ. of Ariz. Press: Tucson, AZ), p. 275.

Havnes, O., Morfill, G.E. and Goertz, C.K. 1984, *J. Geophys. Res.* **89**, 10999.

Havnes, O., Goertz, C.K., Morfill, G.E., Grün E. and Ip, W. 1987, *J. Geophys. Res.* **92**, 2281.

Havnes, O. and Morfill, G.E. 1984, *Adv. Space Res.* **4**, 85.

Hill, J.R. and Mendis, D.A. 1980, *Astrophys. J.* **242**, 395.

Horanyi, M. and Goertz, C.K. 1990, *Astrophys. J.* **361**, 155.

Kadomtsev, B.B. and Kontorovich, V.M. 1974, *Sov. Radiophys. and Quantum Electron.* **17**, 511 (*in Russian*).

Kats, A.V. and Kontorovich, V.M. 1990, *Sov. Phys.—JETP*, **97**, 3 (*in Russian*).

Landau, L.D. and Lifshitz, E.M. 1982. Electrodynamics of Solids (Nauka: Moscow), p. 621 (*in Russian*).

Mendis, D.A. and Rosenberg, M. 1992, *IEEE Trans. on Plasma Science* **20**, 929.

Mendis, D.A. and Rosenberg, M. 1994, *Ann. Rev. Astron. Astrophys.* **32**, (*in press*).

Mendis, D.A. 1989. In H. Kikuchi (Ed.) *Laboratory and Space Plasmas*, (Springer Verlag), p. 51.

Meyer-Vernet, N. 1982, *Astron. Astrophys.* **105**, 98.

Mukai, T. 1981, *Astron. Astrophys.* **99**, 1.

Nitter, T. and Havnes, O. 1992, *Earth, Moon, and Planets* **56**, 7.

Northrop, T.G. 1992, *Physica Scripta* **45**, 475.

Northrop, T.G., Mendis, D.A. and Schaffer, L. 1989, *Icarus* **79**, 101.

Öpik, E.J. 1956, *Irish Astron J.* **4**, 84.

Rosenberg, M. and Mendis, D.A. 1992, *J. Geophys. Res.* **97**, 14773.

Sekamina, Z. and Farrell, J.A. 1980. In J. Halliday and B.A. McIntosh (Eds.) *Solid Particles in the Solar System*, p. 267.

Summers, D. and Thorne, R.M. 1991, *Phys. Fluids.* **83**, 1835.

Whipple, E.C. 1981, *Rep. Prog. Phys.* **44**, 1197.

Whipple, E.C., Northrop, T.G. and Mendis, D.A. 1985, *J. Geophys. Res.* **90**, 7405.

# 2 Gravitoelectrodynamics

The hybrid term of gravitoelectrodynamics has appeared in the literature quite recently, after the missions of *Pioneer-10* and *Pioneer-11* (and *Voyager-1* and *2* some time later) investigated the spatial distributions and time behavior of micron and submicron-size grains in the interplanetary medium. The dynamical patterns of such particles suggested the action of electromagnetic forces along with gravitation. In fact, that was something to be expected as dust grains in space are virtually always electrically charged, while the space is filled with electric and/or magnetic fields. Traditionally, it was believed that the spheres of prevailing influence of the gravitational or electromagnetic forces could be easily demarcated. The motion of large celestial bodies was assumed to be governed by forces of gravitation, while trajectories of charged microparticles (i.e. electrons and ions) were controlled by electromagnetic fields. This is certainly true, but the situation is different with electrically charged macroscopic particles of micron or submicron size for which the forces of the two kinds are equipotent, such that the introduction of the new term is justified. Curiously enough, this is a rare case where the linguistic process of term formation can be traced back. First, two distinct terms were used in the literature, i.e. gravitation and electrodynamics. At a later stage, Mendis *et al.* (1982) introduced the half-integral, hyphenated "gravito-electrodynamics", its two parts finally merging into gravitoelectrodynamics (Goertz, 1989).

## 2.1 Forces Acting on an Isolated Charged Particle in the Vicinity of a Planet

### 2.1.1 Gravitational and Electromagnetic Forces

This Chapter is chiefly an analysis of the dynamics of individual charged particles moving in a planetary magnetosphere. In the close vicinity of a massive celestial body (ranges within a few planetary radii) the particle motion is largely controlled by the gravitational force which will be assumed spherically symmetric,

$$\mathbf{F}_G = -GM_p m\mathbf{e}_r/r^2. \quad (1)$$

Here $G = 6.672 \times 10^{-8}\,(\text{dyne/g}^2)\text{cm}^2$ is the gravitation constant; $M_p$ the planet mass; $r$ the distance from the gravitation center and $\mathbf{r}_0$ the unit vector along the radial. By assuming this representation for the gravitation force, we certainly lose the possibility of analyzing the effects associated with deviations from spherical symmetry, such as oblate geometry or nonuniformity of the planet itself that give rise to multipole moments in the gravitation field, or the influence of large satellites on the motion of small-size particles. Because of satellites, the gravitation field may be locally perturbed both in space and in time. However these questions belong to "pure" celestial mechanics rather than gravitoelectrodynamics. We will not treat them here, just mentioning a few literature references (Franklin *et al.*, 1984; Dermott, 1984) relating to the dynamics of small neutral grains in planetary rings. The effect of plasma "self-gravitation", i.e. grain–grain interaction through the gravitation fields they produce, will not be analyzed either, since at this time we will consider isolated solitary grains.

The force of gravity, $\mathbf{F}_G$, influences charged and uncharged particles in an equal manner. However, a dust grain carrying an electric charge $Q$ is also affected by the forces

$$\mathbf{F}_E = Q\mathbf{E} \tag{2}$$

and

$$\mathbf{F}_B = (Q/c)[\mathbf{V} \times \mathbf{B}] \tag{3}$$

produced by the electric field $\mathbf{E}$ and magnetic field $\mathbf{B}$. (Here $\mathbf{V}$ is the grain velocity and $c = 3 \times 10^8\,\mathrm{ms}^{-1}$ the speed of light). The planetary magnetic field $\mathbf{B}$ will be subject to further simplifying assumptions. First, we consider it as the field of a magnetic dipole of moment $\mathcal{M}$ located at the gravity center and oriented along the rotation axis of the planet,

$$\mathbf{B} = 3\mathbf{r}(\mathcal{M}\mathbf{r})r^{-5} - \mathcal{M}r^{-3} \tag{4}$$

Within the equatorial plane, this becomes

$$\mathbf{B} = -\mathcal{M}/r^3 \quad \text{and} \quad B = B_0 R_p^3 r^{-3} \equiv B_0 L^{-3}, \tag{5}$$

where $B_0$ is the magnetic induction at the planet surface and $L$ the magnetic shell parameter ($L = r/R_p$, where $R_p$ is the planet radius).

As for the interior, the planet will be represented by a uniformly magnetized sphere with

$$\mathbf{B}(r \leq R_p) = 2\mathcal{M}R_p^{-3} \quad \text{and} \quad \mathbf{H} = -\mathcal{M}R_p^{-3}.$$

(The induction $\mathbf{B}$ and the field strength $\mathbf{H}$ inside the planet are different owing to magnetization. The magnetic moment per unit volume is $\mathbf{m} = 3\mathcal{M}/4\pi R^3$. Outside the planet, $\mathbf{B} = \mathbf{H}$).

Table 1: Gravitation and magnetic field parameters of Jupiter, Saturn and Earth.

| | Mass of the planet, $M_p$[kg] | Magnetic moment, $\mathcal{M}$[Gs $\cdot$ cm$^3$] | Magnetic induction at the equator, $B_0$[Gs] | Equatorial radius, $R_p$[cm] | Corotation radius, $R_{co}$[cm] |
|---|---|---|---|---|---|
| Jupiter | $1.9 \times 10^{27}$ | $1.6 \times 10^{30}$ | 4.2 | $7.14 \times 10^9$ | $1.59 \times 10^{10}$ |
| Saturn | $5.68 \times 10^{26}$ | $4.4 \times 10^{28}$ | 0.2 | $6.03 \times 10^9$ | $1.12 \times 10^{10}$ |
| Earth | $5.98 \times 10^{24}$ | $8 \times 10^{25}$ | 0.5 | $6.38 \times 10^8$ | $4.37 \times 10^9$ |

The origin of the planetary magnetic field is not discussed here. We simply take the planet being magnetized for granted, considering the magnetic moment $\mathcal{M}$ among its basic parameters. The situation is different for the electric field about the planet, **E**. It results from rotation of the magnetized planet about its axis and can be calculated from the knowledge of $\mathcal{M}$ and the rotation frequency $\Omega_p$.

### 2.1.2 Unipolar Induction

The appearance of an electric field near a rotating magnetized sphere, known as unipolar induction, is a purely relativistic effect. In fact, separation of an electromagnetic field into an electric and magnetic component is not absolute but depends on the frame of reference.

The unipolar induction field in a planetary magnetosphere can be evaluated quite simply if the plasma is of sufficiently high conductivity and corotates rigidly with the magnetized planet. The medium would be at rest in the frame $K'$ moving at the velocity $\mathbf{V} = [\mathbf{\Omega}_p \times \mathbf{r}]$ with respect to a (fixed) observation point. Ohm's law in $K'$ can be written in the standard form $\mathbf{j} = \sigma\mathbf{E}'$, $\sigma$ being the plasma conductivity and $\mathbf{E}'$ the electric field vector. The latter is related to the field reported by the fixed observer as $\mathbf{E}' = \mathbf{E} + [\mathbf{V} \times \mathbf{B}]/c$. Hence, the current observed in the fixed frame $K$ is $\mathbf{j} = \sigma(\mathbf{E} + [\mathbf{V} \times \mathbf{B}]/c)$ (transformation of the value $\mathbf{j}$ itself results in corrections of the order $v^2/c^2$ which are disregarded). By demanding that the

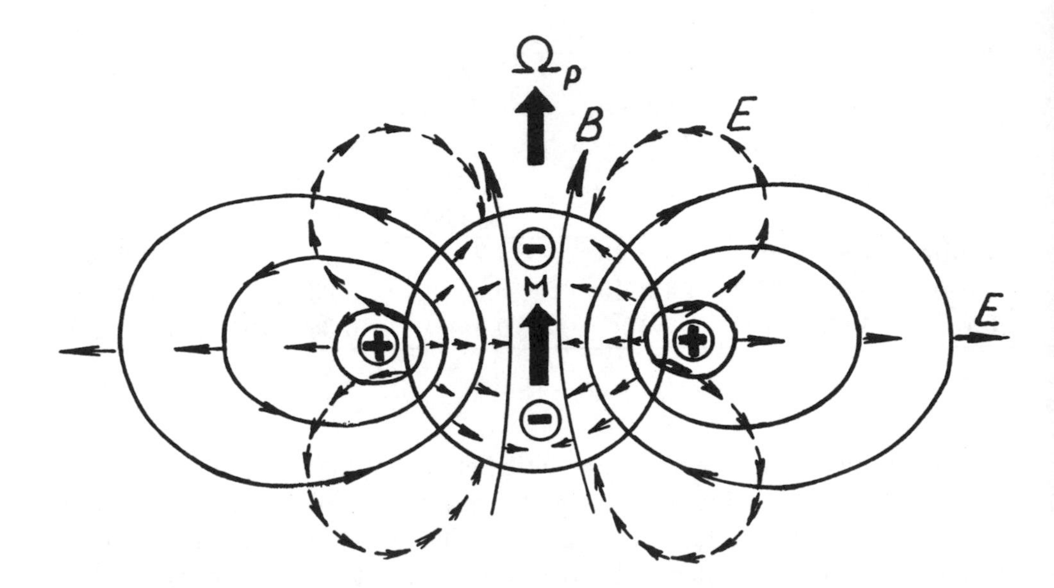


Figure 1: The Unipolar Induction Field **E**: A Magnetized Sphere of Magnetic Moment $\mathcal{M}$ Rotates in a Conducting Medium at an Angular Velocity $\mathbf{\Omega}_p$.

current should remain finite at $\sigma \to \infty$ we arrive at

$$\mathbf{E} = -\frac{1}{c}\left[[\mathbf{\Omega}_p \times \mathbf{r}] \times \mathbf{B}\right]. \tag{6}$$

Unipolar induction field lines are shown schematically in Figure 1. Sometimes it proves convenient to express this field in terms of the potential

$$\Phi = -\frac{1}{cr}\Omega_p \mathcal{M} \sin^2\theta, \tag{7}$$

$\mathbf{E} = -\nabla\Phi$, where $\theta$ is the angle between the rotation axis and the radius-vector $\mathbf{r}$.

The Lorentz force acting on a particle of charge $Q$ orbiting around the planet at an angular velocity $\Omega$ is given by

$$\mathbf{F}_L = \frac{Q}{c}\left[[(\mathbf{\Omega} - \mathbf{\Omega_p}) \times \mathbf{r}] \times \mathbf{B}\right]. \tag{8}$$

Within the equatorial plane, $\mathbf{F}_L$ is radially oriented and equals

$$F_L = \frac{QB}{c}(\Omega - \Omega_p)R = \frac{QB_0}{c} \cdot \frac{R_p^3}{R^2}(\Omega - \Omega_p) \tag{9}$$

at a distance $R$ from the axis of rotation. The force is zero ($F_L = 0$) at the synchronous orbit, where $\Omega = \Omega_p$. Recalling that the particle moves in a gravitation field, hence its orbital frequency depends upon $R$ according to the Kepler law

$$\Omega = \Omega_K = \left(\frac{GM_p}{R^3}\right)^{1/2}, \tag{10}$$

we can find the radius of the synchronous orbit, $R_{co} = [(GM_p)/\Omega_p^2]^{1/3}$, known as the corotation radius. The value is listed in Table 1 for a number of planets.

In a sense, Equation (8) concludes our consideration of unipolar induction, providing a description of the electromagnetic force involved. However, it has been derived under certain limiting assumptions (in particular, that of a high conductivity of the plasma around the planet). It seems worthwhile to analyze the changes in $\mathbf{E}$ that might follow if we refused this simplification. Let the plasma environment (and the planet itself) be characterized by certain electric parameters, $\varepsilon$ and $\sigma$. The electric field $\mathbf{E} = -\nabla\Phi$ can be conveniently considered as one produced by an ensemble of sources (electric charges $\rho$) distributed in space. Once the source density $\rho$ has been determined, the problem reduces to the Poisson equation

$$\Delta\Phi = -4\pi\rho \tag{11}$$

with boundary conditions reflecting some simple physics. Evidently, the source density $\rho$ is concentrated at the planet or in its close vicinity, therefore $\Phi(r \to \infty) = 0$. On the other hand, the potential $\Phi$ should assume finite values everywhere, including the point $r = 0$, hence $\Phi(r = 0) \neq \infty$.

A rather general solution of the electrostatic problem in a two-layer model of the planetary magnetosphere with differentially rotating layers has been given by Bliokh and Yaroshenko (1983). Here, we shall confine ourselves to a simpler case which is in a sense opposite to the previously considered model of an infinitely conducting medium.

Consider a uniformly magnetized sphere of radius $R_p$, characterized by a dielectric constant $\varepsilon$ and rotating in a vacuum at an angular velocity $\Omega_p$ (Landau and Lifshitz, 1982). Free charges are absent both within and without the sphere, hence $\nabla\mathbf{D} = 0$ (where $\mathbf{D}$ is the electric induction). On the other hand, the density of bound charges can be found from $\nabla \mathbf{E} = 4\pi\rho$ and the $\mathbf{D}-\mathbf{E}$ relation for the moving dielectric (the Minkowski equation),

$$\mathbf{D} = \varepsilon\mathbf{E} + \frac{\varepsilon}{c}[\mathbf{V}\times\mathbf{B}] - \frac{1}{c}[\mathbf{V}\times\mathbf{H}]. \tag{12}$$

This can be written in a simpler form if the magnetic induction, $\mathbf{B}$, and magnetic field, $\mathbf{H}$, are expressed in terms of the magnetization vector $\mathbf{m}$, *viz.* $\mathbf{B} = 8\pi\mathbf{m}/3$ and $\mathbf{H} = -4\pi\mathbf{m}/3$,

$$\mathbf{D} = \varepsilon\mathbf{E} + \frac{4\pi(2\varepsilon+1)[\mathbf{V}\times\mathbf{m}]}{3c}. \tag{13}$$

Recalling that $\mathbf{V} = \mathbf{\Omega}_p \times \mathbf{r}$ and $\mathbf{m}$ and $\mathbf{\Omega}$ are parallel vectors, we arrive at $\nabla\cdot[\mathbf{V}\times\mathbf{m}] = -2\Omega_p m$ and $\rho = -2(2\varepsilon+1)m\Omega_p/3\varepsilon c$. These sources are concentrated inside the planet where the Poisson equation takes the form

$$\Delta\Phi^{(i)} = 8\pi(2\varepsilon+1)m\Omega_p/3\varepsilon c \qquad (r \le R_p). \tag{14}$$

Outside the sphere, $\varepsilon = 1$, $\mathbf{B} = \mathbf{H}$ and $\mathbf{D} = \mathbf{E}$, such that Equation (12) and $\nabla\cdot\mathbf{D} = 0$ yield $\nabla\cdot\mathbf{E} = 0$, whence $\rho = 0$ and

$$\Delta\Phi^{(e)} = 0 \qquad (r \ge R_p). \tag{15}$$

The radial component of $\mathbf{D}$ should be continuous at the boundary, which condition allows matching the potentials $\Phi^{(i)}$ and $\Phi^{(e)}$ at $r = R_p$,

$$\left[-\varepsilon\frac{\partial\Phi^{(i)}}{\partial r} + 4\pi(2\varepsilon+1)R_p\Omega_p m\frac{\sin^2\theta}{3c}\right]_{r=R_p} = -\frac{\partial\Phi^{(e)}}{\partial r} \qquad (r = R_p),$$

(the polar angle $\theta$ is with respect to the $z$-axis which is along the axis of rotation).

As can be shown, a rotating magnetized sphere possesses a quadrupole electric moment with the components $D_{zz} = -4R_p^2\Omega_p\mathcal{M}\cdot(2\varepsilon+1)/3c(2\varepsilon+3)$ and $D_{xx} = D_{yy} = -D_{zz}/2$. The potentials $\Phi^{(e)}$ and $\Phi^{(i)}$ are

$$\begin{aligned}\Phi^{(e)} &= \frac{D_{zz}(3\cos^2\theta-1)}{4r^3};\\ \Phi^{(i)} &= \frac{D_{zz}r^2(3\cos^2\theta-1)}{4R_p^5} + 4\pi(2\varepsilon+1)m\Omega_p\frac{r^2-R_p^2}{9\varepsilon c}.\end{aligned}$$

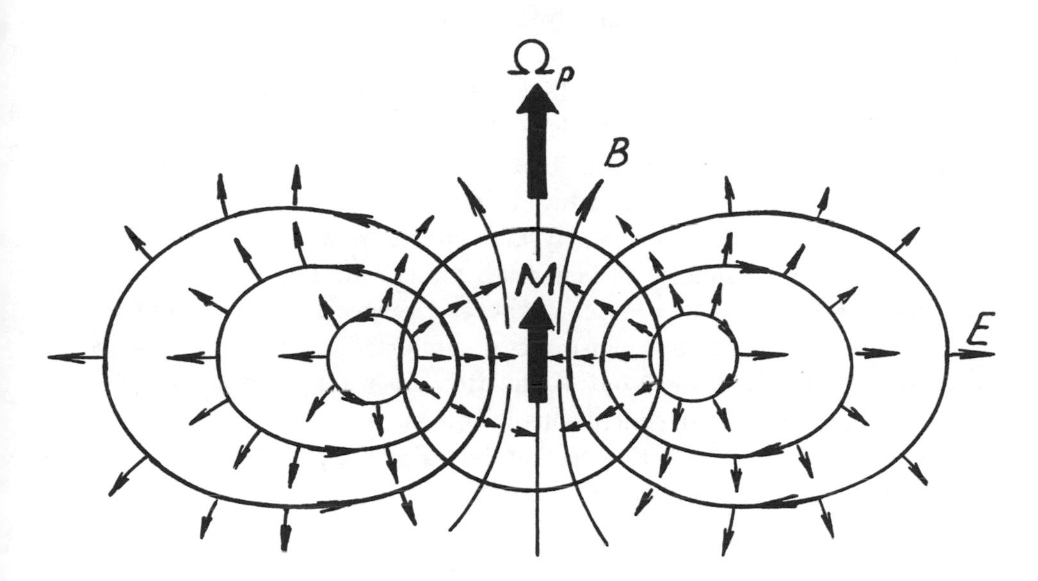


Figure 2: The Unipolar Induction Field **E**: A Magnetized Sphere of Magnetic Moment $\mathcal{M}$ Rotates an Angular Velocity $\mathbf{\Omega}_p$ in a Vacuum.

For a metal sphere, $\varepsilon$ should be tended to infinity. Then the potential in the domain of interest, i.e. outside the sphere would be

$$\Phi^{(e)} = -\frac{\Omega_p \mathcal{M} R_p^2}{3cr^3}(3\cos^2\theta - 1) = -\frac{B_0 \Omega_p R_p^5}{6cr^3}(3\cos^2\theta - 1). \tag{16}$$

The electric unipolar induction field appearing in the vacuum is shown schematically in Figure 2.

From comparing Equations (16) and (7), it can be seen how different the two limiting cases are (i.e. the highly conducting and the vacuum magnetosphere), not in the quantitative way alone. To mention just two distinctions, note that Equation (7) implies each magnetic line of force (with an arbitrary McIlwain parameter $L$, $r = R_p L \sin^2\theta$) to be an equal-potential line as $\Phi_L = -\Omega_p \mathcal{M}/cR_pL = \text{const}$. Indeed, according to Equation (6), $\mathbf{E} \perp \mathbf{B}$. Whereas in the vacuum magnetosphere the **E**-component along **B** is distributed along a line of force symmetrically with respect to the apex point ($\theta = \pi/2$), with the maximum potential difference existing between the apex and those points where the line hits the planet surface (i.e. $\theta_L = \pm \arcsin L^{-1/2}$).

The other distinction is that the Lorentz force turns to zero at the synchronous orbit if $\mathbf{E}$ is determined from Equation (7) but does not for the vacuum magnetosphere.

This analysis, however brief, demonstrates the necessity of making a choice between the two models discussed. Otherwise we would have to consider a self-consistent problem of the electric field in a plasma whose dynamics is determined by the induced field. That would be strictly necessary if we tried to analyze extremely high magnetic and electric fields, like those in the magnetospheres of neutron stars (Goldreich and Julian, 1969; Lyubarsky, 1994). For our present purposes, however, it is sufficient to notice that the magnetospheric plasma can support d.c. currents, and hence the vacuum approximation is out of question. In what follows, we shall use only Equation (7) and the conclusions it brings.

The numerical values of the electric field $\mathbf{E}_K = \mathbf{F_L}/Q$ acting on a Keplerian particle within the planet's equatorial plane can be found from Equations (9) and (10) (the field is purely radial),

$$E_K = E_0 L^{-2}(L^{-3/2}\Omega_{K,0}/\Omega_p - 1), \tag{17}$$

with $E_0 = B_0 R_p \Omega_p / c$ and $\Omega_{K,0} = \Omega_K(r = R_p)$. The values assumed by $E_0$ in the case of Jupiter, Saturn and Earth are, respectively, $5.4 \times 10^{-2}$ V/cm; $1.9 \times 10^{-3}$ V/cm and $2.2 \times 10^{-4}$ V/cm. The $E_K = E_K(L)$ curves are shown for different planets in Figure 3.

## 2.2 Radiation Pressure and Drag

### 2.2.1 Radiation Pressure

Dust grains in space are affected not only by d.c. force fields but by electromagnetic waves as well that are radiated by sources of various kind. An irradiated particle intercepts a fraction of the wave energy, $\mathcal{E}$, and the associated momentum $p = \mathcal{E}/c$. That is how the force acting on the particle appears. We shall consider the electromagnetic wave as a photon beam, with each photon characterized by an energy $\mathcal{E}_1 = \hbar\omega_0$ and momentum $p_1 = \hbar\omega_0/c$ ($\hbar$ being Planck's constant and $\omega_0$ the radiation frequency). If the photon number density is $N_{ph}$, then the momentum transferred per unit time (1 s) to a particle of radius $a$ is $\pi a^2 N_{ph} \hbar\omega_0$, which value is the radiation pressure force, $F_{ph}$.

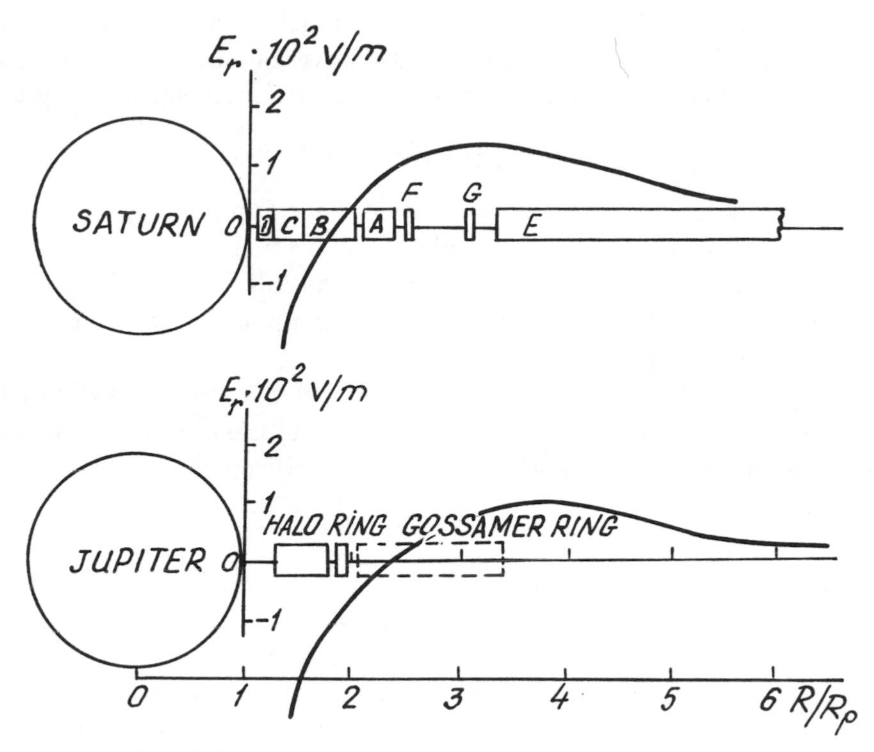


Figure 3: The Unipolar Induction Field $E_r$ Acting upon Grains that Move at a Keplerian Velocity along Circular Orbits within the Equatorial Plane.

Generally, the radiation intensity is characterized by the power flux $I_0$ across a unit area, rather than the photon density. Since $I_0$ can be written as $I_0 = N_{ph}\omega_0\hbar c$, the pressure force becomes

$$\mathbf{F}_{ph} = \pi a^2 I_0 \mathbf{e}_r / c, \tag{18}$$

which vector is oriented along the incident wave vector $\mathbf{k}$. If the radiation source is so far away as to appear point-size, then $\mathbf{F}_{ph}$ is oriented along the radius-vector connecting it with the dust grain (the unit vector $\mathbf{e}_r$ in Equation (18)).

In fact, the equation is rather approximate. First, it is only valid for particles that can absorb all the photons incident on their surface and whose radius $a$ is much greater than the radiation wavelength (ray optical approximation). To allow for a fraction of the electromagnetic power to be reflected and scattered into different directions, $I_0$ in Equation (18) should be replaced by $I_0 - I_{sc}(\theta, \varphi)$, where $\theta$ is the polar angle of scattering taken from $\mathbf{r}_0$, and $\varphi$ the azimuthal angle. Besides, the particle is heated by the absorbed power and hence emits electromagnetic waves of flux density $I_e(\theta, \varphi)$. As a result, Equation (18) should be substituted by a more general formula

$$F_{ph}(\theta, \varphi) = A\left[I_0(\theta, \varphi) - I_{sc}(\theta, \varphi) - I_e(\theta, \varphi)\right]/c, \tag{19}$$

where $\mathbf{F}_{ph}$ has been replaced by its angular density function $F_{ph}(\theta, \varphi)$ and $A$ is an area factor instead of the circular cross-section $\pi a^2$, as the particles generally are not spherical. To determine the full pressure force, it is necessary to estimate its radial and tangential components by integrating Equation (19) over angles.

It may be convenient to represent Equation (19) in a somewhat different form. Since both $I_{sc}$ and $I_e$ are proportional to $I_0$, all the re-radiated modes can be taken into account through a single factor $\gamma(\theta, \varphi)$,

$$F_{ph}(\theta, \varphi) = A I_0 \gamma(\theta, \varphi)/c. \tag{20}$$

Within the ray optical approximation, $\gamma = 1$ in the case of total absorption of the incident power and $\gamma = 2$ if the radiation is totally reflected toward the source. The product $A\gamma$ is the effective cross-section area that may greatly exceed the geometrical area of the dust particle, especially if its size is comparable with the radiation wavelength $\lambda$. E.g., a needle-like particle of thickness $\delta \ll \lambda$ and length $l \sim \lambda/2$ produces resonance scattering with $A\gamma \sim \lambda^2 \gg A \sim \delta l$.

### 2.2.2 Radiation Drag. The Poynting-Robertson Effect

In the case of a spherically symmetric particle the $\gamma(\theta, \varphi)$ factor is independent of $\varphi$, such that the net pressure force will have, upon integration over $\theta$ and $\varphi$, only a radial component along the incidence direction. For a particle of an irregular geometry, a tangential force component will be also present. However, a lateral force can appear in the radiation field even for a spherical particle. This occurs if the dust particle moves with respect to the radiation source at a velocity $\mathbf{w}$ that is not purely radial (the Poynting-Robertson effect).

The effect can be best understood if emission and scattering of the radiation are analyzed in the frame of reference moving with the particle. In that frame, Equations (19) and (20) that were derived for a particle at rest remain valid. While both equations remain formally the same, two circumstances should be taken into account:

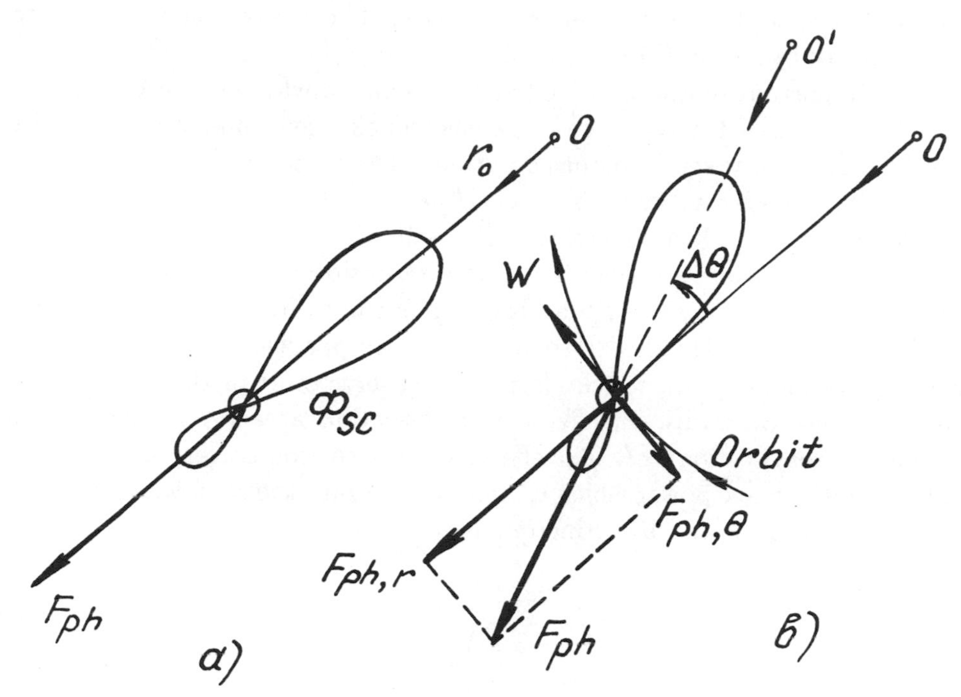


Figure 4: The Radiation Pressure on a Particle at Rest ($a$) and one Moving at a Velocity **w**. The Tangential Component $\mathbf{F}_{ph,\theta}$ Decelerates the Moving Particle (the Poynting-Robertson Effect).

— because of the Doppler effect, the frequency of the radiation incident on the particle will be $\omega = \omega_0(1 - w_r/c)$, where $w_r$ is the radial component of **w**, and the photon momentum $p_1 = \hbar\omega/c$ will change in magnitude;

— owing to aberration, the wave will propagate at an angle $\Delta\theta = w_\theta/c$ to the radial, where $w_\theta$ is the tangential component of **w**.

As a result, the axial symmetry with respect to the radial will be violated and a tangential force component will appear (Figure 4). To terms of order $w/c$, the radiation force will be

$$\mathbf{F_{ph}} = I_0(A\gamma/c)[(1 - 2w_r/c)\mathbf{e}_r - \mathbf{e}_\theta \cdot w_0/c], \tag{21}$$

where $\mathbf{e}_\theta$ is the tangential unit vector, $\mathbf{e}_\theta \perp \mathbf{e}_r$.

Owing to the tangential component of the radiation pressure, a particle orbiting around the attraction center which at the same time is the source of radiation (e.g., the Sun), is continuously slowed down. Because of this braking force (known as "radiation drag") the particle loses some of its orbital energy and the orbit radius decreases. The particle moves closer to the source of radiation (Mignard, 1984).

The important question arising at this point is whether the forces that have been discussed will change if the electric charge $Q$ possibly carried by the dust grain is taken into consideration. The force acting on a charged particle in an electromagnetic field is $Q\mathbf{E}_0 e^{i(\omega t - \mathbf{kr})}$ ($E_0$ is the amplitude of the electric vector). It produces periodic oscillations of the dust grain which thus becomes a source of electromagnetic radiation of the same frequency. The power radiated by the vibrating particle comes from the primary electromagnetic wave. Along with some power, the particle receives a fraction of the wave momentum, which effects ultimately result in the same Equation (18), however with an effective cross-section area $\sigma_Q$ in lieu of the geometric cross-section $\pi a^2$. The effective cross-section corresponds to that fraction of the wave power which is transformed into the radiation scattered by the charge $Q$. It can be estimated as

$$\sigma_Q = \frac{8\pi}{3} \cdot \left(\frac{Q^2}{mc}\right)^2, \tag{22}$$

($m$ is the mass of the dust grain) which formula is recognizable as the Thomson scattering cross-section. It involves the grain radius only implicitly, through $Q \simeq \psi a$ and $m \simeq (4\pi/3)\rho a^3$, where $\rho$ is the material mass density. Omitting numerical factors and assuming $\rho \simeq 1\,\mathrm{g/cm^3}$ we can write the estimate $\sigma_Q \sim \psi^4/a^2c^4$. As can be seen, the effective cross-section $\sigma_Q$ decreases as $a$ is increased, whereas the geometric cross-section increases as $\sim a^2$. Comparing the "charge-dependent" cross-section area with the geometric, $\sigma_Q/a^2 \sim \psi^4/(a^4c^4) \sim 10^{-36}\psi^4\,[\mathrm{V}]/a^4\,[\mu]$, one can easily see that $\sigma_Q \ll a^2$ with all reasonable values of $\psi$ and $a$, hence the effect of the grain charge on the radiation pressure is negligible.

However, this is only true for an isolated grain in a vacuum. In the case of a grain in a plasma environment allowance should be made for the motion of charged particles in the Debye layer. Electromagnetic wave scattering by charged dust grains in a plasma will be discussed below in more detail (see Chapter 3).

### 2.2.3 Plasma Drag

A charged dust grain moving through the magnetospheric plasma is subject to power losses due to collisions with the heavy plasma components (ions and neutral atoms). The drag forces in such plasmas have been analyzed in many papers (e.g., Grün, Morfill and Mendis, 1984). Basically, the physics can be described as follows.

The drag force is determined as the amount of momentum transferred per unit time from the moving grain to plasma atoms and ions (collisions with electrons can be neglected). The relative grain–plasma velocity appears because the plasma is involved in a rigid corotation with the planet, while the grain motion is governed mainly by Kepler's laws. The relative velocity, and hence the plasma drag, vanish at the synchronous orbit.

The derivations are particularly simple for collisions with neutral atoms (molecules). The momentum transferred in one collision is $p_1 = m_n w$, where $m_n$ is the atom mass and $w$ the relative grain velocity. The number of collisions per unit time is equal to $\sim \pi a^2 w n_{0,n}$ with $w \gg v_{T,n}$ or $\pi a^2 v_{T,n} n_{0,n}$ with $w \ll v_{T,n}$, where $v_{T,n}$ is the thermal velocity of the atoms and $n_{0,n}$ their number density. Therefore, the drag force owing to collisions of the dust grain with neutrals can be written as

$$F_n^{(d)} \simeq \begin{cases} \pi a^2 n_{0,n} m_n w^2 & w \gg v_{T,n} \\ \pi a^2 n_{0,n} m_n w v_{T,n} & w \ll v_{T,n}. \end{cases} \tag{23}$$

The situation with ions is somewhat more complex, since a moving charged particle can transfer its momentum to an ion not through direct collisions alone but through Coulombian long range interaction as well. The impact collisions are described by the same equations as the grain–neutral collisions, with $n_{0,n}$ replaced by $n_{0,i}$ and $m_n$ by $m_i$. Besides, the grain radius $a$ should be replaced by the impact parameter $b$ determined from the moment of momentum conservation condition for the centrally symmetric field of a negatively charged particle. The moment of ion momentum at a great distance from the dust grain is $m_i v_\infty b$, while at the impact point it is $m_i v_i a$, where $v_i$ is the velocity of the ion that has passed through the Coulombian field of the dust grain. The energy conservation law, $(m v_i^2/2) + Q\psi = m v_\infty^2/2$, and $m_i v_\infty b = m_i v_i a$ yield

$$b = a\left[1 - \frac{2Q\psi}{m_i v_\infty^2}\right]^{1/2}. \tag{24}$$

Like in the case of neutral atoms, $v_\infty$ should be understood as $w$ if $w \gg v_{T,i}$ or $v_\infty = v_{T,i}$ with $w \ll v_{T,i}$. Thus, the drag force resulting from impact collisions with ions can be written as

$$F_i^{(d)} \simeq \begin{cases} \pi b^2 n_{0,i} m_i w^2 & w \gg v_{T,i} \\ \pi b^2 n_{0,i} m_i v_{T,i} w & w \ll v_{T,i}. \end{cases} \tag{25}$$

Formal application of this formula to distant Coulombian collisions would result in an infinite increase of $b$ accompanied by a reduction of the transferred momentum.

Actually, the dust grain–ion interaction occurs in a plasma medium rather than in a vacuum, and hence the local impact parameter is limited at least by the Debye length $\lambda_D$. In addition, another impact parameter is often introduced, namely $b_{\pi/2}$. The ion trajectory for $b = b_{\pi/2}$ (in the frame of reference associated with the dust grain) is deflected by $\pi/2$ from its initial direction. (According to the Rutherford formula, $b_{\pi/2} = Qe/(m_i v_\infty^2)$). As a result, the drag force produced by distant Coulombian collisions is

$$F_{i,c}^{(d)} \simeq \gamma \, F_i^{(d)}\Big|_{b=\pi/2}, \tag{26}$$

where $\gamma = \log\left(\lambda_D^2 + b_{\pi/2}^2\right) / \left(b^2 + b_{\pi/2}^2\right)$ is the Coulomb logarithm. In many cases it can be assumed that $b_{\pi/2} \simeq b \simeq a \ll \lambda_D$, such that $\gamma \simeq \log \lambda_D^2/a^2$.

### 2.2.4 Comparison of the Different Forces

If all the forces acting on a charged particle were of comparable magnitude, its motion would be extremely difficult to analyze. In fact it is often possible to indicate one (or two) dominant forces, treating the rest as perturbations. The comparative analysis of different forces is rather complicated because of the great number of parameters involved, however the dependence on the particle size (radius) can be followed quite easily. According to Equation (1), $F_G \sim m \sim a^3$, while Equations (18) and (9) suggest $F_{ph} \sim a^2$ and $F_L \sim Q \sim a$. The latter dependence though can be accepted as a rough approximation only, since $F_L$ happens to depend on the dust grain velocity which is generally different for grains of different size.

The general tendency revealed however is insufficient for indicating the principal forces under specific conditions. It seems reasonable to believe that in the vicinity of a planet the major role belongs to the planet's gravity and the Lorentz force, whereas the dominant forces near a comet would be solar gravitation, radiation pressure and forces of electromagnetic origin.

Since our interest is concentrated mostly on the dynamics of charged particles in planetary magnetospheres, let us consider the $F_L/F_G$ ratio. For grains within the equatorial plane, it can be represented, with the use of Equations (1) and (9), as

$$\frac{F_L}{F_G} = (\Omega - \Omega_p)\frac{\Omega_B}{\Omega_K^2}, \tag{27}$$

where $\Omega_B = QB/(mc)$ is the grain's gyrofrequency and $\Omega_K = (GM_p/r^3)^{1/2}$ the Kepler frequency. We would like to know the dependence of $F_L/F_G$ upon the grain size. When trying to answer this question with the aid of Equation (8) one should remember that the mass and charge are not the only size-dependent grain parameters. The orbital frequency $\Omega$ also depends on the grain radius and may prove different from the Kepler frequency if the ratio $F_L/F_G$ is not low. In either case, the ratio depends on the distance from the axis of rotation which precludes indicating a single "critical" size $a_{cr}$ for which $F_L$ and $F_G$ are comparable in magnitude.

In fact, the condition $F_L/F_G \sim 1$ is excessive. The effect of electromagnetic forces is sufficiently important even with $F_L/F_G \ll 1$. Along with the principal force $F_G$, the moving grain is influenced by various perturbing factors that may often prove essential. For $F_L$ to be taken into account, it suffices that it be comparable in magnitude with the perturbing agents as the grain trajectory would be greatly altered even with small $F_L/F_G$.

Let us re-write Equation (27) so as to emphasize the values depending on the planet parameters,

$$\frac{F_L}{F_G} = K_p[\Omega(R,a) - \Omega_p]\frac{\psi}{a^2}, \tag{28}$$

with $K_p \simeq 3B_0R_p^3/(4M_pG\pi\rho c)$ and $m = (4\pi/3)a^3\rho$. Limiting the analysis to $F_L/F_G \ll 1$ we have $\Omega(R,a) \simeq \Omega_K(R)$. Then the critical size $a_{cr}$ can be easily determined for the magnetospheres of different planets, given the $F_L/F_G$ ratio. The derivation is much more complex if the limitation is not imposed. Calculated values of $a_{cr}$ (without the $F_L/F_G \ll 1$ limitation) for different dust grain potentials are shown in Figure 5 (Grün *et al.*, 1984; Mendis, 1984). Assuming $F_L/F_G \sim 10^{-2}$ as a conventional boundary to the influence of electromagnetic forces, we can see that the motion of dust grains $a \leq 1\,\mu$ under the potential of a few volts should be considered as a problem belonging to gravitoelectrodynamics. This estimate relates to Saturn's E-ring ($L = 5$). It would be different for other distances or other planets, however always remaining in the range of micron or submicron particles.

The effect of different forces on the particle dynamics can be estimated for the interplanetary space where gravity fields of the planets are of little importance. In Figure 6, lifetimes of different size grains in heliocentric orbits are shown (Mukai, 1986). Once again, Lorentzian forces are predominant for particles of submicron size, while the orbits of larger grains are perturbed by collisions and radiation drag.

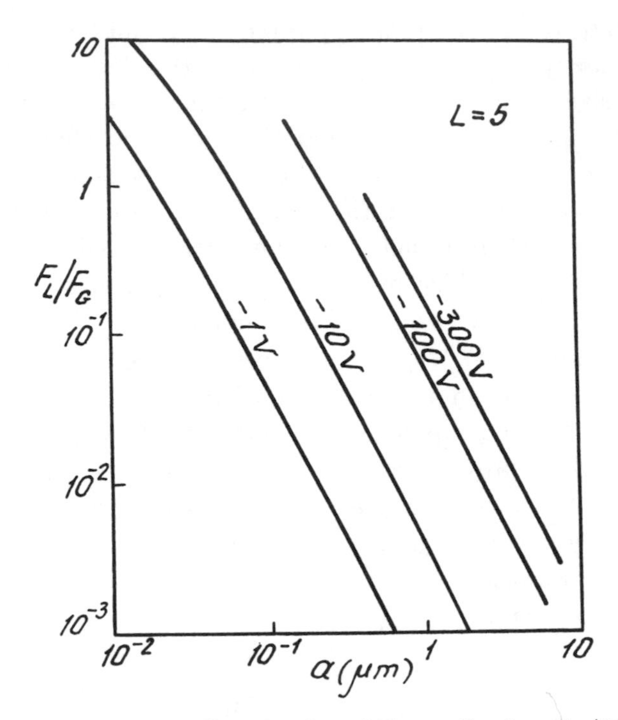


Figure 5: The Electric-to-Gravitational Force Ratio, $F_L/F_G$ as a Function of the Grain Size and Potential $\psi$ within Saturn's Magnerosphere at $L = 5$ (the Broad E-Ring) (after Mendis *et al.*, 1982).

## 2.3 Orbital Motion of Charged Grains

In this paragraph, we will consider the motion of the mass center of dust grains, without touching on the question of their orientation. This formulation of the problem is typical for gravitoelectrodynamics where the trajectories of "point-size" charged grains in planetary magnetospheres were first calculated by Mendis and Axford (1974) more than twenty years ago. Since that time, a great number of publications have appeared, especially after the *Pioneer* and *Voyager* space missions that collected ample material on the distribution of dust grains near Jupiter and Saturn.

Consider the equation of motion of a charged grain subjected to the above mentioned forces,

$$\frac{d^2\mathbf{r}}{dt^2} = -\frac{\mu\mathbf{r}}{mr^3} + \frac{Q(\mathbf{r})}{mc}\left[\frac{d\mathbf{r}}{dt} \times \mathbf{B}(\mathbf{r}) - [\Omega_p \times \mathbf{r}] \times \mathbf{B}(\mathbf{r})\right] + \mathbf{F}(\mathbf{r}), \tag{29}$$

where $\mathbf{F}(\mathbf{r})$ is the sum of all the forces except gravitation and the Lorentz

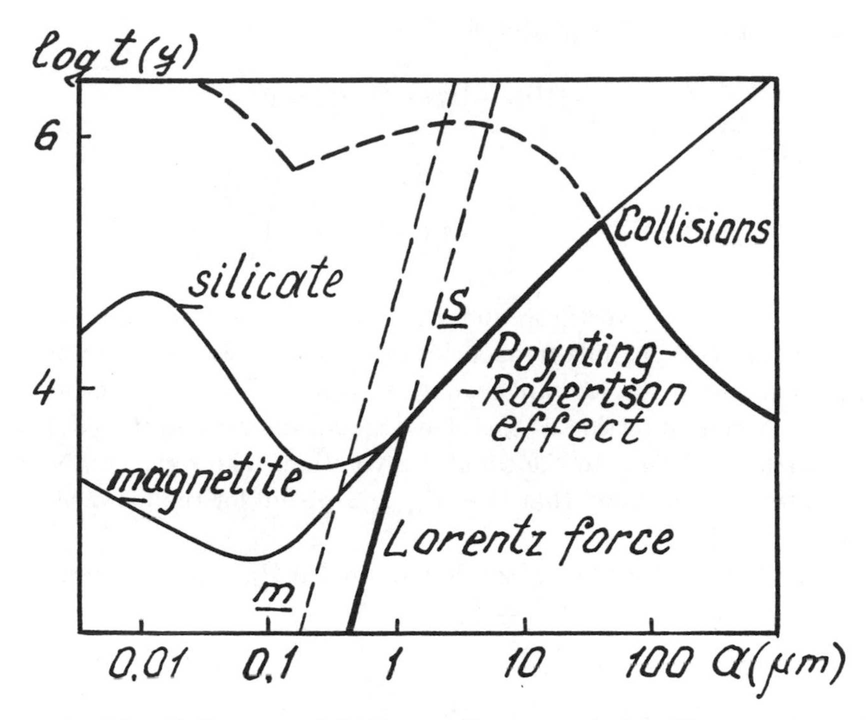


Figure 6: The Influence of Different Factors on the Characteristic Time Scale (in Units of Years) of Grains of Different Size (after Mukai, 1986).

force. Note that the grain charge generally depends on the grain position, and hence on time, as the particle is in motion. The physics of this dependence and the associated effects will be discussed below. Analyzing Equation (29) in its complete form is too complicated a task to be even attempted here. As has been shown in this Section, the principal forces acting on charged grains in planetary magnetospheres are $F_G$ and $F_L$, which alone will be taken into account. Even with $\mathbf{F} = 0$, the equation remains quite complex; therefore we begin the analysis with the simplest case of circular orbits within the equatorial plane. This simplified problem is directly related to the dynamics of grains in planetary rings.

### 2.3.1 Equatorial Plane: Circular Orbits

The orbital frequency $\Omega$ for a trajectory of radius $R$ can be found from the balance condition involving all forces, including the centrifugal force $m\Omega^2 R$.

The result is a quadratic equation for $\Omega$,

$$\Omega^2 - \Omega\Omega_B - \Omega_K^2 + \Omega_B\Omega_p = 0, \tag{30}$$

which yields

$$\Omega = \frac{\Omega_B}{2}\left\{1 \pm \left[1 + 4\left(\frac{\Omega_K^2}{\Omega_B^2} - \frac{\Omega_p}{\Omega_B}\right)\right]^{1/2}\right\} \tag{31}$$

(Mendis *et al.*, 1982; Northrop and Hill, 1983). The two signs in Equation (31) imply two possible frequencies of revolution along the same orbit. One corresponds to the forward revolution (the "+" sign, the orbital motion is in the same direction as the planet's rotation around its axis), while the other (the "−" sign) to the counter revolution. The signs can be easily identified from the demand that $\Omega \to \Omega_K$ at $\Omega_B \to 0$ for the forward orbital motion.

The $\Omega = \Omega(R)$ dependence can be analyzed if $\Omega_K$ and $\Omega_B$ are replaced by their respective values for $R = R_p$,

$$\Omega = \frac{\Omega_{B0}}{2L^3}\left\{1 \pm \left[1 + 4L^3\left(\frac{\Omega_{K0}^2}{\Omega_{B0}^2} - \frac{\Omega_p}{\Omega_{B0}}\right)\right]^{1/2}\right\} \tag{32}$$

$\Omega_{B0} = \Omega_B(R_p)$ and $\Omega_{K0} = \Omega_K(R_p)$ $(L = 1)$.

We shall be mostly interested in small deviations of $\Omega$ from $\Omega_K$, therefore Equation (31) can be re-written as

$$\Omega = \Omega_K\left\{\frac{\Omega_B}{2\Omega_K} \pm \left[1 + \frac{\Omega_B^2}{4\Omega_K^2} - \frac{\Omega_p\Omega_B}{\Omega_K^2}\right]^{1/2}\right\}.$$

or, with the "+" sign in front of the square bracket and small $\Omega_B/\Omega_K$

$$\Omega \simeq \Omega_K + \frac{\Omega_B}{2\Omega_K}(\Omega_K - \Omega_p). \tag{33}$$

As can be seen, the principal small parameter of gravitoelectrodynamics is the ratio $|\Omega_B|/\Omega_K = (|\Omega_{B0}|/\Omega_{K0})L^{-3/2}$. The influence of electromagnetic forces drops off at greater distances from the planet. The micron-size particles ($a \simeq 1\,\mu$) of mass density $\rho \simeq 2\,\mathrm{g/cm^3}$ near the surface of Saturn are characterized by $|\Omega_{B0}|/\Omega_{K0} \sim 5 \times 10^{-3}$, while smaller particles ($a \sim 0.1\,\mu$) have $|\Omega_{B0}|/\Omega_{K0} \sim 5 \times 10^{-1}$. Similar estimates for the vicinity of Jupiter

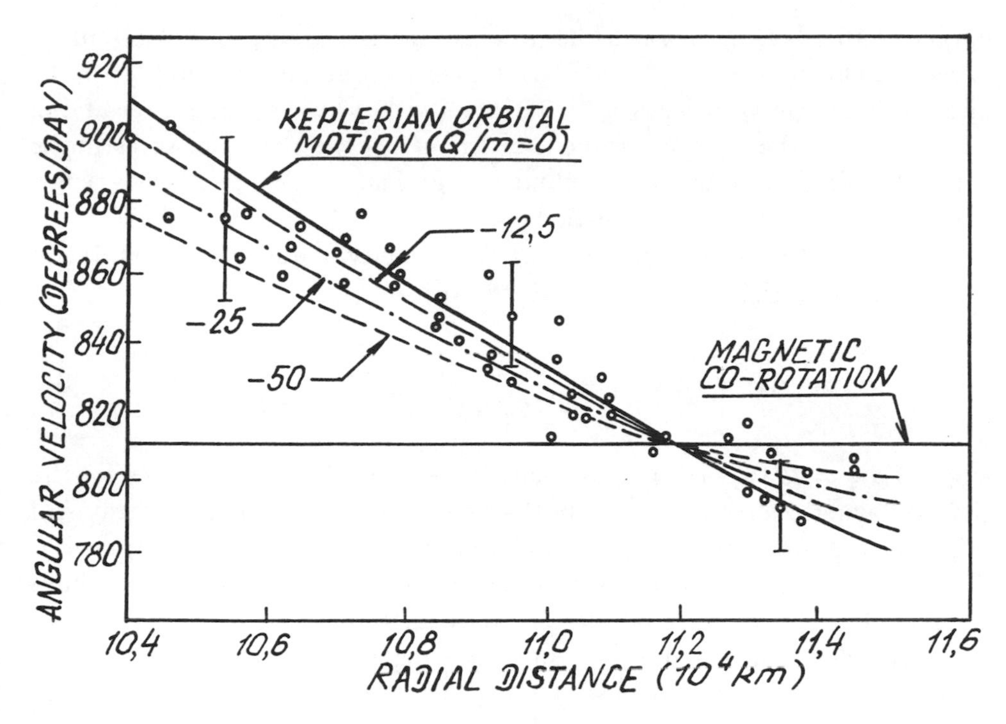


Figure 7: The Angular Velocity of Spoke Features in Saturn's B-Ring. The Kepler Velocity and the Planet Rotation Rate Are Indicated. The Dashed Lines Show the Angular Velocities of the Particles with the Charge-to-Mass Ratio (Coulomb/kg) as Indicated (after Thomsen *et al.*, 1982).

increase by more than one order of magnitude. Thus, the approximate formula of Equation (33) can be applied to micron and submicron grains near Saturn and to micron grains only near Jupiter.

Knowledge of the particle revolution frequency $\Omega$ allows determining, with the aid of Equation (33), $\Omega_B$ and the parameter as important as the charge-to-mass ratio, $Q/m$. This has actually become possible as a result of the *Voyager-1* space mission. Shown in Figure 7 are measured values of $\delta\Omega(R) = \Omega(R) - \Omega_K$ for the region of "spokes" in Saturn's rings (the Figure is after Thomsen *et al.* (1982)). Inside the synchronous orbit ($\Omega_K > \Omega_p$), negative values of $\delta\Omega(R)$ are nearly twice as numerous as positive. Whereas outside the synchronous orbit ($\Omega_K < \Omega_p$) positive deviations, $\delta\Omega > 0$, are more frequent, which implies $\Omega_B < 0$, i.e. the charge (and potential) of the dust grains is negative.

The average charge-to-mass ratio following from the data of Figure 7 is

$|Q|/m \simeq 10\,\mathrm{C/kg}$, from which the potential at the grain, $|\psi| \simeq (|Q|/m) \times 4\pi\rho a^2/3$, can be estimated. The grain size (radius) corresponding to the minimum charge $Q = e$, is $a = 0.016\,\mu$ (with $\rho \simeq 1\,\mathrm{g/cm^3}$). Assuming dust grains in the "spokes" to be $a \simeq 1\,\mu$ in size we arrive at an estimate $|\psi| \geq 10^2\,\mathrm{V}$. The potential of submicron grains, $a \sim 0.1\,\mu$, reaches a few volts, which seems a more realistic value.

Equation (33) permits determining the velocity of dust grains relative to the plasma in a rigid corotation with the planet,

$$V_G = V_{G,K}\left(1 + \frac{\Omega_B}{2\Omega_K}\right) = V_{G,K}\left(1 + \frac{a_0^2}{a^2}\right), \tag{34}$$

where $V_{G,K} = R(\Omega_K - \Omega_p)$ is the relative velocity without allowance for the electromagnetic force and $a_0^2 = 3\psi B/(8\pi\rho c\Omega_K)$. As can be seen, grains of different size are characterized by different velocities along the same orbit. While the velocity difference is small (provided the assumptions that have been made are true), it may be of importance for grain coagulation and collective effects in the dusty plasma. Owing to the fact that $V_G$ is size dependent, dust grains moving along the same orbit can collide, giving rise to larger grains. The electrostatic repulsion of likely charged particles cannot prevent the grains from coagulation if their relative velocity is sufficient for overcoming the Coulombian force. For neutral particles, collisions at the same orbit are out of question, all of them having the same Kepler velocity. It should be noted that, along with charged particles, the induction field can affect the motion of neutrals as well. A neutral grain placed in the electric field $\mathbf{E}$ gets polarized, acquiring the dipole moment $\mathbf{p} = a^3(\varepsilon - 1)/(\varepsilon + 2)\mathbf{E}$. The polarized particle is acted upon by the force $\mathbf{F}_p = (\mathbf{p}\nabla)\mathbf{E}$ whose orientation within the equatorial plane is radial and the magnitude is $F_p = (a^3/2)(\varepsilon - 1)/(\varepsilon + 2)\partial(E^2)/\partial r$. Since $E^2$ increases in both directions from the synchronous orbit $R_{co}$, the force $\mathbf{F}_p$ goes against the gravity force at $R > R_{co}$ and along it at $R < R_{co}$. Accordingly, the rotation frequency of the neutral particle somewhat decreases at $R > R_{co}$ and increases at $R < R_{co}$. The effect is *a priori* very weak. It should have been allowed for if neutral grains had really existed. Actually, as soon as the grain acquires an electric charge (which is always the case), the Lorentz force $F_L$, greatly exceeding $F_p$, comes into play. A complete analysis of the variations in the rotation frequency of a neutral grain subjected to the polarization force $\mathbf{F}_p$ can be found in Bliokh and Yaroshenko (1983) and Yaroshenko (1985), while we restrict the presentation here to order of magnitude estimates of $F_p/F_L$.

The scale length of variations in $E$ being of the order of the planet radius, $F_p$ can be estimated as $F_p \sim a^3 \partial(E^2)/\partial r \sim a^3 E^2/R_p$, whence

$$\frac{F_p}{F_L} \sim \frac{a^3 E^2}{R_p Q E} = \frac{\delta\psi}{|\psi|} \cdot \frac{a}{R_p},$$

where $\delta\psi = aE$ is the potential difference of the unipolar induction field across the grain. At this point we can make use of the above given estimates of $E$ for different planets. Taking $E \sim E_0 \sim 10^{-1}$ V/cm, one obtains for a micron-size grain $\delta\psi \sim 10^{-5}$ V. With $|\psi| \sim 10$ V and $R_p \simeq 10^9$ cm, this gives $F_p/F_L \sim 10^{-19}$, i.e. the effect is negligible. The ratio $F_p/F_L$ increases with the grain radius; however, the principal dimensionless parameter is $a/R_p$, which is small as long as $a \ll R_p$.

### 2.3.2 Equatorial Plane: Non-Circular Orbits

The circular orbits predicted by Newtonian mechanics for neutral particles are a particular case of elliptical trajectories. It seems natural to expect elliptical orbits in gravitoelectrodynamics with $F_L/F_G \ll 1$, at least orbits with a small eccentricity $e$. As we have seen, circular orbits ($e = 0$) do exist with arbitrary $F_L/F_G$ ratios.

Note however that a charged particle moving in a planetary magnetosphere is subjected, along with the radial force $\mathbf{F}_G(r)$, to tangential forces acting along the trajectory. These arise from the plasma and radiation drag and the Lorentz force when variations in $Q(\mathbf{r})$ are allowed for. Owing to the tangential force $\mathbf{F_T}$, the elliptical or nearly circular trajectory transforms into a spiral. If the deformation is sufficiently slow, it can be described in terms of variations of the semi-axes, $A$ and $B$, and eccentricity of the elliptical trajectory. The moment of the particle momentum changes according to $d(mRV_G)/dt = RF_T$. With slow variations of the radius the orbit is locally close to circular, and hence $V_G^2 \simeq GM_p/R$, which yields (Northrop *et al.*, 1989)

$$\frac{dR}{dt} = \pm \frac{2R^{3/2}F_T}{m(GM_p)^{1/2}}.$$

The "+" sign corresponds to a tangential force directed along $\mathbf{V}_G$, while the "−" to an oppositely directed force. Different kinds of drag can produce differently oriented $\mathbf{F}_t$s. The effect of radiation drag is the simplest (see Subsection 2.2.2) as it always tends to slow the particle down and reduce its orbit radius.

The force produced by plasma drag (Subsection 2.2.3) may have different orientation, depending on the orbit radius. At $R > R_{co}$, the rigidly corotating plasma moves faster than the dust grain with its practically Keplerian velocity, since $\Omega_K < \Omega_p$ outside the synchronous orbit. Contrary to this, $\Omega_K > \Omega_p$ for $R < R_{co}$. In this region, the effect of plasma drag is to decelerate the dust grain and reduce the size of its orbit. In either case, the radial plasma-induced drift carries the particle away from the synchronous orbit. The situation is different with the gyrophase drift. It can be shown to occur in any direction, in particular toward the synchronous orbit.

This gave Northrop, Mendis and Schaffer (1989) grounds to suggest an explanation to the sizable increase in the density of Jupiter's "gossamer" ring in the vicinity of the synchronous orbit ($L \simeq 2.24$). The source of the submicron dust grains of which the "gossamer" ring consists admittedly is the satellite Amalthea, closest to the synchronous orbit ($L_A \simeq 2.54$). However, Amalthea lies beyond the synchronous orbit, and hence plasma drag cannot be responsible for particle migration toward the region of their enhanced number density. The writers have given a detailed analysis to the three types of drag capable of producing the drift of dust grains of different size, with different ambient plasma parameters. Their result has been that the radiation drag is of negligible importance, while the other two mechanisms generally should be taken into account simultaneously. In particular, the gyrophase drift may prove more important than the effect of plasma drag, and capable of transporting dust grains from Amalthea toward the synchronous orbit where the "gossamer" ring reaches its highest density.

The evolution of an elliptic orbit is determined not by the ambient plasma parameters alone but initial values of the semi-minor axis $A_0$ and eccentricity $e_0$ of the ellipse. These initial parameters are controlled by the conditions of dust grain injection into the magnetosphere. Havnes, Morfill and Melandsø (1992) analyzed the effects for Saturn's diffuse E-ring spanning from $L \simeq 3$ to $L \simeq 8$. The source of dust grains admittedly was the satellite Enceladus at $L \simeq 4$. The orbits of sufficiently large particles (like ice grains of $a \sim 1\,\mu$ and $\rho \sim 1\,\mathrm{g/cm^3}$) were gravity controlled, while the electromagnetic forces represented perturbations. The equations by which the deforming elliptic orbit was described were integrated numerically and averaged over a few revolutions of the particle. The resultant average values $\langle \dot{A}/A_0 \rangle$ and $\langle \dot{e} \rangle$ were analyzed as functions of $e_0$ for a variety of plasma parameters and $L$ values of 4; 6 and 8. The plasma models discussed included radial gradients of density and temperature, with allowance for suprathermal electrons that can essentially affect the grain charge through secondary emission. For sufficiently small initial eccentricities ($e_0 \leq 0.3$) $\langle \dot{e} \rangle$ can be shown to be negative, i.e. the evolution of such orbits ends in circularization. The calculated values of $\langle \dot{A}/A_0 \rangle$ allow estimating the lifetimes of the grains orbiting in the E-ring.

The averaging over a few revolutions is equivalent to separating variations of $A$ and $e$. Along with the secular variations, the orbit elements experience periodic oscillations. The oscillation amplitudes are very small in the case under discussion (i.e. comparatively large particles and great separations from the gravitation center).

The importance of the Lorentz force increases for dust grains of smaller size or those located closer to the planet. The result may be cycloidal trajectories. The perturbed motion can be described rather conveniently as elliptic gyration around a leading center whose orbit is circular and the frequency of revolution is given by Equation (31). For the case of small perturbations, the elliptic gyration was discussed in detail by Mendis *et al.* (1982). Their result for the gyration frequency $\Omega_r$ (which is also the variation frequency for the grain's radial distance) is

$$\Omega_r^2 = \Omega_B^2 - 4\Omega_B \Omega + \Omega^2. \tag{35}$$

The semi-axes $A$ and $B$ of the gyration ellipse are related as

$$\frac{A}{B} = -\frac{\Omega_r}{2\Omega - \Omega_B}, \tag{36}$$

with the semi-minor axis oriented radially. Equations (30), (35) and (36) specify the frequencies $\Omega$ and $\Omega_r$ and the ratio $A/B$ as functions of the known magnitudes $\Omega_K$, $\Omega_p$ and $\Omega_B$, thus providing an opportunity to follow the transition from Newtonian (neutral-particle) mechanics to electromagnetically controlled motion. Equation (30) can be re-written as

$$\Omega_B = \frac{\Omega^2 - \Omega_K^2}{\Omega - \Omega_p}. \tag{37}$$

In the case of neutral particles or "very large" grains characterized by minor variations of the specific charge $Q/m$, the equations hold $\Omega_B \simeq 0$ and $\Omega = \Omega_K$, i.e. the angular velocity of the leading center is equal to the Kepler frequency. While $\Omega_r = \pm\Omega$ (according to Equation (35)), the $A/B$ ratio should be positive by the definition of $A$ and $B$. Hence, the correct result is $\Omega_r = -\Omega$, and $A/B = 1/2$. Since $\Omega_r$ is negative, this value of $A/B$ implies a retrograde epicyclic rotation known from the theory of motion of neutral particles in a central gravitation field.

The other limiting case of very large $|Q|/m$ ratios, when $|\Omega_B| \gg \Omega_p, \Omega_K$, gives $\Omega \simeq \Omega_p$ or $\Omega \simeq \Omega_B$. The latter solution corresponds to $\Omega_r^2 < 0$, or imaginary $\Omega_r$'s and hence should be discarded as unstable. The remaining equation $\Omega \simeq \Omega_p$ implies corotation with the planet, with $\Omega_r = \Omega_B$ and $A/B = 1$ (the root $\Omega_r = -\Omega_B$ is meaningless because of $A/B < 0$). Thus, the "electromagnetic limit" corresponds to circular motion of the leading center at the frequency $\Omega_p$ (corotation with the planet). The gyration frequency is equal to the Larmor $\Omega_B$, and the ellipse of gyration has degenerated into a circle. Since $\Omega_B = QB/(mc)$, the negatively charged particles perform retrograde rotation around the leading center ($\Omega_r < 0$), while the positively charged grains move in the prograde mode ($\Omega_r > 0$). These results are well known for electrons and ions in a magnetic field. In the general case of arbitrary $Q/m$ ratios, the values of $A/B$ lie between 1/2 and 1, i.e. $1 \geq A/B \geq 1/2$.

If all the parameters, except grain radius, were fixed, it might be possible to watch gravitoelectrodynamics change from electrodynamics over to Newtonian mechanics as the grains increased in size. This transition is illustrated in Figure 8 taken from Mendis *et al.* (1982). The calculations were for Saturn's magnetosphere in the vicinity of F-ring, dust grains of mass density $\rho \simeq 1\,\mathrm{g/cm^3}$ at a potential of $\psi = -38\,\mathrm{V}$. As the radius $a$ was changed from $0.01\,\mu$ to $0.9\,\mu$, the $A/B$ ratio decreased rapidly from 0.998 to about 0.5. Shown in the same Figure are variations of the other parameters demonstrating transitions from gravitoelectrodynamics to specific limiting cases. We have already mentioned that $\Omega_r^2 > 0$ is in fact the condition for stability of the grain's orbit. This important problem has been given a detailed analysis by Mendis *et al.* (1982). They analyzed Equations (30), (35) and (36) to identify stable and unstable orbits at various distances from the center of attraction, considering for definiteness Saturn's magnetosphere over the range from the planet surface to the corotation boundary ($L \simeq 1.86$). In the $(\Omega, \alpha)$ plane (where $\alpha = \Omega_K/\Omega_p$ is a measure of separation from the axis of rotation), domains of orbit stability ($\Omega_r^2 > 0$) and instability ($\Omega_r^2 < 0$) were determined for dust grains of different charge signs, performing direct or retrograde orbital motion.

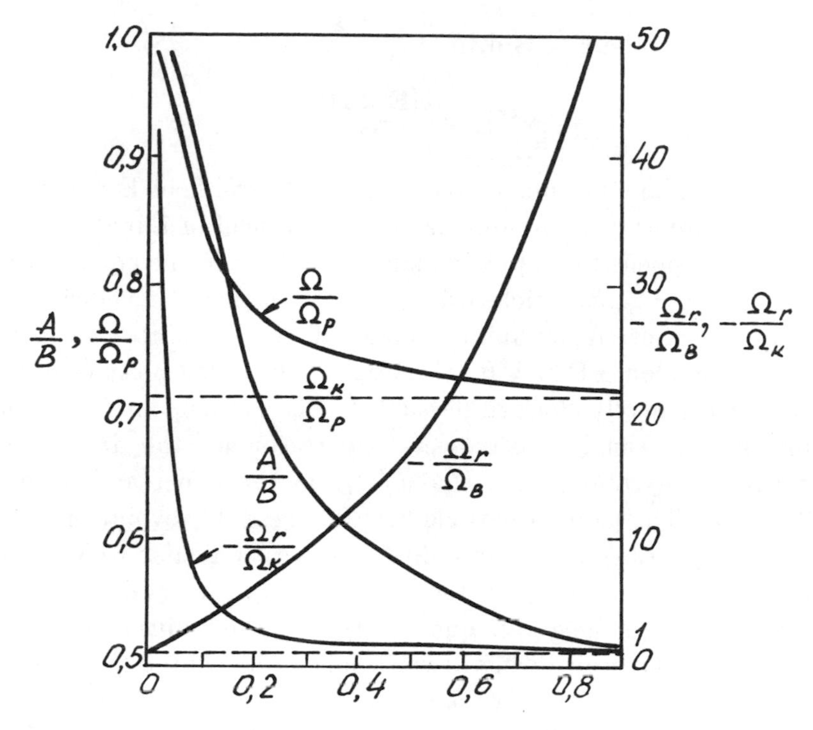


Figure 8: The Variations of $\Omega/\Omega_p$, $-\Omega_r/\Omega_B$, $-\Omega_r/\Omega_K$ and $A/B$ as Functions of the Grain Radius $a$ within Saturn's F-Ring ($\rho = 0.9\,\mathrm{g/cm^3}$, $\psi = -38\,\mathrm{V}$) (after Mendis *et al.*, 1982).

### 2.3.3 Gyrophase Drift

As has been shown, the motion of dust grains of large charge-to-mass ratio becomes magnetic-field-dominated, similar to electrons and ions. Yet the dust particles retain a distinction responsible for the above-mentioned gyrophase drift.

The concept of drift motion is widely used to describe the kinematics of electrons and ions. It proves particularly convenient when the particle in a magnetic field is subject to the action of other force fields, or when the magnetic field itself is nonuniform. If the additional forces vary but slightly over one revolution of the particle around the magnetic line of force, then the trajectory can be averaged over the rotation of frequency $\Omega_B$. The resultant velocity of the leading center is what is known as the drift velocity $\mathbf{V_D}$. The familiar example is the $\mathbf{E} \times \mathbf{B}$ drift,

$$\mathbf{V_D} = \frac{c[\mathbf{E} \times \mathbf{B}]}{B^2}. \tag{38}$$

In the case of a planetary magnetosphere, an electric field $\mathbf{E}$ perpendicular to $\mathbf{B}$ is also present; it is the unipolar induction field of Equation (6). The unipolar drift happens to be purely azimuthal. Within the equatorial plane its velocity is $V_D = \Omega_p R$, which corresponds to the angular velocity $\Omega = \Omega_p$ of the leading center. A nonuniform magnetic field produces an azimuthal drift in the direction $\pm\mathbf{B} \times \nabla B$ (the upper sign is for positively and the lower one for negatively charged particles). The gradient drift is explained in Figure 9a. As can be easily seen, the reason for the drift motion is the difference of gyration radii, $l = w/\Omega_B$, in the upper and lower semicircles ($l_1 > l_2$). Therefore, a particle performing a full revolution over time $T = 2\pi/\Omega_B$ becomes displaced in a direction perpendicular to $\nabla B$.

In the case of electrons or ions the gyroradius $\rho_B$ can change solely due to changes of $\mathbf{B}$, whereas the gyrofrequency $\Omega_B$ (and the radius $l$) of dust grains also varies along the orbit owing to variations of its charge $Q(\mathbf{r})$. The drift velocity in this general case takes the form $\mathbf{V}_D \sim \pm\mathbf{B} \times \nabla\Omega_r$. The frequency $\Omega_B$ has been replaced by the gyration frequency $\Omega_r$ of Equation (35) which coincides with $\Omega_B$ for large values of $Q/m$.

Along with the radial dependence of $B(r)$, the magnetospheric plasma may be characterized by radial gradients of $T$ and $n_0$, and possibly a gradient of composition, i.e. alteration with altitude of the dominant sort of ions. All of these variations can influence the grain charge $Q(\mathbf{r})$, so that the direction of $\nabla\Omega_B = \nabla[Q(r)B(r)]/(mc)$ may either coincide with or be opposite to $\nabla B$. When determining the azimuthal drift direction, one should be aware that the variations of $Q(\mathbf{r})$ are not instantaneous but are rather characterized by a certain delay $t_0$ (cf. Subsection 1.1). Therefore, the extremum values of $Q$ are shifted from the apices of the gyrocycle semicircles by $\theta = \Omega_r t_0$ and the vector $\nabla\Omega_B$ is deviated from the radial by the same angle. The drift velocity $\mathbf{V}_D$ acquires a radial component $V_D \sin\theta$ (Figure 9b). Calculating the trajectory of a grain with a variable charge (drift motion in particular) obviously is much more complicated than analyzing the motion of electrons or ions.

Northrop and Hill (1983) developed a generalized adiabatic theory allowing for arbitrary relations between the drift velocity and rotation around the leading center. The conventional adiabatic theory for the motion of electrons and ions through a nonuniform magnetic field $\mathbf{B}(\mathbf{r})$ requires that the

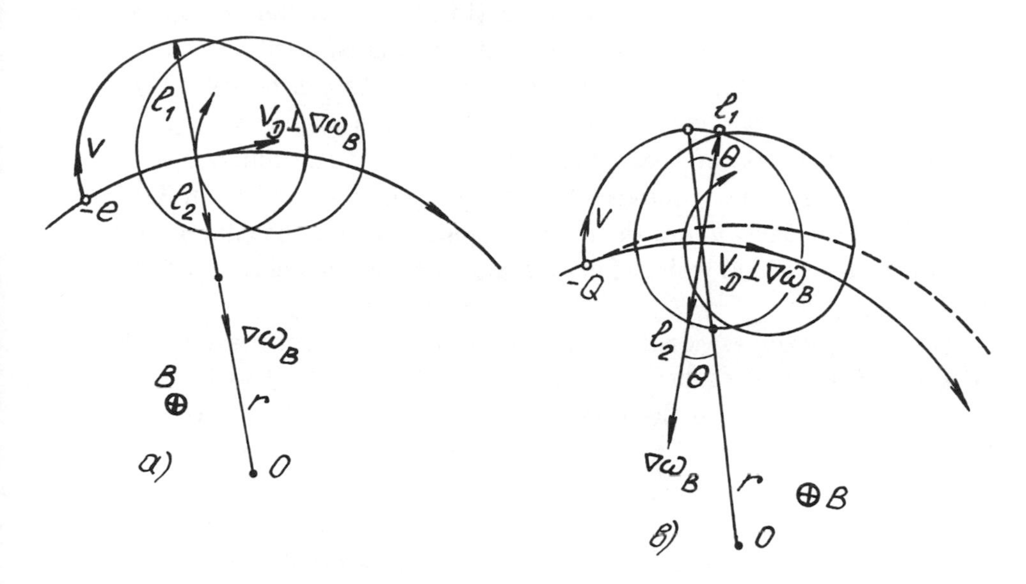

Figure 9: The Gradient Drift of an Electron ($a$) and Gyrophase Drift of a Negatively Charged Grain ($b$). Owing to the Delay by an Angle $\theta$ of the Grain Recharging, the Drift Acquires a Radial Component and the Grain Moves along a Spiral Trajectory.

drift velocity be much lower than the gyrovelocity. The Northrop – Hill theory allows for variability of the grain charge $Q(t)$. While we have discussed above variations of $Q = Q(\mathbf{r})$ in space, the motion about the leading center may be considered known, as the variations of $Q$ are small, and hence $Q(\mathbf{r})$ may be replaced by $Q(t)$. The latter is a periodic function representable as a Fourier series. To analyze the gyrophase drift, it is sufficient to retain the fundamental harmonic only, i.e. $Q(t) \simeq Q_0 + Q_1 \cos(\Omega_r t - \theta)$. Numerical estimates of Hill and Mendis (1980) show the amplitude of the second harmonic to be very small ($Q_2 \ll Q_1$) under realistic conditions.

The variability of $Q(t)$ violates the invariance of the magnetic moment owing to rotation about the leading center. Yet exact invariants of motion still exist with $Q = Q(t)$ (Northrop and Hill, 1983).

A detailed analysis of the gyrophase drifts of different origin in Jupiter's magnetosphere was given by Northrop *et al.* (1989). The drift produced by a radial gradient of plasma temperature is directed so as to move the particle to the area of higher $T$.

The composition gradient gives rise to a modulation of $Q(t)$, since the particle potential (and charge, see Subsection 1.1.1) can depend on the magnitude of the $m_i/m_e$ ratio. E.g., the dust grain is charged more negatively in a plasma with $O^+$ ions than in one with $H^+$. As can be seen, the radial displacement of particles owing to the composition gradient occurs toward greater concentration of heavy ions. Variations of the number density of plasma particles cannot produce changes in $Q(t)$; however they influence directly the phase delay $\theta$, i.e. the higher is $n_0$, the faster the charge-discharge cycle of the dust grain, which effect tends to reduce the delay $\theta$ and diminish the drift.

Note the gyrophase drift to also occur in a uniform plasma, when all the above described gradients are absent. It is caused by variations of the particle velocity relative the plasma during cycloidal motion of frequency $\Omega_r$. The relative velocity $w$ can affect the particle potential, as has been discussed in Subsection 1.1.1. Consider the question in more detail, always remembering that the velocities $w$ obey $w \ll v_{T,e}$. That is why the electronic current flowing to the particle from the plasma is practically independent of $w$. As for the ion current, it may change because of two reasons. If the relative velocity of the ions incident on the front hemisphere increases, the current to that hemisphere increases too, while decreasing in the back hemisphere. With sufficiently small values of $w/v_{T,i}$ the latter effect prevails. In this case $|\psi|$ should increase, compared with the particle at rest, to such a value as to restore the balance of currents $J_i = J_e$. At still greater values of $w$ the current to the back hemisphere nearly vanishes (at $w > v_{T,i}$) which results in a four-fold decrease of the current to the grain. Indeed, the effective area of ion interception reduces from $4\pi a^2$ to $\pi a^2$. However, this effect can be compensated by a simultaneous increase of the current to the front hemisphere if $\pi a^2 w > 4\pi a^2 v_{T,i}$, or $w/v_{T,i} \geq 4$. Further growth of the $w/v_{T,i}$ ratio leads to an increase of the ion current and reduction of $|\psi|$. The analysis of Northrop *et al.* (1989) actually shows $|\psi|$ to grow for $0 < w/v_{T,i} \leq 2$, dropping down at $w/v_{T,i} > 2$. The $|\psi|$ value at $w/v_{T,i} \sim 4$ is the same as for the fixed grain. A calculated curve $|\psi| = f(w/v_{T,i})$ taken from Northrop *et al.* (1989) (also *cf.* Mendis and Rosenberg, 1994) is shown in Figure 10.

The drift direction depends on whether the dust grain moves on the left or on the right side of the maximum of $|\psi| = f(w/v_{T,i})$. On the left slope, the grain is shifted radially away from the synchronous orbit, while

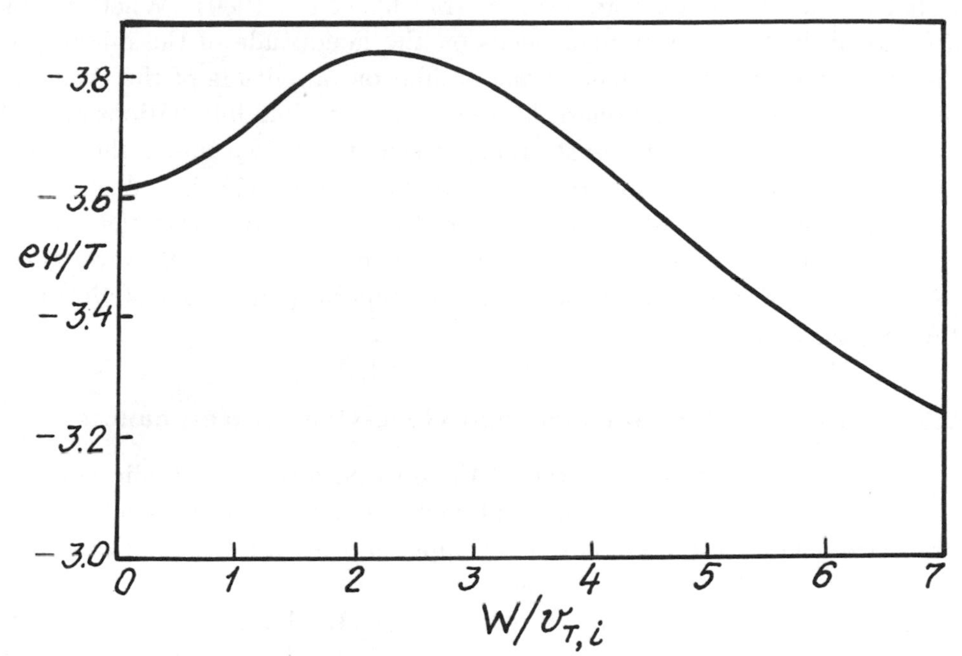


Figure 10: Normalized Grain Potential as a Function of the Ratio of the Grain Velocity to the Ion Thermal Velocity (after Northrop *et al.*, 1989).

on the right slope it moves in the opposite direction. In case $w(r)$ varies near the maximum of the curve in Figure 10, the Fourier spectrum of $Q(t)$ contains an intense second harmonic. This situation has not been studied analytically; however numerical calculations do not show any effects of the second harmonic on the gyrophase drift (Northrop *et al.*, 1989). Northrop and Hill (1983) analyzed the orbit evolution in the equatorial plane, with account of the charge variations $Q(t)$ owing to the radial dependence of the relative velocity $w(r)$, for the case where the radial drift was toward the synchronous orbit. While the trajectory approached to the synchronous orbit, the cycloidal oscillations reduced in amplitude, such that the orbit could become circular well before reaching the synchronous radius.

Along with the regular factors producing radial displacements of the orbit, the process is subject to random influences. In fact, the above calculated magnitudes of the dust grain potential are some average values, against which background the electric parameters fluctuate. The smaller is the grain size, the lower the number of electrons or ions it can intercept, hence the higher its relative charge fluctuation level $\delta Q$. This is how the diffusional radial motion of the dust grain arises (Morfill *et al.*, 1980). Whether this diffusional drift is important depends on the magnitude of the radial gradients in the plasma parameters and oscillation amplitude of the cycloidal motion. As follows from numerical analysis, the random fluctuations of $Q(t)$ may become dominant for dust grains of size $a \sim 0.25\,\mu$ under the conditions of a weak temperature gradient (Northrop *et al.*, 1989). If dust grains are emanated as a cloud, all at the same Kepler velocity, from the surface of a large satellite, they start drifting as a compact group. However their radial distribution is soon smeared out by random fluctuations of $Q(t)$ and the resultant diffusion.

### 2.3.4 Magneto-Gravitational and Gyro-Orbital Resonances

The evolution of orbits of the celestial bodies performing periodic motions is greatly influenced by resonance interactions between the bodies. Gravitoelectrodynamics offers many more opportunities for the appearance of resonances than "pure" gravitation can.

Indeed, while in mechanics there is just one (Keplerian) frequency corresponding to each circular orbit, an epicyclic equatorial orbit in gravitoelectrodynamics is characterized by the frequencies $\Omega$ and $\Omega_r$. Besides, the two frequencies depend, via $\Omega_B$, on the charge-to-mass ratio of the grains $Q/m$, which dependence is particularly important.

For a medium of dust grains with a certain spectrum of masses (sizes) one can indicate the corresponding spectrum of frequencies $\Omega(a)$ and $\Omega_r(a)$. The probability for some of these frequencies to coincide with the perturbation frequency is much higher than for a small number of discrete frequencies.

A characteristic example is the so-called magneto-gravitational resonance first predicted by Mendis *et al.* (1982). The orbital frequency $\Omega$ of a dust grain lies, according to Equation (31), between $\Omega_p$ and $\Omega_K$, where the inequality outside the synchronous orbit is $\Omega_K < \Omega < \Omega_p$ and at closer ranges to the central body $\Omega_p < \Omega < \Omega_K$. The motion of dust grains can be perturbed by the gravitation field of a massive satellite, provided its orbit lies fairly close to the grain's. The pertubation would be particularly strong for coincident frequencies $\Omega_s = \Omega$, where $\Omega_s$ is the satellite's orbital frequency. This is the orbital 1:1 resonance, impossible for Keplerian motion where different orbits are characterized by different revolution frequencies $\Omega_K$. The writers noticed the magnetogravitational resonance condition to be satisfied by the particles of Saturn's F-ring and the internal "shepherding" satellite

S-27 whose orbit is only 1000 km away from the F-ring. The ring lies beyond the synchronous orbit, therefore the condition is met $\Omega > \Omega_K$, and the 1:1 resonance is only possible with the S-27 but not the outer shepherding satellite, S-26.

The resonance conditions are met by grains of a definite size that can be found from Equation (37) by setting $\Omega = \Omega_s$ in the right-hand side, *viz.*

$$\Omega_{B,res} = \frac{\Omega_s^2 - \Omega_K^2}{\Omega_s - \Omega_p}. \tag{39}$$

Now, if $\Omega_B$ is expressed in terms of the grain parameters, $\Omega_B \simeq 4.4 \times 10^{-8}\psi/\rho a^2$, and $\Omega_s$ is taken as $\Omega_s \simeq 1.83 \times 10^{-4}\,\mathrm{s}^{-1}$ (S-27), then the size for $\rho \sim 1\,\mathrm{g/cm}^3$ and $\psi = -38\,\mathrm{V}$ (i.e. the data of Mendis *et al.*, 1982) is $a_{res} \simeq 0.5\,\mu$. Obviously enough, the grains of size not greatly different from $a_{res}$ would be also influenced by the satellite, staying near it for quite a long time. With $\Omega \neq \Omega_s$, the grain performs an extra full revolution, compared with the satellite, over the time $\tau_s = 2\pi/|\Omega - \Omega_s|$. The epicyclic gyration period is $\tau_g = 2\pi/\Omega_r$; hence the number of gyrations during the time of one full revolution with respect to the satellite would be $\tau_s/\tau_g = \Omega_r/|\Omega - \Omega_s|$. The arc length passed by the particle over that time is $2\pi R_F$ (in the frame of reference associated with the satellite; $R_F$ is the F-ring radius). The radial oscillation wavelength is $\lambda = 2\pi R_F|\Omega - \Omega_s|/\Omega_r$. Possibly, this is an explanation for the wavelike structure detected in the F-ring, although it is not free from contradictions (*cf.* the review of Goertz (1989)). First, the charge of dust grains in $F$-ring seems much lower than should be supposed for fitting the calculations to experiment. The reason for the reduced charge magnitude might be the too great concentration of dust, $n_d^{1/3}\lambda_D \geq 10^{-3}$, preventing the grains from performing as isolated (*cf.* Section 1.2). Besides, the wavelike structure would be smeared if the spectrum of grain sizes in the $F$-ring were sufficiently wide.

Another reason for the appearance of resonances might be the periodical changes of grain charge when the particle orbit crosses the planet's shadow. Transitions from the illuminated part of the orbit into the shadow and vice versa modulate the photocurrent and the grain charge $Q(t)$ at a frequency equal to the orbital. The modulation frequency coincides with the orbital frequency automatically, therefore noticeable variations of the orbit radius and eccentricity can be produced even by weak electromagnetic forces (Mendis, 1984; Hill and Mendis, 1982). The effect would be particularly bright for small-size (submicron) particles if the conditions for the so-called gyro-orbital resonance were met, i.e. the ratio of $\Omega$ and $\Omega_r$ were integer. Modulation of $Q(t)$ can also result in a parametric instability of the flux of charged particles. The effect will be discussed later in Chapter 4.

### 2.3.5 Particles Leaving the Equatorial Plane

A natural sequence to the consideration of equatorial orbits is the analysis of their stability against perturbations pushing the particle off the plane of the equator. Of interest is the particle's reaction to the perturbing force directed along the rotation axis. Two situations can be imagined. One is that a particle receiving an initial momentum leaves the equatorial plane along a magnetic line of force and follows that until colliding with the planet. The other is an oscillatory motion along the magnetic line, characterized by multiple crossings of the equatorial plane. In the first case, the equatorial orbit should be considered unstable, while in the second it is stable unless the oscillation amplitude increases with time. We have only discussed the motion along the magnetic line of force, since motions across the magnetic field may be considered forbidden for particles of a relatively high charge-to-mass ratio.

The problem of stability of the equatorial orbits is closely associated with the question of whether gravitoelectrodynamics permits circular orbits whose plane does not include the center of attraction. That is where we begin the analysis. Obviously, such orbits cannot be realized in "pure" Newtonian mechanics, since the gravity force and the centrifugal force would be oriented at an angle to one another and hence would not be mutually compensated. The situation is different for the above discussed approximation where the allowed motion is along the line of force only. The equilibrium condition then does not require that the sum of gravity and the centrifugal force be zero but reduces to compensation of the components of the two forces along the line of force. The gravity force tends to tear the grain off the equatorial plane and bring it closer to the planet, while the centrifugal force acts in the opposite direction (Figure 11). Meanwhile, a rigorous analysis should not be based on this simple concept but proceeds from the complete equation of motion Equation (29) accounting automatically for the prohibition of motions transverse to the magnetic lines of force. Let Equation (29) be re-written as

$$\frac{d^2\mathbf{r}}{dt^2} = -\frac{\mu\mathbf{r}}{mr^3} + \frac{Q\mathbf{E}}{m},$$

with $\mathbf{E} = 1/c\{[(\mathbf{\Omega} - \mathbf{\Omega}_p) \times \mathbf{r}] \times \mathbf{B}\}$, where the additional forces, $\mathbf{F}$, are

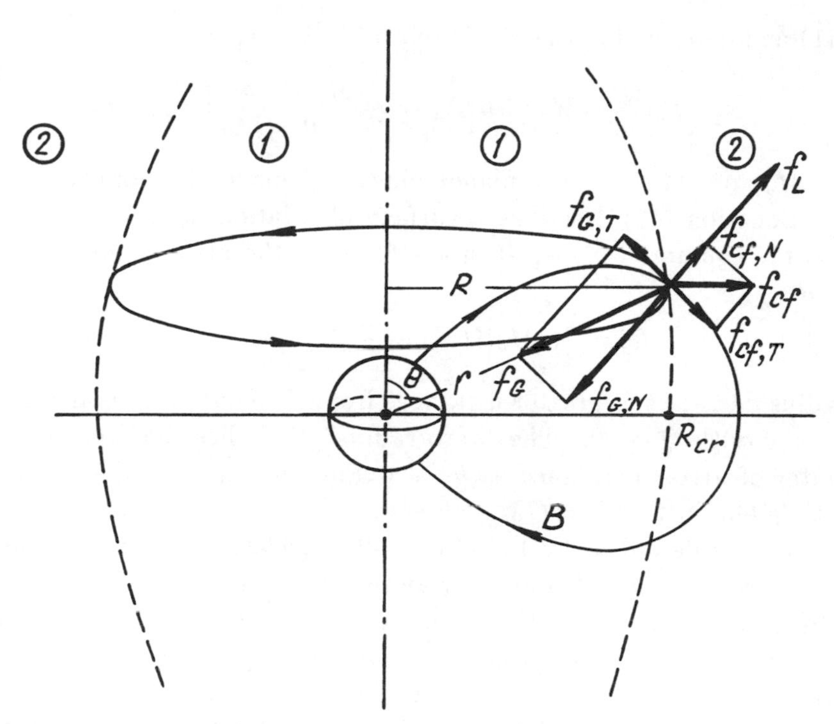


Figure 11: The Balance of Forces in an Off-Equatorial Elevated Orbit. The Tangential Components $f_{G,T}$ of Gravity and $f_{cf,T}$ of the Centrifugal Force Are Mutually Compensated along **B**. The Normal Components Are Compensated with Account of the Lorentz Force. Region 1 is where Gravity Prevails, while in Region 2 Dominates the Centrifugal Force.

not allowed for as before. By projecting this Equation onto the $\mathbf{e}_r$ and $\mathbf{e}_\theta$ directions of the polar spherical coordinate system (with the polar angle $\theta$ counted from the rotation axis), one can obtain the set

$$[\Omega^2 + \Omega_B(\Omega_p - \Omega)] \sin^2\theta - \Omega_K^2 = 0 \tag{40}$$

$$[\Omega^2 - 2\Omega_B(\Omega_p - \Omega)] \sin 2\theta = 0. \tag{41}$$

As long as equatorial orbits alone were considered (i.e. $\theta = \pi/2$), Equation (41) was satisfied identically and Equation (40) proved sufficient for finding $\Omega = \Omega(R)$ (the equation coincided with Equation (30)). In the general case, we are interested first in obtaining the existence condition for off-equatorial orbits. This can be found by excluding $\Omega$ from Equations (40) and (41) and determining the $R - \theta$ relation,

$$R^3 = \frac{2}{3} G M_p \sin\theta \left[\Omega_p - \frac{G M_p}{3 \sin^2\theta} \cdot \frac{mc}{Q\mathcal{M}}\right]^{-2}. \tag{42}$$

(Recall $M_p$ and $\mathcal{M}$ to be the planet mass and magnetic moment, respectively.) Equation (42) describes a surface of rotation, symmetric with respect to the equatorial plane. It intersects with the plane along a circle of radius $R_c$,

$$R_c^3 = \frac{2}{3} G M_p \left[\Omega_p - \frac{G M_p}{3} \cdot \frac{mc}{Q\mathcal{M}}\right]^{-2}. \tag{43}$$

The radius decreases at greater distances from the equatorial plane, as well as $R_c \to 0$ with $Q \to 0$. The "zero-radius" orbit lies infinitely far from the center of attraction, since at $\theta \to 0$ Equations (40) and (41) can be compatible only with $\Omega_K^2 = GM_p/r^3 \to 0$.

The surface described by Equation (42) separates the circumplanetary space into two domains. The inner volume (region 1 in Figure 11) is dominated by gravitation, while in the outer (region 2 in Figure 11) the centrifugal force prevails. Since the role of gravitational forces is to take the grain away from the equatorial plane, the critical radius $R_c$ of Equation (43) can be regarded as the lower boundary to the domain of stable equatorial orbits (marginal stability radius). The question was given a detailed analysis by Northrop and Hill (1982) who calculated particle trajectories for different orbital radii and different initial axial momenta launching the particles off the equatorial plane. The equatorial orbits with $R < R_c$ were indeed shown to be unstable (see Figure 12).

The writers attempted to use this result for interpreting a spectacular structural feature in Saturn's B-ring. The inner and outer part of the ring are separated quite sharply at a distance about 1.63 $R_p$ from the planet. The inner ring is more transparent and of lower density than the outer part. To compare the radius of this boundary with the $R_c$ of Equation (43), one would need to know the charge-to-mass ratio of the dust grains in the ring. However the value can be assumed sufficiently large to allow taking the limit $R_c(Q/m \to \infty) \simeq (2GM_p/3\Omega_p^2)^{1/3}$ which in the case of Saturn yields $R_c \simeq 98049\,\text{km}$ or $R_c = 1.625\,R_p$. Apparently, the coincidence is excellent. The reduced number density of dust grains in the inner B-ring thus finds a natural explanation. The particles subject to small perturbations are swept out of the ring plane.

The gravitoelectrodynamic effects of charged particles in the magnetospheres of Saturn and Jupiter, as well as of other celestial bodies like the

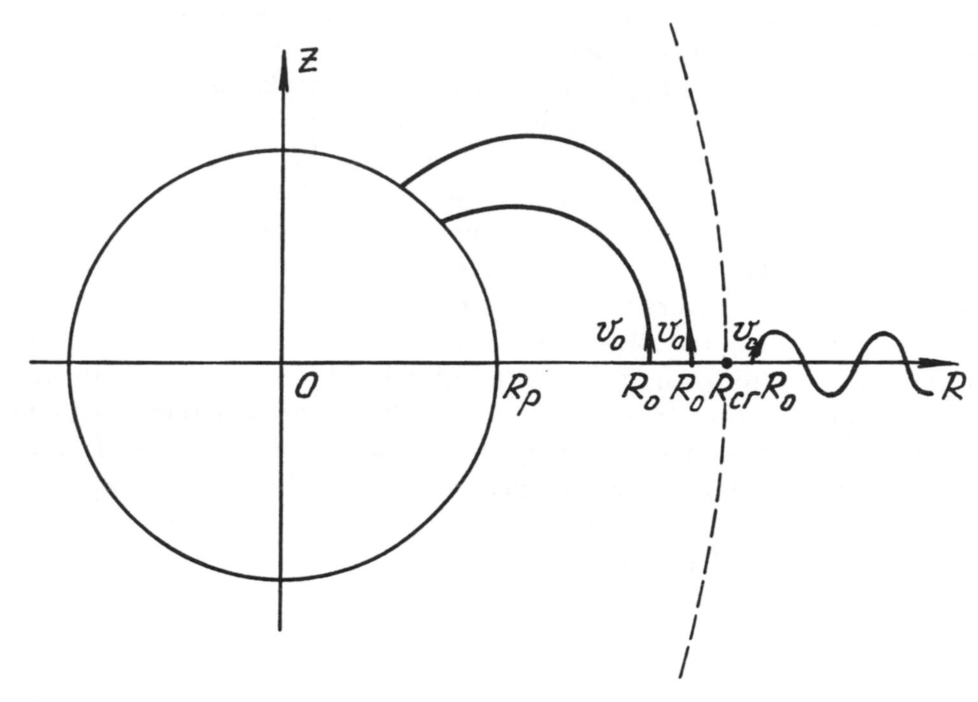


Figure 12: The Trajectories of Particles with the Initial Velocity $\mathbf{V}_0 \parallel OZ$ Launched from Different Off-Axis Distances $R_0$. The Particles with $R_0 < R_{cr}$ Hit the Planet (Instability), while with $R_0 > R_{cr}$ the Particle Oscillates about the Equatorial Plane (Stability). The Dashed Line Shows the Critical Surface Trace.

Earth, smaller planets and comets, are abundantly described in the literature. The reviews of Mendis and Rosenberg (1994) and Goertz (1989) and the bibliographies are of note.

## 2.4 Polarized Particles Rotating around the Mass Center. Binary Interactions

### 2.4.1 Rotation around the Mass Center

Until now, the charged grains have been considered as "point-size" particles, although the finite dimension was taken into account to describe the charge per unit mass. Meanwhile, the motion of a solid body is fully determined by the motion of its mass center and rotation about that point. Obviously, the rotational motion (orientation) is essential only for non-spherical particles.

The torque owing to gravitation tends to align a grain of oblong geometry with the gravitational lines of force, i.e. orientate it along the radius-vector of the point of orbit. The electric field acts in a similar way, namely its associated torques tend to orientate the particle along the lines of force. However, the force lines of the unipolar induction field generally may not coincide with the radial direction, hence orientation of particles at arbitrary orbits requires special consideration. The equatorial plane is an exception and the analysis here is much simpler, for the electric field is radial and the two torques coincide in direction. We have only to compare their magnitudes to understand where the role of electric forces can be essential.

Let us start from the analysis of gravitational forces. Seeking to obtain just order of magnitude estimates, we will not specify either the grain geometry or precise numerical factors. Consider an oblong grain of mass $m \sim da^2\rho$, where $d$ is length and $a$ the cross-section radius. The force acting across the grain owing to the gravitation gradient is $\Delta F_G \sim (\partial F_G/\partial r)\,\Delta r$. It is radially oriented and, with the grain's "axis" deviated from the radial by an angle $\alpha$ ($\Delta r = d\cos\alpha$), becomes

$$\Delta F_G \sim \rho\Omega_K^2 a^2 d^2 \cos\alpha. \tag{44}$$

In celestial mechanics, $\Delta F_G$ is known as the tidal force. It tends to tear the grain apart with $\alpha = 0$, while with $\alpha \neq 0$ also tries to orientate it along the radial. The torque is $K_G \sim \Delta F_G d \sin\alpha$,

$$K_G \sim \Omega_K^2 a^2 d^3 \sin 2\alpha. \tag{45}$$

A dielectric particle placed in the unipolar induction field becomes polarized and the corresponding torque is $\mathbf{K}_E = \mathbf{p}\times\mathbf{E}$, where $\mathbf{p}$ is the particle's dipole moment. Proceeding from the exact solution for a highly anisotropic ellipsoid (Landau and Lifshitz, 1982), $K_E$ can be estimated as

$$K_E \sim E^2 da^2(\varepsilon-1)^2 \sin 2\alpha. \tag{46}$$

As the grain length $d$ is increased, the gravitational torque $K_G$ grows at a higher rate than $K_E$. Accordingly, the electric forces can only play an essential role for small-size particles. The orientating effect of the unipolar induction field prevails for $d_{cr} \sim E(\varepsilon-1)/(\Omega_K\rho^{1/2})$. In the case of Saturn's rings with $E \sim 10^{-4}\,\mathrm{V/cm}$ and $\Omega_K \sim 10^{-4}\,\mathrm{s}^{-1}$, we have $d_{cr} \sim 100\,\mu$ for $\varepsilon \sim 3$ and $\rho \sim 1\,\mathrm{g/cm}^3$. Thus, allowance for electromagnetic forces should be essential for the analysis of orientation of micron-size dust grains. The question is of importance for wave scattering and diffraction by the dust grains.

### 2.4.2 Binary Interactions of Polarized Particles

In contrast to the pure gravitational interaction, electric forces may result in mutual repulsion, as well as attraction, of the particles. Such effects underlie wavelike disturbances of the particle number density, quite similar to those occurring in conventional plasmas. In this Chapter, however the analysis will be restricted to binary interactions.

Without electric forces, any two particles are subject to attraction alone. Under the action of this force, particles moving along the same orbit may merge to form stable orbit-aligned aggregates of nominally unlimited length. The merging of particles from close, however different, orbits should give rise to chains extended along the radial. These can be stable to a certain limit, because of the tidal forces tending to tear them apart. Here are some estimates for tne grains of Saturn's rings.

The highest tensile force can be shown to appear in the middle of a radial chain of a few particles. Its magnitude can be roughly estimated from Equation (44) with $\alpha = 0$ if $d$ is understood as the chain length. Meanwhile, the force of attraction between the two halves of the chain is $f_G \sim G(\rho a^2 d)^2 d^{-2} - G\rho^2 a^4$. The radial chain would be stable if the tidal force remained lower than the mutual attraction, whence the limitation on the chain length is $d \leq 1.5a$ (for the case of Saturn's rings). Thus, radial chains may incorporate just 2 or 3 grains.

The situation is greatly different if we allow for electric forces. First, consider the interaction of two polarized particles at a relatively great separation ($d > a$). The dipole moments $\mathbf{p}_1$ and $\mathbf{p}_2$ are both parallel to the external electric field vector $\mathbf{E}$ which is perpendicular to the orbit (see Figure 13). The force component along $\mathbf{d}$ is

$$f_p = \frac{3p_1p_2}{d^4}(1 - 3\sin^2\chi), \tag{47}$$

where the angle $\chi$ has been defined in the Figure. The force $\mathbf{f}_p$ can be either attraction or repulsion, depending on the mutual position of the dipoles. In Figure 13 showing the $f_p(\chi)$ dependence, the domain of angles corresponding to attraction has been hatched. Dust grains moving along the same orbit are repulsed upon approaching one another and hence orbit-aligned chains are impossible. The particles moving along different orbits can be attracted to form a radially extended chain, since the electric and gravitation forces act in the same direction. The highest value of $f_p$ for spherical grains corresponds to $\chi = \pi/2$ and equals $6E^2a^6d^{-4}(\varepsilon-1)^2/(\varepsilon+2)^2$, while $f_G = G(4\pi R^3\rho/3d)^2$.

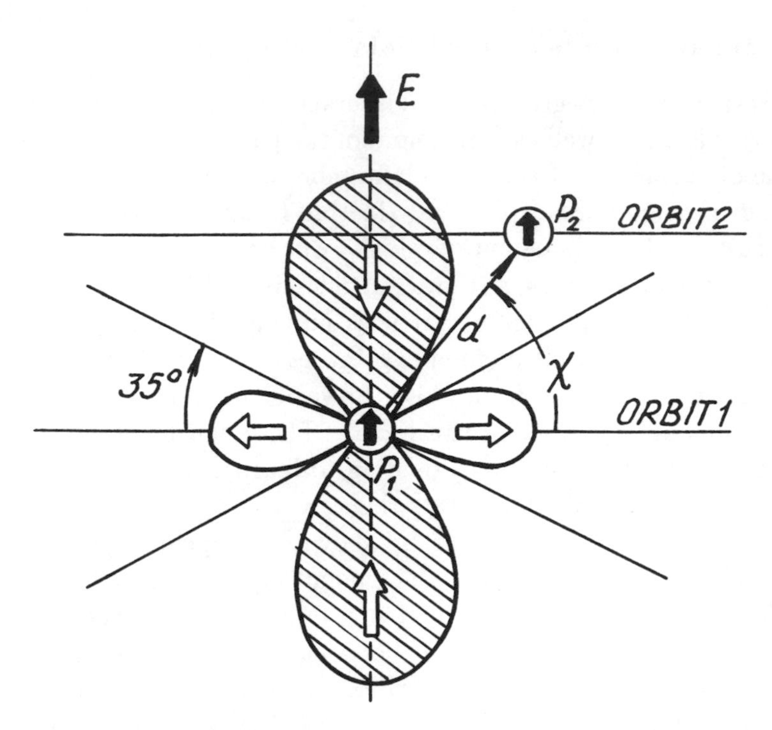


Figure 13: The Interaction of Polarized Particles of Dipole Moments $\mathbf{p}_1$ and $\mathbf{p}_2$. The Hatched Angular Domain $\chi_1$ Is where the Particles Are Mutually Attracted.

The electric forces are characterized by a shorter range of action, and hence can dominate at $d \le E(\varepsilon - 1)/(\varepsilon + 2)\rho\sqrt{G}$, which for the case of Saturn's rings is $d \le 50\,\mu$. The particle size apparently should be even smaller.

Had it not been for tidal forces, the polarized particles could aggregate into long chains oriented along the orbit radius. In fact such configurations can only be of limited length, which scale size should be estimated. Equation (47) cannot be used for the purpose as it is valid for $d > a$, i.e. does not permit particle merging. Consider a chain model in the form of a prolate ellipsoid of revolution with the symmetry axis parallel to the external field. Imagine the ellipsoid to be separated into halves by a plane perpendicular to $\mathbf{E}$ and evaluate the force of attraction between the two halves. The total force acting on a body on the part of an external field is known to be (Landau and Lifshitz, 1982)

$$\mathbf{F}_E = \frac{1}{4}\oint \left[\mathbf{E}(\mathbf{nE}) - \frac{1}{2}E^2\mathbf{n}\right] dS, \tag{48}$$

where $\mathbf{n}$ is the outward normal to the integration surface and $dS$ the surface element. The attraction force is found through integration of Equation (48) along the surface of one half of the ellipsoid,

$$F_E = \frac{E^2 a^2 (\varepsilon - 1)^2}{16[1 + (\varepsilon - 1) n_E]^2 e^2} \left[ 1 + \frac{1 - e^2}{e^2} \log(1 - e^2) \right],$$

where $e = (1 - a^2/d^2)^{1/2}$ is the eccentricity and $n_E = (1 - e^2)[\log(1 + e) \times (1 - e)^{-1} - 2e]/2e^3$ the depolarization factor along the external field. Assuming $a \ll d$, $e \simeq 1$ and $n_E \ll 1$ one obtains

$$F_E \simeq \frac{E^2 a^2 (\varepsilon - 1)^2}{16}. \tag{49}$$

Meanwhile, the tidal force trying to break the ellipsoid is $3\pi\Omega_K^2 \rho a^2 d^2/4$. By comparing the two magnitudes, we can estimate the "critical" length $d_{cr}$ for the chain to remain stable,

$$d_{cr} \sim (6\pi\rho)^{-1/2} \frac{E(\varepsilon - 1)}{\Omega_K}.$$

In the case of Saturn's rings this becomes $d_{cr} \sim 300\,\mu$, while on Jupiter $d_{cr} \sim 10^3\,\mu$. Since $a \ll d$, formation of radial chains is possible under the action of electric forces. Such chains might incorporate a few tens of particles. An anisotropy of the kind might affect light scattering by the rings and probably have relation to the effects discovered recently, like depolarization of the radiation scattered by internal Saturn rings (Bugaenko and Morozhenko, 1981). It is in the internal regions that the effect should be brighter owing to the relatively high electric field $E$.

In fact, the above given estimates of the chain length are quite rough as electrostatic repulsion of the likely charged particles has not been taken into account. Evaluating the repulsion force as $f_Q \sim (\psi^2 a^2)/a^2 = \psi^2$ we can demand that it should not be greater than the force of attraction, Equation (48), to obtain a lower boundary to the grain size, $a > |\psi|/E$. Taking $|\psi| \sim 1\,\mathrm{V}$ and $E \sim 10^{-4}\,\mathrm{V/cm}$ we have $a > 10^4\,\mathrm{cm}$. In other words, the polarization-induced contraction may be a stronger effect than Coulombian repulsion solely for large and weakly charged fragments. The above estimated scale lengths of grain chains apply to neutral grains only. Besides, the analysis has been done for dielectric particles. If the grain is a fairly good conductor, then polarization effects in an external field result in additional breaking (rather than contracting!) forces (Landau and Lifshitz, 1982).

### 2.4.3 The Roche Limit in Gravitoelectrodynamics

The Roche limit is a parameter to characterize binary stellar systems or planetary satellites orbiting along circular trajectories. The smaller is the orbit radius, the stronger the tidal forces tending to disrupt the satellite. In its turn, the material of the satellite tends to concentrate in a minimum volume owing to self-gravitation. If the tidal force is sufficiently strong, the gravitational condensation with the formation of a single body proves impossible. The radius $R_R$ of the sphere within which the system is unstable in the sense described, is known as the Roche limiting radius. The magnitude of $R_R$ is determined by the ratio of mass densities $\rho$ and $\rho_p$ of the satellite and the planet,

$$R_R \simeq 1.26 R_p \left(\frac{\rho_p}{\rho}\right)^{1/3}, \tag{50}$$

being independent of the satellite radius, $a$.

In the systems governed by the laws of gravitoelectrodynamics, gravitation is supplemented by electrostatic forces which should be allowed for when determining $R_R$. Consider a spherical grain moving along a circular orbit within the equatorial plane. Imagine the grain divided in our minds in two hemispheres, with the cutting plane perpendicular to its radius-vector. Spacing the two halves by an infinitesimal distance, let us estimate the forces of their interaction. First, consider the gravitation. In view of the spherical symmetry assumed, both the contracting force,

$$F_{G,attr} = \frac{\pi^2}{3} G \rho^2 a^4, \tag{51}$$

and the tidal disruptive force,

$$F_{G,dis} \simeq \frac{2\pi^2}{3} \cdot \frac{G \rho \rho_p a^4 R_p^3}{R^3}, \tag{52}$$

can be easily estimated.

The electrostatic forces appearing in the unipolar induction field can be estimated from Equation (49). Note, however, the field across the gap to be $3\varepsilon E/(2+\varepsilon)$ in the case of a dielectric or zero for a conductor. The result is the contracting force

$$F_{E,attr} = \frac{9}{16} \cdot \frac{E^2 a^2 (\varepsilon - 1)^2}{(2+\varepsilon)^2} \tag{53}$$

in the dielectric or disruptive force,

$$F_{E,dis} = \frac{9}{16} E^2 a^2 \tag{54}$$

in the conductor. Additional disruptive forces appear in either case if the grain carries a charge $Q$. E.g., for the charge on the spherical surface the force is

$$F_{Q,dis} = \frac{Q^2}{8a^2}. \tag{55}$$

The Roche limit is where the contracting and disruptive forces are equal in magnitude,

$$R_R \simeq 1.26 R_p (\rho_p/\rho)^{1/3} (1+\delta)^{-1/3}, \tag{56}$$

with

$$\delta = \left[\frac{9}{16} E^2 a^2 \left(\frac{\varepsilon - 1}{\varepsilon + 2}\right)^2 - \frac{Q^2}{8a^2}\right] \left(\frac{\pi^2}{3} G \rho^2 a^4\right)^{-1} \tag{57}$$

for the case of dielectric material or

$$\delta = -\left(\frac{9}{16} E^2 a^2 + \frac{Q^2}{8a^2}\right) \left(\frac{\pi^2}{3} G \rho^2 a^4\right)^{-1} \tag{58}$$

for a conductor. Setting $E = 0$ and $Q = 0$ one obtains the familiar value of Equation (50), while generally the $R_R$ radius is different, being dependent on the grain radius, electric properties of its material and charge magnitude. Conductors are characterized by a negative $\delta$, and hence the Roche radius $R_R$ of Equation (56) increases, tending to infinity with $\delta \to -1$.

The $\delta$ of a dielectric grain may be either positive or negative, with $R_R$ decreasing or increasing accordingly. The parameter changes its sign at a definite size of the particle, roughly coincident with the estimate of the preceding Section, $a \sim \psi/E$, which is $a \sim 10^4$ cm in the vicinity of Saturn's or Jovian rings. The estimate can be brought to the form $\delta\psi \sim \psi$, where $\delta\psi$ is the potential difference, across the grain size, of the unipolar induction field and $\psi$ the grain potential.

Similar calculations present no difficulties of principle for grains with off-equatorial orbits, since the field vector $\mathbf{E}$ is known everywhere. However the corresponding formulas are too cumbersome, for the field vector $\mathbf{E}$ is no more aligned with the radius-vector of the grain. The Roche sphere is replaced by another "critical" surface.

## References

Bliokh, P.V. and Yaroshenko, V.V. 1983, *Kosmich. Issledovaniya* **21**, 940 (*in Russian*).

Bugaenko, O.I. and Morozhenko, A.V. 1981, *Adv. Space Res.* **1**, 183.

Dermott, S.F. 1984. In R. Greenberg and A. Brahic (Eds.) *Planetary Rings*, (Univ. of Ariz. Press: Tucson, AZ), p. 589.

Franklin, F., Lecar, M. and Wiesel, W., 1984. In R. Greenberg and A. Brahic (Eds.) *Planetary Rings*, (Univ. of Ariz. Press: Tucson, AZ), p. 562.

Goertz, C.K. 1989, *Rev. Geophys* **27**, 271.

Goldreich, P. and Julian, W. 1969, *Asroph. J.* **157**, 869.

Grün, E., Morfill, G.E. and Mendis, D.A. 1984. In R. Greenberg and A. Brahic (Eds.) *Planetary Rings*, (Univ. of Ariz. Press: Tucson, AZ), p. 275.

Havnes, O., Morfill, G.E. and Melandsø, F. 1992, *Icarus* **98**, 141.

Hill, J.R. and Mendis, D.A. 1980, *Moon and Planets* **23**, 53.

Hill, J.R. and Mendis, D.A. 1982, *Moon and Planets* **26**, 217.

Landau, L.D. and Lifshitz, E.M. 1982. Electrodynamics of Solids (Nauka: Moscow), p. 306 (*in Russian*).

Lyubarsky, Yu. 1994, *Astrophys. and Space Sci.*, (*in press*).

Mendis, D.A. and Axford, W.I. 1974, *Rev. Earth and Planet. Sci.* **2**, 163.

Mendis, D.A., Harry, L.F. Houpis and Hill, J.R. 1982, *J. Geophys. Res.* **87**, 3449.

Mendis, D.A. 1984, *Proc. Indian Acad. Sci.* **93**, 177.

Mendis, D.A. and Rosenberg, M. 1994, *Ann. Rev. Astron. Astrophys.* **32**, (*in press.*).

Mignard, F. 1984. In R. Greenberg and A. Brahic (Eds.) *Planetary Rings*, (Univ. of Ariz. Press: Tucson, AZ), p. 313.

Morfill, G.E., Grün, E. and Johnson, T.V. 1980, *Planet. Space Sci.* **28**, 1087.

Mukai, T., 1986. In A. Bonetti and S. Aiello (Eds.) *Lectures at the Enrico Fermi Summer School* (Enrico Fermi Institute Publ.).

Northrop, T.G. and Hill, J.R. 1982, *J. Geophys. Res.* **87**, 6045.

Northrop, T.G. and Hill, J.R. 1983, *J. Geophys. Res.* **88**, 1.

Northrop, T.G., Mendis, D.A. and Schaffer, L. 1989, *Icarus* **89**, 101.

Thomsen, M.F., Goertz, C.K., Northrop, T.G. and Hill, J.R. 1982, *Geophys. Res. Lett.* **9**, 423.

Yaroshenko, V.V. 1985. Electrostatic Waves in Planetary Rings, PhD Thesis (Institute of Radiophysics and Electronics: Kharkov, Ukraine) (*in Russian*).

# 3 Cooperative Effects in Self-Gravitational and Dusty Plasmas

The preceding Chapter, dedicated to gravitoelectrodynamics, treated the motion of charged particles through external fields of force. Now we are starting an analysis of cooperative effects owing to interactions of the particles via long-range Coulombian forces. Speaking of the Coulombian fields in a mixture of charged particles, it is tacitly agreed they are electric fields. In fact, the particles are always subject to mutual attraction due to gravitation, which effect also follows Coulomb's law. The question whether or not the gravitational interaction is significant can be resolved numerically. Consider, for instance, two electrons at a distance $d$. They are mutually repulsed with the force $f_E^e = e^2/d^2$. Meanwhile, as material particles of mass $m_e$, the same electrons are attracted with the force $f_G^e = Gm_e^2/d^2$. The ratio of $f_E^e$ and $f_G^e$ is $f_E^e/f_G^e \sim 10^{42}$, which value is independent of range. Evidently, the enormous domination of electric forces over the gravitation makes it absolutely senseless to take the gravitational electron-electron interaction into account, which is indeed never done. In the case of ions, the $f_E^i/f_G^i$ ratio is lower by several orders of magnitude but still remains far too high for any gravitational effects to be noticeable.

The situation may prove essentially different with macroscopic particles. We will estimate the ratio $f_E/f_G = Q^2/(Gm^2)$ (with $Q$ and $m$ being the particle charge and mass, respectively) on the assumption that the particles have a simple spherical geometry (the sphere radius is $a$) and consist of a material of mass density $\rho$. Since $m = 4/3\pi a^3\rho$, the force ratio becomes

$$f_E/f_G = 9Q^2/[(4\pi\rho a^3)^2 G].$$

The two forces are roughly equal, provided $a \simeq a_{cr} = [9Q^2/(16\pi^2\rho^2 G)]^{1/6}$.

Assuming the minimum possible charge $Q = e$, we see that the electric forces prevail as before for sufficiently small macroscopic particles ($a \leq 10^{-2}$ cm, with $\rho \leq 1\,\mathrm{g/cm^3}$). For a realistic value of $Q = 10^3 e$ the critical radius $a_{cr} \sim Q^{1/3}$ increases to $a_{cr} \simeq 10^{-1}$ cm. This value can be assumed for the limiting size at which the prevailing force that controls cooperative processes in the plasma is altered. Self-gravitation is not essential for small-size grains $a \ll a_{cr}$, so the previous term of a "dusty plasma" remains valid. In a mixture of larger grains, $a \simeq a_{cr}$, the gravitation forces become important and the name of a self-gravitational plasma is more appropriate. It should be noted, however that the gravitational attraction dominates in the latter case solely with respect to large-size grains and does not include their interaction with microscopic dust particles or electrons and ions. This implies that electromagnetic forces should always be included in the analysis of cooperative processes, especially because many astrophysical objects demonstrate a broad spectrum of grain sizes. (E.g., Saturn's rings contain both submicron grains and fragments of a few meters in size).

For these reasons, we will first consider a rather generally formulated problem, to be further able to analyze some limiting cases of interest. The eigenvalue spectrum of the self-gravitational plasma will be seen to be quite complex, with the great variety of grain sizes and variability of their electric charge resulting in peculiar effects that are never observed in "pure" plasmas. Among these, there is excitation of electromagnetic waves by a massive neutral particle moving through the plasma, special effects in wave scattering and nonlinear effects involving macroscopic grains.

## 3.1 Waves in the Self-Gravitational Plasma

### 3.1.1 Problem Formulation. Basic Equations

The electromagnetic fields existing in a plasma medium and the ensemble of its charged particles are a self-consistent system that can be described by an equation set involving Maxwell's equations, some equations of the medium dynamics and constituent relations to express the charge and current densities in terms of the number of particles. If gravitational interaction between the particles is taken into account, then the electromagnetic field vectors **E** and **H** must be supplemented by the gravitation field $\nabla\psi_G$ (where $\psi_G$ is the gravitation potential) produced by all the particles of the system. The equation sets corresponding to rigorous theoretical approaches, such as the kinetic theory, are quite complex. Meanwhile, the majority of plasma effects of importance for physical applications can be analyzed in a relatively simple model known as the hydrodynamic (or gas dynamic) approximation (Krall and Trivelpiece, 1973; Akhiezer *et al.*, 1974). Essentially, it consists of replacing the analysis of the particle behavior by consideration of mutually penetrating, electrically charged or neutral "liquids", each corresponding to a specific sort of particles. The major part of this Chapter will be devoted to such effects that permit hydrodynamic description, before all to wavelike disturbances in self-gravitational and dusty plasmas.

We begin the analysis with the simpler case of a dusty plasma in which the gravitation interaction between particles can be neglected. The only distinction from the common plasma lies in the presence of relatively massive charged dust grains. From a formal point of view, we are discussing a multicomponent plasma whose particles are characterized by the charge $Q_\alpha$ and mass $m_\alpha$ ($\alpha$ is the subscript for the particle species). Therefore, we can make use of the standard set of hydrodynamic equations.

Maxwell's equations will relate the electromagnetic field vectors $\mathbf{E}$ and $\mathbf{H}$ to the charge density $\rho$ and electric current $\mathbf{j}$ in the plasma,

$$\nabla \times \mathbf{E} = -\frac{1}{c}\frac{\partial \mathbf{H}}{\partial t} \tag{1}$$

$$\nabla \cdot \mathbf{E} = 4\pi\rho \tag{2}$$

$$\nabla \times \mathbf{H} = \frac{1}{c}\cdot\frac{\partial \mathbf{E}}{\partial t} + \frac{4\pi}{c}\mathbf{j} \tag{3}$$

$$\nabla \cdot \mathbf{H} = 0 \tag{4}$$

with

$$\rho = \sum_{\alpha=1}^{N} Q_\alpha n_\alpha \tag{5}$$

and

$$\mathbf{j} = \sum_{\alpha=1}^{N} Q_\alpha n_\alpha \mathbf{v}_\alpha \tag{6}$$

(the summation concerns all sorts of micro- and macroscopic particles; $N$ denotes the number of particle species, $\mathbf{v}_\alpha$ and $n_\alpha$ are the velocity and number density, respectively, for the $\alpha$-species particles). The response of the medium to the electromagnetic disturbances $\mathbf{E}$ and $\mathbf{H}$ can be derived from the equations of motion

$$\frac{\partial \mathbf{v}_\alpha}{\partial t} + (\mathbf{v}_\alpha\nabla)\mathbf{v}_\alpha = \frac{Q_\alpha}{m_\alpha}\left(\mathbf{E} + \frac{1}{c}\mathbf{v}_\alpha \times \mathbf{H}\right) - \frac{v_{T\alpha}^2}{n_\alpha}\nabla n_\alpha, \tag{7}$$

and continuity equations for all the "liquids",

$$\frac{\partial n_\alpha}{\partial t} + \nabla(n_\alpha \mathbf{v}_\alpha) = 0, \tag{8}$$

where $v_{T,\alpha} = (T_\alpha m_\alpha^{-1})^{1/2}$ is the thermal velocity of the particles of species $\alpha$ and $T_\alpha$ the corresponding temperature.

By solving the sets Equations (1-6) and Equations (7-8) simultaneously we can obtain a self-consistent set of field vectors, particle densities and velocities in the hydrodynamic approximation.

Now, the question is how to modify Equations (1-8) to allow for the self-gravitation in the ensemble of particles. The obvious answer is to include a supplementary term in the right hand part of the equations of motion Equation (7), namely the acceleration owing to the gravitation force, $f_{G,\alpha}/m_\alpha = -\nabla\psi_G$. In addition, the set should be supplemented with another Poisson equation relating the gravitation potential $\psi_G$ to the particle densities $n_\alpha$, *viz.*

$$\Delta\psi_G = -4\pi G \sum_{\alpha=1}^{N} m_\alpha n_\alpha. \tag{9}$$

Thus we arrive at a complete set of hydrodynamic equations to describe the self-gravitational plasma.

### 3.1.2 Dispersion Laws for Transverse and Longitudinal Waves

Consider free oscillations of the plasma, assuming all the external fields and average velocities of the particles to be zeros in the unperturbed (i.e. equilibrium) state. The equation set Equations (1-9) can be linearized with respect to small perturbations. Let

$$\mathbf{E} = \mathbf{E}', \quad \mathbf{H} = \mathbf{H}', \quad \mathbf{v}_\alpha = \mathbf{v}'_\alpha, \quad \psi_G = \psi'_G \quad \text{and} \quad n_\alpha = n_{0,\alpha} + n'_\alpha \tag{10}$$

where the values with the subscript "0" relate to the equilibrium state of the system (i.e. the "background"), while primed quantities characterize deviations from the equilibrium. The linearized Equations (1-9) in which quadratic and higher order terms in the perturbation amplitude, like $\mathbf{v}'\times\mathbf{H}'$, $(\mathbf{v}'\nabla)\mathbf{v}$, *etc.* have been neglected, take the form

$$\frac{\partial \mathbf{v}'_\alpha}{\partial t} = \frac{Q_\alpha}{m_\alpha}\mathbf{E}' - \nabla\psi'_G - v_{T,\alpha}^2 \nabla n'_\alpha / n_{0,\alpha} \tag{11}$$

$$\frac{\partial n'_\alpha}{\partial t} + \nabla\cdot(n_{0,\alpha}\mathbf{v}'_\alpha) = 0 \tag{12}$$

$$\nabla\cdot\mathbf{E}' = 4\pi\sum_{(\alpha)} Q_\alpha n'_\alpha \tag{13}$$

$$\nabla\cdot\mathbf{H}' = 0 \tag{14}$$

$$\Delta\psi'_G = -4\pi G\sum_{(\alpha)} m_\alpha n'_\alpha \tag{15}$$

$$\nabla\times\mathbf{E}' = -\frac{1}{c}\frac{\partial \mathbf{H}'}{\partial t} \tag{16}$$

$$\nabla\times\mathbf{H}' = \frac{1}{c}\frac{\partial \mathbf{E}'}{\partial t} + \frac{4\pi}{c}\sum_{\alpha} Q_\alpha n_{0\alpha}\mathbf{v}'_\alpha. \tag{17}$$

We shall consider a uniform self-gravitational plasma, assuming all the quantities to vary as $\exp(i\mathbf{kr} - i\omega t)$. Then the operator $\partial/\partial t$ becomes $-i\omega$ and $\nabla \rightarrow i\mathbf{k}$. It is also convenient to decompose all the vectors in Equations (11-17) into a longitudinal (with respect to $\mathbf{k}$) and transverse part, i.e.

$$\mathbf{E}' = \mathbf{E}'_{\parallel} + \mathbf{E}_{\perp}'; \quad \mathbf{H}' = \mathbf{H}'_{\parallel} + \mathbf{H}'_{\perp} \quad \text{and} \quad \mathbf{v}' = \mathbf{v}'_{\parallel} + \mathbf{v}_{\perp}, \tag{18}$$

with $\mathbf{E}_{\parallel} = (\mathbf{k}/k^2)(\mathbf{k} \times \mathbf{E})$, $\mathbf{E}_{\perp} = (\mathbf{k}/k) \times \mathbf{E}$, *etc.* Then Equations (11-17) become a set of algebraic equations splitting into two independent homogeneous sets, one for the longitudinal and the other for the transverse perturbations,

$$-i\omega \mathbf{v}_{\alpha_{\parallel}} = \frac{Q_\alpha}{m_\alpha}\mathbf{E}_{\parallel} - i\mathbf{k}\psi_G - i\mathbf{k}v_{T,\alpha}^2 n_{0,\alpha} \tag{19}$$

$$-\omega n_\alpha + k v_{\alpha_{\parallel}} n_{0,\alpha} = 0 \tag{20}$$

$$ikE_{\parallel} = 4\pi \sum_\alpha Q_\alpha n_\alpha \tag{21}$$

$$k^2\psi_G = 4\pi G \sum_\alpha m_\alpha n_\alpha \tag{22}$$

and

$$-i\omega \mathbf{v}_{\alpha_{\perp}} = \frac{Q_\alpha}{m_\alpha}\mathbf{E}_{\perp} \tag{23}$$

$$\mathbf{k} \times \mathbf{E}_{\perp} = \frac{\omega}{c}\mathbf{H}_{\perp} \tag{24}$$

$$\mathbf{k} \times \mathbf{H}_{\perp} = -\frac{\omega}{c}\mathbf{E}_{\perp} - \frac{4\pi i}{c}\sum_\alpha Q_\alpha n_{0,\alpha}\mathbf{v}_{\alpha_{\perp}}. \tag{25}$$

(The primes have been omitted both here and below). By equating the determinants of the two equation sets to zeros we arrive at characteristic or dispersion relations whose solutions relate the perturbation frequencies $\omega$ to the corresponding wave vectors $\mathbf{k}$.

The set Equations (23-25) for transverse perturbations does not involve terms owing to the self-gravitation or density perturbations (i.e., both $\psi_G$ and $n_\alpha$ may be set equal to zeros). Hence, the dispersion law for the transverse waves is unaffected by the gravitational attraction and remains exactly the same as for electromagnetic waves in a conventional multicomponent plasma,

$$\omega^2 = k^2c^2 + \sum_\alpha \omega_{p,\alpha}^2. \tag{26}$$

As for the longitudinal disturbances, the structure of the corresponding equation set Equations (19-22) is such that the gravitational and plasma interactions are involved in an equipotent manner, which is in contrast to transverse waves. This equality becomes quite evident if we set $\mathbf{E}_{\|} = -\nabla\psi_E$ (the waves in question are potential waves in view of $\mathbf{H}_{\|} = 0$). Then the set Equations (19-22) can be reduced to two coupled equations with respect to the electric, $\psi_E$, and gravitation, $\psi_G$, potentials, *viz.*

$$\begin{aligned} \psi_E \left(1 - \sum_\alpha \frac{\omega_{p,\alpha}^2}{\omega^2 - k^2 v_{T,\alpha}^2}\right) - \psi_G \cdot 4\pi \sum_\alpha \frac{Q_\alpha n_{0,\alpha}}{\omega^2 - k^2 v_{T,\alpha}^2} &= 0 \\ \psi_E \cdot 4\pi G \sum_\alpha \frac{Q_\alpha n_{0,\alpha}}{\omega^2 - k^2 v_{T,\alpha}^2} + \psi_G \left(1 + \sum_\alpha \frac{\omega_{G,\alpha}^2}{\omega^2 - k^2 v_{T,\alpha}^2}\right) &= 0. \end{aligned} \tag{27}$$

The corresponding dispersion relation for the longitudinal waves is (Bliokh and Yaroshenko, 1985; 1991)

$$\left(1 - \sum_\alpha \frac{\omega_{p,\alpha}^2}{\omega^2 - k^2 v_{T,\alpha}^2}\right)\left(1 + \sum_\alpha \frac{\omega_{G,\alpha}^2}{\omega^2 - k^2 v_{T,\alpha}^2}\right) + K_T^2 = 0, \tag{28}$$

where $\omega_{p,\alpha} = (4\pi Q_\alpha^2 n_{0,\alpha} m_\alpha^{-1})^{1/2}$ and $\omega_{G,\alpha} = (4\pi G m_\alpha n_{0,\alpha})^{1/2}$ are the plasma and the Jeans frequency, respectively, for the particles of species $\alpha$, and

$$K_T = 4\pi G^{1/2} \sum_\alpha Q_\alpha n_{0,\alpha} (\omega^2 - k^2 v_{T,\alpha}^2)^{-1}$$

is the coupling factor owing to the different thermal velocities. The frequencies $\omega_{p,\alpha}$ and $\omega_{G,\alpha}$ characterize the time scales of cooperative effects that are associated, respectively, with the electric and gravitational interaction. Let us recall their physical meaning.

The principal cooperative effect in a cold electron-ion plasma consists of charge density variations known as the Langmuir oscillations. Suppose the plasma electrons have been displaced with respect to the homogeneous ionic background in such a way that the number density of electrons has increased at some point. As a result, an electric field will arise, oriented so as to restore the electric neutrality of the plasma, i.e. to move the electrons back to their initial position. While moving towards their point of equilibrium, the electrons will certainly overshoot it because of their inertia. Hence, they are bound to start vibrating about the equilibrium at the characteristic frequency $\omega_{p,e}$ known as the electron Langmuir, or plasma frequency ($n_{0,e}$ is the electron number density in the unperturbed plasma).

Now, consider cooperative effects in a simple self-gravitating model. The simplest system of the kind would be an infinite homogeneous medium. That was the model first analyzed in the classical paper by Jeans (1929). Strictly speaking, the problem that he formulated, i.e. the behavior of small perturbations in density should be considered for a time-dependent background. (In contrast to the electron-ion plasma, a uniform infinite self-gravitational medium actually is time dependent (Fridman and Polyachenko, 1983).) However, the principal qualitative results obtained by Jeans in a steady-state model are correct, in particular the conclusion that small perturbations in a self-gravitating system increase with time exponentially, at a rate $\omega_G^{-1}$. The underlying physics of this instability is quite simple, i.e. an ensemble of sufficiently massive particles is compressed under the action of its own gravity. This principal instability of gravitational media is known as the Jeans instability and $\omega_G$ is called the Jeans frequency.

Since the particles of a self-gravitational plasma interact through both electrical and gravitation forces, the possible cooperative effects are eventually conditioned by a competition of two tendencies, one for the development of Langmuir oscillations or waves of frequency $\omega_{p,\alpha}$ and the other for the growth of aperiodic perturbations characterized by the scale time $\omega_{G,\alpha}^{-1}$. As a result, the dielectric constant $\varepsilon$ of such a medium is markedly different from that of the common plasma, including the frequency dependence $\varepsilon = \varepsilon(\omega)$. New branches can appear in the spectrum of free oscillations, and the criteria for the stability of wavelike perturbations, as well as their propagation conditions, become altered (Bliokh and Yaroshenko, 1988; 1989).

### 3.1.3 Electrostatic Waves

In this Section, we will consider the characteristic Equation (28) in more detail. Its first factor,

$$\varepsilon_p(\omega, k) = 1 - \sum_\alpha \frac{\omega_{p,\alpha}^2}{\omega^2 - k^2 v_{T,\alpha}^2}, \tag{29}$$

arises from the "purely electric" interaction between particles, and hence it coincides with the longitudinal dielectric constant of the common plasma. The inside of the second brackets,

$$\varepsilon_G(\omega, k) = 1 + \sum_\alpha \frac{\omega_{G,\alpha}^2}{\omega^2 - k^2 v_{T,\alpha}^2}, \tag{30}$$

is associated with gravitation forces alone, and the equation $\varepsilon_G(\omega, k) = 0$ determines free waves in a self-gravitating electrically neutral (i.e. $Q_\alpha = 0$) medium (Zeldovich and Novikov, 1975; Fridman and Polyachenko, 1981). Finally, the last term in the left-hand side of Equation (28) that can be written as

$$K_T^2(\omega, k) = \left( \sum_\alpha \frac{\omega_{p,\alpha}\omega_{G,\alpha}}{\omega^2 - k^2 v_{T,\alpha}^2} \right)^2 \tag{31}$$

is responsible for coupling of the gravitational and plasma effects. The coupling coefficient turns into zero if either the plasma frequencies $\omega_{p,\alpha}(\alpha = 1, 2, ...N)$ or the Jeans frequencies $\omega_{G,\alpha}(\alpha = 1, 2, ...N)$ are zeros for all the species. $K_T^2$ can also vanish with $\omega_{p,\alpha} \neq 0$ if thermal velocities of the particles are the same for all the species, i.e. $v_{T,\alpha} = v_T$ (in particular, if $v_{T,\alpha} = 0$, i.e. in a cold plasma). Certainly, the latter is true only for quasineutral media, $\sum_{(\alpha)} Q_\alpha n_{0,\alpha} = 0$.

With $K_T^2 = 0$, the electrical and the gravitational disturbances described by the equations $\varepsilon_p(\omega, k) = 0$ and $\varepsilon_G(\omega, k) = 0$, respectively, can develop independently. Otherwise, eigenwaves of the medium obey the characteristic equation of the general form

$$\varepsilon_p(\omega, k)\varepsilon_G(\omega, k) + K_T^2(\omega, k) = 0 \tag{32}$$

whose solutions $\omega = \omega(\mathbf{k})$ can be represented as complex combinations of the Jeans and plasma frequencies for individual components and show a marked dependence on the thermal velocities. To analyze Equation (28) with $K_T^2 \neq 0$, it proves convenient to bring it to the form

$$1 - \sum_\alpha \frac{\omega_{p,\alpha}^2 - \omega_{G,\alpha}^2}{\omega^2 - k^2 v_{T,\alpha}^2} - \frac{1}{2} \sum_{\substack{\alpha,\beta \\ \alpha \neq \beta}} \frac{(\omega_{p,\alpha}\omega_{G,\beta} - \omega_{p,\beta}\omega_{G,\alpha})^2}{(\omega^2 - k^2 v_{T,\alpha}^2)(\omega^2 - k^2 v_{T,\beta}^2)} = 0. \tag{33}$$

In a few special cases, this structure can be simplified. First, this is the case of equal plasma and Jeans frequencies, i.e. $\omega_{p,\alpha} = \omega_{G,\alpha}(\alpha = 1, 2, ...N)$. Then the gravitational and the electric interaction forces get "subtracted" and the characteristic equation yields a set of eigenwave solutions whose phase velocities are equal to the thermal velocities for individual components of the medium. The dispersion law is linear,

$$\omega = k v_{T,\alpha}, \qquad \alpha = 1, 2, ...N, \tag{34}$$

which resembles acoustic waves. Note, however, that similar "sound waves" can only exist in a self-gravitational plasma obeying the "compensation"

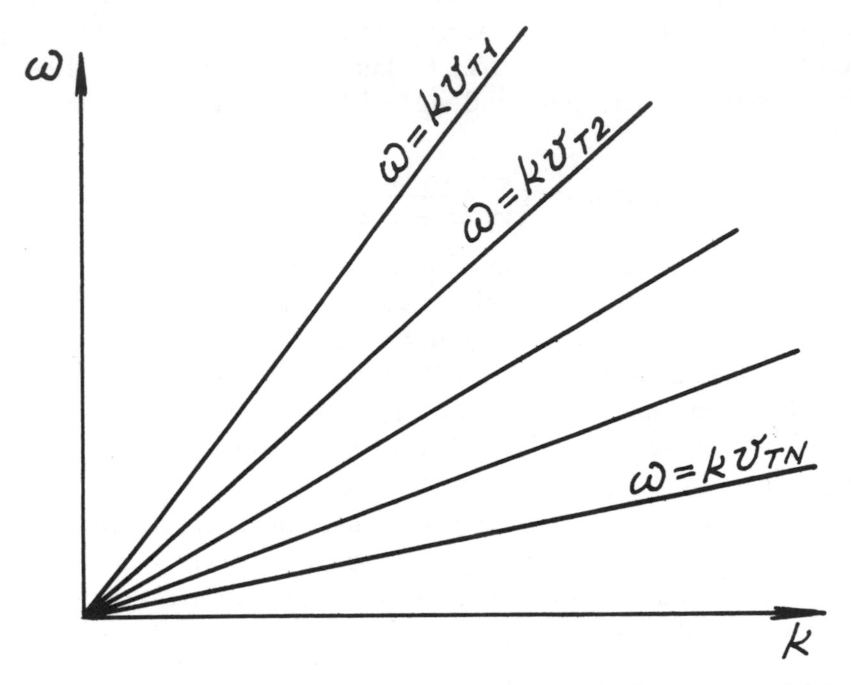


Figure 1: The Acoustic Wave Family in a Self-Gravitational Plasma with $\omega_{p,\alpha} = \omega_{G,\alpha}$ $(\alpha = 1, 2, \ldots N)$ and $v_{T,1} > v_{T,2} > \cdots > v_{T,N}$.

condition $\omega_{p,\alpha} = \omega_{G,\alpha}$, while they cannot develop either in conventional plasmas or in purely gravitational media (Figure 1). Another interesting case is $\omega_{p,\alpha}\omega_{G,\beta} = \omega_{p,\beta}\omega_{G,\alpha}$, with $\alpha \neq \beta$. The characteristic Equation (33) takes the simpler form

$$1 = \sum_{\alpha} \frac{\omega_{p,\alpha}^2 - \omega_{G,\alpha}^2}{\omega^2 - k^2 v_{T,\alpha}^2}. \tag{35}$$

Here, a variety of qualitatively different solutions are possible, depending on the relative magnitude of $\omega_{p,\alpha}$ and $\omega_{G,\alpha}$. With $\omega_{p,\alpha} > \omega_{G,\alpha}$ Equation (35) reduces to the familiar dispersion relation for longitudinal waves in a conventional multicomponent plasma, i.e. $\varepsilon_p(\omega, \mathbf{k}) = 0$, where $\varepsilon_p(\omega, \mathbf{k})$ is given by Equation (29), however with $\omega_{p,\alpha}^2$ replaced by $\omega_{p,\alpha}^2 - \omega_{G,\alpha}^2$. The spectrum of longitudinal plasma waves is quite well known (Krall and Trivelpiece, 1973; Akhiezer *et al.*, 1974). We shall not discuss it here, just noting that disturbances of that plasma type are always stable. To check the assertion, consider a graphical representation of Equation (35). Shown in Figure 2 are

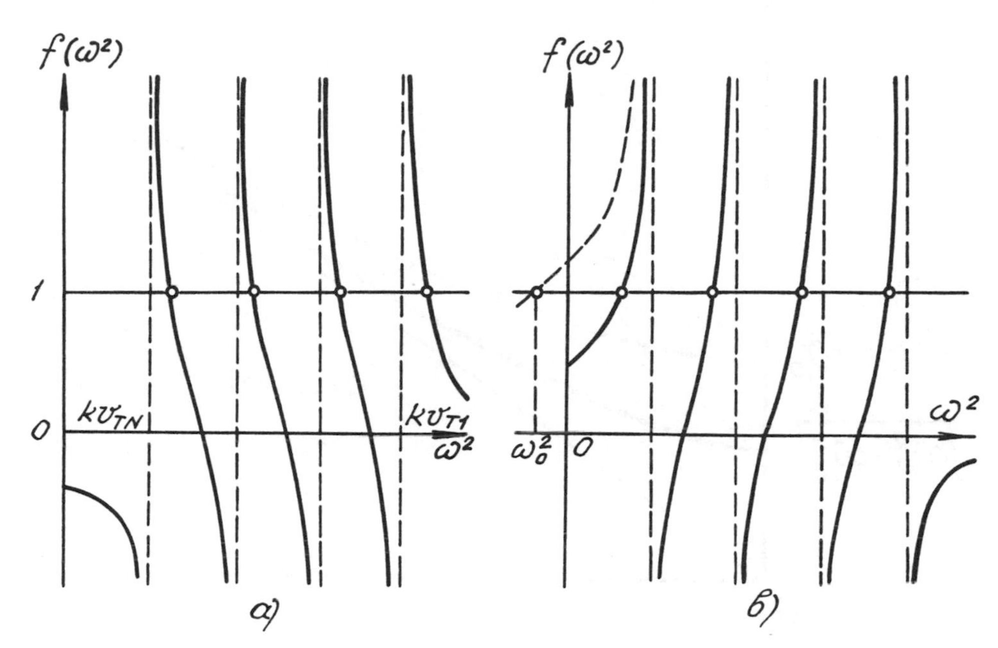


Figure 2: Roots of the Dispersion Relation Equation (35) with $\omega_{p,\alpha}\omega_{G,\beta} = \omega_{p,\beta}\omega_{G,\alpha}$ ($\alpha \neq \beta$): a) $\omega_{p,\alpha} > \omega_{G,\alpha}$; b) $\omega_{p,\alpha} < \omega_{G,\alpha}$ and $\sum_\alpha(\omega^2_{G,\alpha} - \omega^2_{p,\alpha})/k^2v^2_{T,\alpha} > 1$ (Broken Line) or $\sum_\alpha(\omega^2_{G,\alpha} - \omega^2_{p,\alpha})/k^2v^2_{T,\alpha} < 1$ (Solid Line).

plots of $y = f(\omega^2)$ and $y = 1$, $f(\omega^2)$ being the right-hand side of Equation (35) (in terms of $\omega^2$ rather than $\omega$). Figure 2a is for $\omega_{p,\alpha} > \omega_{G,\alpha}$, while 2b for $\omega_{p,\alpha} < \omega_{G,\alpha}$. As can be seen, with $\omega_{p,\alpha} > \omega_{G,\alpha}$ all the roots are positive, hence $\mathrm{Im}\omega = 0$ and the oscillations are stable. If the characteristic frequencies $\omega_{p,\alpha}$ and $\omega_{G,\alpha}$ obey the opposite inequality, i.e. $\omega_{p,\alpha} < \omega_{G,\alpha}$, then the dispersion relation can be written as $\varepsilon_G(\omega, k) = 0$, however with the part of the Jeans frequency for the $\alpha$-species played by $(\omega^2_{G,\alpha} - \omega^2_{p,\alpha})^{1/2}$. As is seen from Figure 2b, the equation may have a single negative root ($\omega_0^2 < 0$) if

$$\sum_\alpha \frac{\omega^2_{G,\alpha} - \omega^2_{p,\alpha}}{k^2 v^2_{T,\alpha}} > 1.$$

The disturbances with $\omega_0^2 < 0$ will be unstable. Indeed, their time dependence is described by a combination of terms $\exp(|\omega_0|t)$ and $\exp(-|\omega_0|t)$ of which the first one grows with $t$ exponentially. This instability of the self-

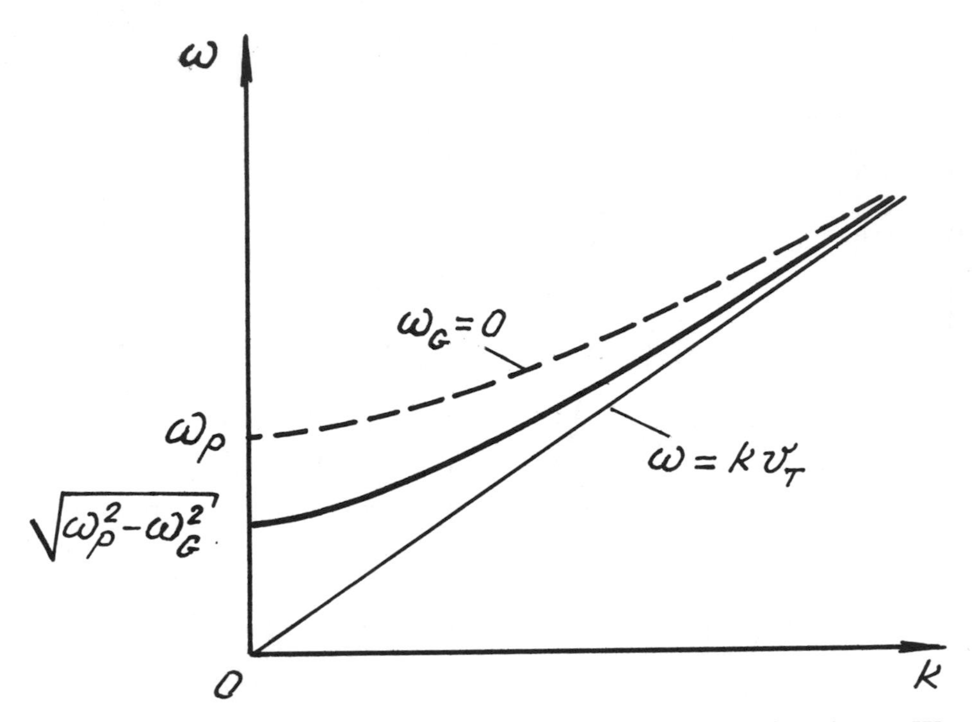


Figure 3: Effect of Self-Gravitation on the Dispersion Law for Plasma Waves ($\omega_G = 0$, Broken Line and $\omega_G \neq 0$, $\omega_p > \omega_G$ Solid Line).

gravitational plasma is physically identical to the Jeans instability (Fridman and Polyachenko 1981; 1983).

Thus, even the simpler particular versions of the characteristic Equation (33) show that the spectrum of free oscillations in the self-gravitational plasma is much richer than both in "pure" ion-electron plasmas or media composed of gravitating particles alone. By way of example, consider some practically important cases of single- and bicomponent plasmas.

1. Single-component medium ($\omega_{p,\alpha} = \omega_p$; $\omega_{G,\alpha} = \omega_G$ and $v_{T,\alpha} = v_T$). The dispersion law following from Equation (33) is

$$\omega = (\omega_p^2 - \omega_G^2 + k^2 v_T^2)^{1/2}. \tag{36}$$

With $\omega_p > \omega_G$, this represents a simple longitudinal wave in the plasma, characterized by the effective Langmuir frequency $\omega_{p,eff} = (\omega_p^2 - \omega_G^2)^{1/2} < \omega_p$. The appropriate $\omega = \omega(k)$ dependences for $\omega_G = 0$ and $\omega_G \neq 0$ are shown in Figure 3. As can be seen, the effect of the gravitational particle–

particle interaction is to bring the dispersion curve $\omega = \omega(k)$ closer to the asymptote $\omega = kv_T$.

With $\omega_p < \omega_G$, the dispersion law Equation (36) suggests the existence of a critical wavenumber $k_{cr}$ (or wavelength $\lambda_{cr} = 2\pi/k_{cr}$) separating the domains of stable ($\omega^2(k) > 0$, Im$\omega = 0$) and unstable ($\omega^2(k) < 0$) disturbances. The value is given by the condition $\omega^2 = 0$, i.e.

$$k_{cr} = \frac{(\omega_G^2 - \omega_p^2)^{1/2}}{v_T}. \tag{37}$$

The "shortwave" disturbances, $k > k_{cr}$, propagate in almost the same manner as the "thermal" sound waves (with a small correction to the phase velocity owing to the "effective Jeans frequency" $\omega_{G,eff} = (\omega_G^2 - \omega_p^2)^{1/2} < \omega_G$), *viz.*

$$\omega \approx kv_T \left(1 - \frac{\omega_G^2 - \omega_p^2}{2k^2 v_T^2}\right). \tag{38}$$

Meanwhile, the disturbances of a sufficiently large scale size (i.e. $k < k_{cr}$) can be unstable, for among the roots of the characteristic equation one corresponds to their growth in time,

$$\omega \approx i\omega_G \left(1 + \frac{k^2 v_T^2 + \omega_p^2}{2\omega_G^2}\right). \tag{39}$$

This behavior is quite understandable physically. Indeed, the gravitation potential of a condensation of scale size $\lambda \sim k^{-1}$ is proportional to $\lambda^2$, as $\psi_G \sim Gm/\lambda \sim G \cdot \rho\lambda^3/\lambda$, and hence the gravitation force $F$ is proportional to $\lambda$, $F \sim \partial\psi_G/\partial r \sim \psi_G/\lambda \sim \lambda$. On the other hand, the force owing to the pressure gradient varies as $1/\lambda$. Accordingly, the gravitation force prevails for disturbances of larger wavelengths $\lambda$ (which results in an instability), whereas pressure dominates for smaller $\lambda$ ($k > k_{cr}$). The dispersion curves $\omega = \omega(k)$ for $k > k_{cr}$ and $\omega'' = \text{Im}\omega(\text{k})$, $k < k_{cr}$ are shown in Figure 4 for $\omega_p = 0$ and $\omega_p \neq 0$. As $k$ increases, the curves $\omega = \omega(k)$ approach asymptotically to the dependences $\omega = kv_T$, either from above (with $\omega_p > \omega_G$, Figure 3) or from below (with $\omega_p < \omega_G$, Figure 4). In the limiting case $\omega_p = \omega_G$ the gravitation and electric forces become mutually compensated and only the "thermal" sound wave $\omega = kv_T$ remains.

2. Bicomponent medium. Consider a plasma consisting of two species of particles with sharply different masses (the subscript "1" will relate to the lighter and "2" to the heavier fraction, i.e. $m_1 \ll m_2$).

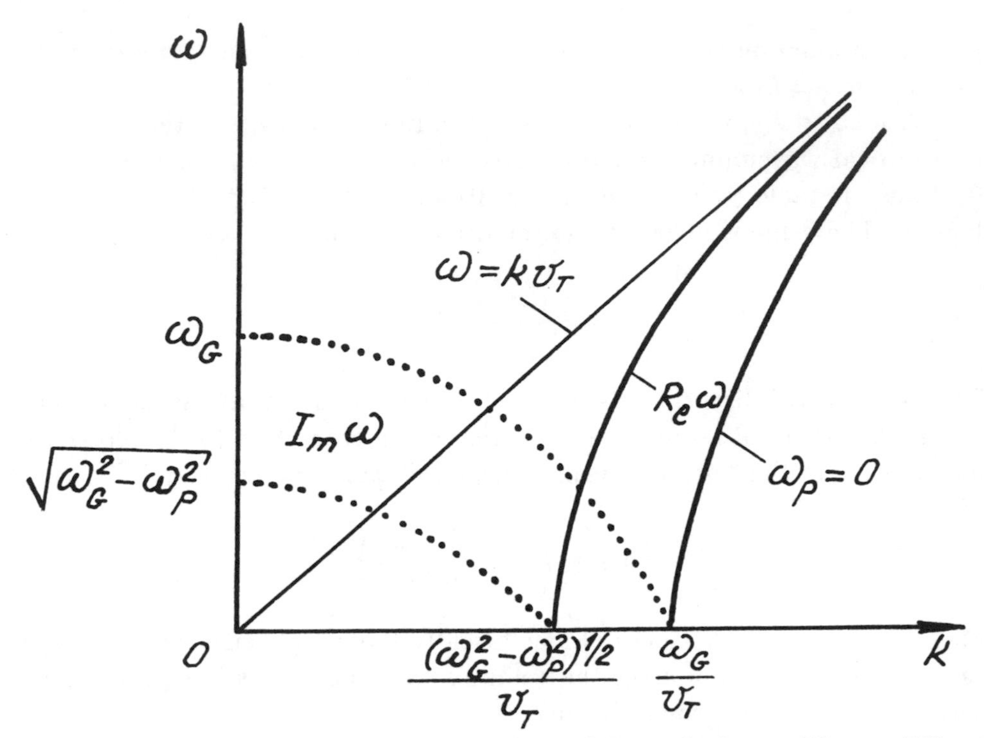


Figure 4: Reduction in the Instability Growth Rate Owing to Plasma Effects in a Gravitating Medium (the Dotted Line is for $\mathrm{Im}\,\omega$ and the Solid Line for $\mathrm{Re}\,\omega$).

For the lighter fraction we shall assume $\omega_{p,1} > \omega_{G.1}$; moreover $\omega_{G,1} \to 0$. It is also possible to neglect the effects associated with thermal motion of the heavier particles, i.e. to assume $v_{T,1} \gg v_{T,2} \to 0$. Then the dispersion relation Equation (33) can be reduced to a biquadratic equation in $\omega$ which is

$$1 \approx \frac{\omega_{p,1}^2}{\omega^2 - k^2 v_{T,1}^2} + \frac{\omega_{p,2}^2}{\omega^2 + \omega_{G,2}^2}. \tag{40}$$

Depending on the relative values of $\omega_{p,1}$ and $\omega_{p,2}$; $\omega_{p,1}$ and $kv_{T,1}$, and $\omega_{p,2}$ and $\omega_{G,2}$, the solutions $\omega^2 = \omega^2(k)$ of Equation (40),

$$\omega_{1,2}^2 \approx \frac{1}{2}(\omega_{p,1}^2 + k^2 v_{T,1}^2 - \omega_{G,2}^2 + \omega_{p,2}^2) \pm \left[\frac{1}{4}(\omega_{p,1}^2 + k^2 v_{T,1}^2 - \omega_{G,2}^2 + \omega_{p,2}^2)^2 - k^2 v_{T,1}^2(\omega_{p,2}^2 - \omega_{G,2}^2) + \omega_{p,1}^2 \omega_{G,2}^2\right]^{1/2} \tag{41}$$

can be either positive or negative and even complex. E.g., with $\omega_{p,1}^2 \gg \omega_{p,2}^2 > \omega_{G,2}^2$ and $k^2 v_{T,1}^2 \gg \omega_{p,1}^2$

$$\omega_1^2 \approx k^2 v_{T,1}^2 > 0 \text{ and } \omega_2^2 \approx \omega_{p,2}^2 - \omega_{G,2}^2. \tag{42}$$

The first of these represents a "sound" wave traveling at the thermal velocity through the lighter component alone, whereas the second corresponds to plasma oscillations of the heavier particles at the frequency $(\omega_{p,2}^2 - \omega_{G,2}^2)^{1/2}$. With $k^2 v_{T,1}^2 \ll \omega_{p,1}^2$, the solutions of Equation (40) become

$$\omega_1^2 \approx \omega_{p,1}^2 \text{ and } \omega_2^2 \approx -\omega_{G,2}^2, \tag{43}$$

hence the second solution describes a Jeans instability with the growth rate determined by the number density of the heavier particles.

In the other limiting case, $\omega_{p,1} > \omega_{G,2} > \omega_{p,2}$, the solutions are

$$\omega_1^2 \approx \omega_{p,1}^2 + k^2 v_{T,1}^2 \text{ and } \omega_2^2 \approx -\omega_{G,2}^2. \tag{44}$$

Apparently, the first of these corresponds to "normal" longitudinal plasma waves in the gas of light-weight charged particles, while the second to the Jeans instability of the macroscopic component.

Without dwelling on Equation (40) any longer, note another interesting feature of the self-gravitational plasma. Let $\omega_{p,1}^2 > \omega_{p,2}^2$, $\omega_{G,2}^2 > \omega_{G,1}^2$ and $v_{T,1}^2 > v_{T,2}^2 \to 0$. While the eigenwave spectrum of the common plasma includes an ion-acoustic wave with the dispersion law

$$\omega \approx k v_{T,1} \omega_{p,2} / \omega_{p,1} = k (T_1/m_2)^{1/2} \qquad (\text{ here } n_{1,0} = n_{2,0}), \tag{45}$$

its analog in the self-gravitational plasma medium obeys

$$\omega \approx \left\{ \left[ (\omega_{p,2}^2 - \omega_{G,2}^2) k^2 v_{T,1}^2 - \omega_{p,1}^2 \omega_{G,2}^2 \right] (\omega_{p,1}^2 - \omega_{G,2}^2)^{-1} \right\}^{1/2}. \tag{46}$$

The phase velocity of the ion-acoustic wave Equation (45) is determined by the temperature of the lighter (electronic) plasma component and mass of the heavier particles. The wave of Equation (46) can propagate only if some conditions are satisfied. It is necessary that either $\omega_{p,1}^2 < \omega_{G,2}^2$ or $\omega_{p,2}^2 > \omega_{G,2}^2$, with $k^2 v_{T,1}^2 > (\omega_{p,1}^2 \omega_{G,2}^2)/\omega_{p,2}^2$.

The corresponding dispersion curves $\omega = \omega(k)$ are shown in Figure 5. If $\omega_{p,2}^2 > \omega_{G,2}^2$, then the wave of Equation (46) can exist only for $k > \omega_{p,1}\omega_{G,2}/v_{T,1}\omega_{p,2}$. Since $\omega_{p,1}\omega_{G,2}/v_{T,1}\omega_{p,2} = \omega_{G,2}/(\omega_{p,2}\lambda_{D,1})$, the wave is

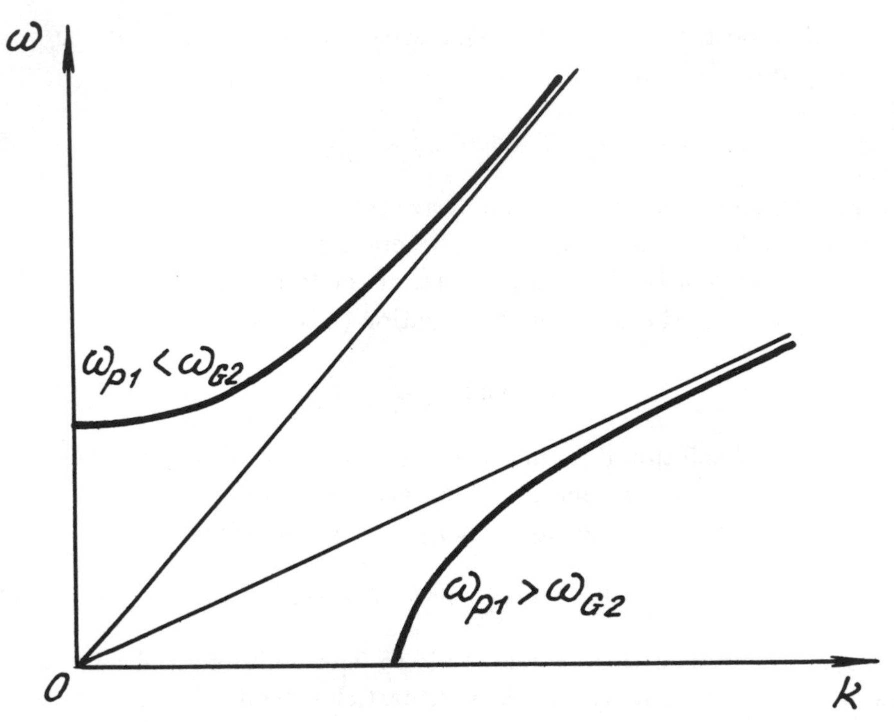


Figure 5: Dispersion Curves for Analogs to Ion-Acoustic Waves in the Self-Gravitational Plasma.

characterized by a spatial scale limited from above by a few Debye lengths of the lighter component, i.e. $\lambda = (2\pi/k) = (\omega_{p,2}\lambda_{D,1})/\omega_{G,2}$. The dispersion law for the case,

$$\omega \approx \frac{\omega_{p,2}kv_{T,1}}{\omega_{p,1}}\left(1 + \frac{\omega_{G,2}^2}{2\omega_{p,1}^2}\right)$$

is only slightly different from Equation (45) characteristic of the common plasma.

In the opposite case, $\omega_{p,1}^2 < \omega_{G,2}^2$, the dispersion law is altered substantially and resembles of the law for plasma waves,

$$\omega \approx (\omega_{p,1}^2 + k^2 v_{T,1}^2)^{1/2}(1 + \omega_{p,1}^2/2\omega_{G,2}^2).$$

### 3.1.4 Dielectric Constant

In contrast to the common ion-electron plasma, charged particles of the self-gravitating medium interact through forces of two different kinds. Accordingly, the frequency dependence shown by the dielectric constant of such a plasma is also different from that of the conventional. We are going to consider it in more detail, using the assumption that components $\varepsilon_{ij}(\omega, \mathbf{k})$ of the dielectric constant tensor can be expressed in a standard way through those of the complex conductivity tensor $\sigma_{ij}(\omega, \mathbf{k})$,

$$\varepsilon_{ij}(\omega, \mathbf{k}) = \delta_{ij} + \frac{4\pi i \sigma_{ij}(\omega, \mathbf{k})}{\omega}. \tag{47}$$

The conductivity tensor sets a relation in the Fourier space between the current density $\mathbf{j}$ and the electric field vector $\mathbf{E}(\omega, \mathbf{k})$,

$$j_i(\omega, \mathbf{k}) = \sigma_{ij}(\omega, \mathbf{k}) E_j(\omega, \mathbf{k}). \tag{48}$$

According to Equation (6), the perturbation in the current density is

$$\mathbf{j} = \sum_\alpha Q_\alpha n_{0,\alpha} \mathbf{v}_\alpha. \tag{49}$$

Making use of the equations of motion for the "liquid particles", the continuity equation and Poisson's equation for the gravitation potential $\psi_G$(*cf.* Equations (11), (12) and (15)), we can obtain the perturbed particle velocity,

$$\mathbf{v}_\alpha = \frac{1}{i\omega} \cdot \frac{Q_\alpha}{m_\alpha} \mathbf{E} - \frac{\mathbf{k}}{k^2 \omega^2} \sum_\beta \omega_{G,\beta}^2 (\mathbf{k}\mathbf{v}_\beta) + \frac{\mathbf{k}(\mathbf{k}\mathbf{v}_\alpha)}{\omega^2} v_{T,\alpha}^2. \tag{50}$$

(As before, all the descriptive variables are assumed to vary as $\exp(i\mathbf{k}\mathbf{r} - i\omega t)$). It is convenient to choose a reference system with the $z$-axis directed along the wave vector $\mathbf{k}$. Then the projections of Equation (50) to the axes $Ox$, $Oy$ and $Oz$ are

$$v_{\alpha,x} = -iQ_\alpha E_x / (\omega m_\alpha) \tag{51}$$

$$v_{\alpha,y} = -iQ_\alpha E_y / (\omega m_\alpha) \tag{52}$$

and

$$v_{\alpha,z} = -(i\omega Q_\alpha E_z / m_\alpha + \sum_\beta \omega_{G,\beta}^2 v_{\beta,z})(\omega^2 - k^2 v_{T,\alpha}^2)^{-1}. \tag{53}$$

Upon multiplying $v_{\alpha,z}$ by $\omega_{G,\alpha}^2$ and performing summation over all particle species we can arrive at

$$\varepsilon_G \sum_\beta \omega_{G,\beta}^2 v_{\beta,z} = -i\omega E_z G^{1/2} K_T, \tag{54}$$

with $\varepsilon_G(\omega, \mathbf{k})$ and $K_T$ given by Equations (30) and (31), respectively. By combining Equations (54) and (53) we obtain an explicit form for $v_{\alpha,z}$, *viz.*

$$v_{\alpha,z} = \left[(-i\omega E_z/(\omega^2 - k^2 v_{T,\alpha}^2)\right](Q_\alpha/m_\alpha - K_T G^{1/2}/\varepsilon_G). \tag{55}$$

Now we are in a position to explicitly write the components of Equation (49) and derive the conductivity tensor $\sigma_{ij}$, *viz.*

$$\sigma_{xx} = \sigma_{yy} = -i \sum_\alpha Q_\alpha^2 n_{0,\alpha}/(\omega m_\alpha) \tag{56}$$

and

$$\sigma_{zz} = -i\omega \sum_\alpha \left( \frac{Q_\alpha^2 n_{0,\alpha}}{m_\alpha} - \frac{Q_\alpha n_{0,\alpha} G^{1/2} K}{\varepsilon_G} \right) /(\omega^2 - k^2 v_{T,\alpha}^2), \tag{57}$$

the rest of $\sigma_{ij}$ components being zeros. By substituting $\sigma_{ij}(\omega, \mathbf{k})$ into Equation (47) we arrive at

$$\varepsilon_{xx} = \varepsilon_{yy} = 1 - \sum_\alpha \omega_{p,\alpha}^2/\omega^2 \tag{58}$$

$$\varepsilon_{zz} = 1 - \sum_\alpha \omega_{p,\alpha}^2/(\omega^2 - k^2 v_{T,\alpha}^2) + K_T^2/\varepsilon_G. \tag{59}$$

The off-diagonal components of $\varepsilon_{ij}$ are all zeros. Thus, the dielectric constant of a self-gravitational plasma is a diagonal second-order tensor, similar to that of the conventional isotropic plasma. If $T_\alpha = 0$, then $K_T$ vanishes owing to the quasineutrality condition, and hence $\varepsilon_{xx} = \varepsilon_{yy} = \varepsilon_{zz}$, or the tensor $\hat{\varepsilon}$ reduces to a scalar. This scalar dielectric constant is exactly the same as for the common plasma. However, the plasma waves with $\varepsilon_p = 1 - \sum_{(\alpha)} \omega_{p,\alpha}^2/\omega^2 = 0$ will not be the only longitudinal mode possible in the self-gravitational plasma. The medium can also support unstable longitudinal disturbances, obeying

$$\varepsilon_G(\omega, \mathbf{k}) = 1 + \sum_\alpha \omega_{G,\alpha}^2/\omega^2 = 0. \tag{60}$$

(Indeed, it can be seen from Equation (54) that $K_T = 0$ can hold only with $\varepsilon_G(\omega, \mathbf{k}) = 0$). These are associated with purely gravitational interaction of the particles. The growth rate is $\mathrm{Im}\omega = \left[\sum_{(\alpha)} \omega_{G,\alpha}^2\right]^{1/2}$. With electromagnetic parameters of the medium expressed in terms of the dielectric constant $\hat{\varepsilon}(\omega, \mathbf{k})$, the dispersion relation can be written immediately

$$\det \left\| k^2 \delta_{ij} - k_i k_j - \frac{\omega^2}{c^2} \varepsilon_{ij}(\omega, \mathbf{k}) \right\| = 0. \tag{61}$$

In the frame of reference with $\mathbf{k} \parallel Oz$ that we have chosen, Equation (61) reduces to two independent equations, namely

$$k^2 = \frac{\omega^2}{c^2}\varepsilon_{xx} = \frac{\omega^2}{c^2}\varepsilon_{yy} \tag{62}$$

and $\varepsilon_{zz} = 0$. It is only natural that these coincide with the above derived Equations (26) and (28) governing transverse and longitudinal waves, respectively.

From the knowledge of $\varepsilon_{ij}(\omega, \mathbf{k})$ it is also possible to estimate the Debye length in the self-gravitational plasma. To that end, let us set $\omega = 0$ in the dispersion relation for longitudinal waves and solve that as an equation for $\mathbf{k}$. A definition of the Debye length is $\lambda_D = [\mathrm{Im}k(\omega = 0)]^{-1}$, from which we can derive a rather simple formula for the Debye length in the above described multicomponent medium with $\omega_{p,\alpha}\omega_{G,\beta} = \omega_{p,\beta}\omega_{G,\alpha} (\alpha \neq \beta)$, *viz.* $\lambda_D^2 = \sum_{(\alpha)}(\omega_{p,\alpha}^2 - \omega_{G,\alpha}^2)/v_{T,\alpha}^2$. As long as the electric interaction between the plasma particles dominates over gravitation forces (i.e. $\omega_{p,\alpha}^2 > \omega_{G,\alpha}^2$), Debye's electrostatic screening remains effective but the scale length $\lambda_D$ is increased. The opposite inequality, $\omega_{p,\alpha}^2 < \omega_{G,\alpha}^2$, indicates a prevailing role of the gravitational attraction of particles, in which case formally $\lambda_D^2 < 0$ and the Debye length concept becomes senseless.

### 3.1.5 Energy Losses of Moving Particles

In the conventional plasma, electromagnetic waves are known to be excited by electric currents, i.e. charged particles moving through the medium. The situation may be different with the inclusion of self-gravitation effects. In particular, electromagnetic waves can be excited in a self-gravitational plasma by a stream of electrically neutral, as well as charged particles. Indeed, a heavy neutral particle moving through the dusty medium acts upon grains of the medium through its gravitation field. The dust grains start moving and produce, owing to their electrical charge, electric currents that excite an electromagnetic field. Contrary to this, a massive charged particle moving through such a medium may fail to excite any perturbations, should the gravitational perturbation be compensated by an electric disturbance. Such a compensation is only possible if the mass $M_0$ of the probe particle and its charge $Q_0$ are related in some special way. To analyze these unusual effects, let us consider energy losses of a particle moving through the self-gravitational plasma at a velocity $\mathbf{v}_0$.

In the conventional plasma, the energy $\mathcal{E}$ spent by a moving charged particle to excite plasma waves can be evaluated as the work done against the braking force owing to the electric field $\mathbf{E}$ at the point where the particle is at the time moment $t$. The change in energy per unit time is

$$d\mathcal{E}/dt = Q_0(\mathbf{v}_0\mathbf{E})_{\mathbf{r}=\mathbf{v}_0 t}. \tag{63}$$

With allowance for gravitation effects, this becomes

$$\frac{d\mathcal{E}}{dt} = Q_0(\mathbf{v}_0\mathbf{E})_{\mathbf{r}=\mathbf{v}_0 t} + M_0(\mathbf{v}_0\mathbf{\Gamma})_{\mathbf{r}=\mathbf{v}_0 t}, \tag{64}$$

where $\mathbf{E} = -\nabla\psi_E$ and $\mathbf{\Gamma} = -\nabla\psi_G$. The two potentials can be found from the basic equation set Equations (11-12), plus Poisson's equations

$$\begin{aligned}
\Delta\psi_E &= -4\pi\left(\sum_\alpha Q_\alpha n_\alpha + Q_0\delta(\mathbf{r}-\mathbf{v}_0 t)\right), \\
\Delta\psi_G &= 4\pi G\left(\sum_\alpha m_\alpha n_\alpha + M_0\delta(\mathbf{r}-\mathbf{v}_0 t)\right),
\end{aligned} \tag{65}$$

with $\delta$ being Dirac's delta. The Fourier components of $\mathbf{E}$ and $\mathbf{\Gamma}$ following from $\psi_E$ and $\psi_G$ are

$$\mathbf{E}_\mathbf{k} = -i\mathbf{k}\frac{4\pi e^{-i\mathbf{k}\mathbf{v}_0 t}}{k^2\varepsilon(\omega,\mathbf{k})}\left[Q_0 - 4\pi G M_0 \frac{K_T^{1/2}(\omega,\mathbf{k})}{1+\sum_{(\alpha)}\omega_{G,\alpha}^2/(\omega^2 - k^2 v_{T,\alpha}^2)}\right]$$

and

$$\mathbf{\Gamma}_\mathbf{k} = -i\mathbf{k}\frac{4\pi e^{-i\mathbf{k}\mathbf{v}_0 t}}{k^2\varepsilon(\omega,\mathbf{k})}\left\{[M_0\varepsilon_p(\omega,\mathbf{k}) + 4\pi Q_0 K_T^{1/2}(\omega,\mathbf{k})]\varepsilon_G^{-1}(\omega,\mathbf{k})\right\}. \tag{66}$$

Applying the inverse Fourier transformation and substituting the result into Equation (64) we can arrive, with an account of $d^3\mathbf{k} = dk_{||}d^2\mathbf{k}_\perp \to 2\pi\mathbf{k}_\perp\, d\mathbf{k}_\perp\, d(\omega/v_0)_{\omega=\mathbf{k}\mathbf{v}_0}$ and $\mathbf{k}_{||} = \mathbf{v}_0(\mathbf{k}\mathbf{v}_0)/v_0^2$, at

$$\frac{d\mathcal{E}}{dt} = -\frac{i}{\pi v_0}\int_0^{k_0} d\mathbf{k}_\perp\,\mathbf{k}_\perp \int_{-\infty}^{\infty}\left\{d\omega\,\omega\left[Q_0^2\left(1+\sum_\alpha\frac{\omega_{G,\alpha}^2}{\omega^2(1-v_{T,\alpha}^2/v_0^2) - k_\perp^2 v_{T,\alpha}^2}\right) - \right.\right.$$

$$\left.\left. -2Q_0M_0G^{1/2}K_T^{1/2}(\omega,\mathbf{k}) - GM_0^2\left(1-\sum_\alpha\frac{\omega_{p,\alpha}^2}{\omega^2(1-v_{T,\alpha}^2/v_0^2) - k_\perp^2 v_{T,\alpha}^2}\right)\right]\right\}\times$$

$$\times \left\{ (k_\perp^2 + \omega^2/v_0^2)\varepsilon(\omega, \mathbf{k}) \left( 1 + \sum_\alpha \frac{\omega_{G,\alpha}^2}{\omega^2(1 - v_{T,\alpha}^2/v_0^2) - k_\perp^2 v_{T,\alpha}^2} \right) \right\}^{-1}. \quad (67)$$

Here $k_0$ is the highest value of $k_\perp$ allowable in the classical approach to the description of collisions between the probe particle and particles of the medium. The integration along the real $\omega$-axis can be performed unambiguously with the use of the standard substitution

$$\frac{1}{\varepsilon(\omega)} = \frac{P}{\varepsilon(\omega)} - i\pi\delta\{\varepsilon(\omega)\},$$

where $P$ denotes the Cauchy principal value and

$$\delta\{\varepsilon(\omega)\} = \sum_s \frac{\delta(\omega - \omega_s)}{\varepsilon'(\omega_s)}$$

with $\varepsilon(\omega_s) = 0$. The latter equation can be recognized as the dispersion relation for longitudinal waves in the plasma and, as we have seen in this Section, that may be rather complex. We shall therefore consider, as before, the simpler particular cases.

For a cold single-component medium, Equation (33) is characterized by two real roots (with $\omega_p^2 > \omega_G^2$). The rate of energy losses in this case is

$$\frac{d\mathcal{E}}{dt} \simeq -\frac{4\pi n_0(Q_0Q - M_0Gm)^2}{mv_0} \log\left(k_0v_0(\omega_p^2 - \omega_G^2)^{-1/2}\right) \quad (68)$$

with $\omega_p^2 > \omega_G^2$ and $k_0^2v_0^2 \gg \omega_p^2 - \omega_G^2$. The roots of the characteristic equation are imaginary if $\omega_p^2 < \omega_G^2$, in which case the integrand in Equation (67) has no poles on the real axis. As a result,

$$\frac{d\mathcal{E}}{dt} = 0, \quad (\omega_p^2 < \omega_G^2). \quad (69)$$

The $|d\mathcal{E}/dt|$ *vs* $\omega_p^2$ dependence for $M = 0$ is shown schematically in Figure 6 (curve 2). At $\omega_r^2 = \omega_G^2$ the function shows a singularity, namely $|d\mathcal{E}/dt| = 0$ at $\omega_p^2 = \omega_G^2 - 0$ and $|d\mathcal{E}/dt| \to \infty$ at $\omega_p^2 = \omega_G^2 + 0$. The singular behavior can be removed by taking wave absorption in the medium into account. With allowance for particle collisions described in terms of a collision frequency $\nu$, the dielectric constant $\varepsilon$ of the single-component medium becomes $\varepsilon = 1 - \omega_p^2/(\omega^2 + i\nu\omega + \omega_G^2)$. If substituted into Equation (67), this yields

$$\frac{d\mathcal{E}}{dt} \simeq -\frac{(Q_0\omega_p - G^{1/2}M_0\omega_G)^2}{v_0} \log\frac{k_0v_0}{\nu}; \qquad \omega_p^2 = \omega_G^2, \qquad k_0v_0 \gg \nu. \quad (70)$$

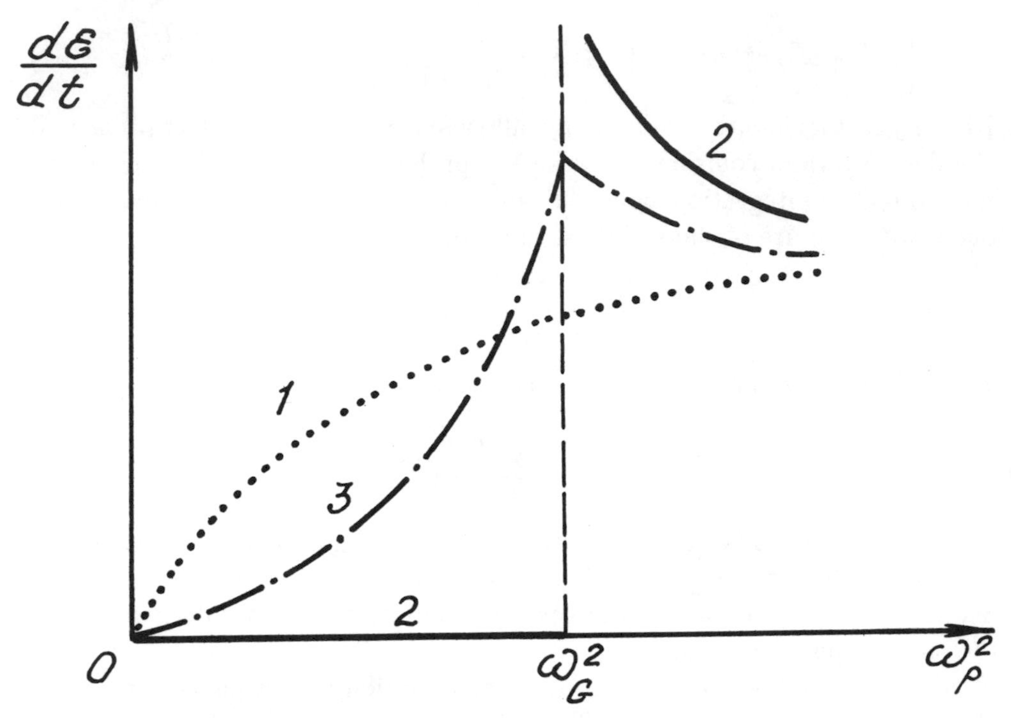


Figure 6: Energy Losses of a Massive Charged Particle Moving through a Plasma: Curve 1 is for the Conventional Plasma; Curve 2 for a Collisionless Self-Gravitational Plasma, and Curve 3 for a Self-Gravitational Plasma with Collisions.

Curve 3 in Figure 6 shows the $|d\mathcal{E}/dt|$ *vs* $\omega_p^2$ dependence for $\nu \neq 0$, $M_0 = 0$. It can be compared with a similar dependence for the common plasma (curve 1) given by (Akhiezer *et al.*, 1974)

$$\frac{d\mathcal{E}}{dt} \simeq -\frac{Q_0^2\omega_p^2}{v_0}\log\frac{k_0 v_0}{\omega_p}. \tag{71}$$

As can be seen from the Figure, the three curves are markedly different even with $M_0 = 0$. The differences are more pronounced with $M_0 \neq 0$. Of greatest interest is the effect shown by self-gravitational plasmas, namely excitation by a neutral particle of longitudinal oscillations involving an electric field component ($\mathbf{E}_k \neq 0$ with $Q_0 = 0$). The energy lost by the neutral particle is given by

$$\frac{d\mathcal{E}}{dt} \simeq -GM_0^2\frac{\omega_G^2}{v_0}\log\frac{k_0 v_0}{(\omega_p^2-\omega_G^2)^{1/2}}, \qquad \omega_p^2 > \omega_G^2. \tag{72}$$

The rate of the energy losses, Equation (68), happens to depend on the sign of the charges $Q_0$ and $Q$; specifically it can be either higher than in Equation (71) ($Q_0$ and $Q$ are of opposite signs) or lower than that ($Q_0$ and $Q$ are of the same sign). Moreover, the polarization losses can vanish if some special relation between $Q_0$, $M_0$ and parameters of the medium holds, *viz.* $Q_0/M_0 = Gm/Q$. This can only occur if the moving probe particle has a charge of the same sign as the ambient plasma particles, and their electrostatic repulsion is balanced by gravitational attraction.

The pattern described will be somewhat different in the case of a bicomponent medium. The rate of energy losses will show a much more complex dependence upon $\omega_{p,1,2}^2$, $\omega_{G,1,2}^2$ and $k^2 v_{T,1,2}^2$. Once again, we shall consider the simpler case of sharply different masses of the two kinds of particles, assuming as before $\omega_{p,1}^2 \gg \omega_{G,1}^2$; $\omega_{p,2}^2 \ll \omega_{G,2}^2$ and $v_0^2 \gg v_{T,1}^2 \gg v_{T,2}^2 \to 0$. Integrating Equation (67) we arrive at

$$\frac{d\mathcal{E}}{dt} \simeq -\frac{\omega_{p,1}^2}{v_0}\left[Q_0^2 - 2G^{1/2}M_0Q_0\frac{\omega_{p,1}\omega_{G,1}+\omega_{p,2}\omega_{G,2}}{\omega_{p,1}^2+\omega_{G,2}^2} - \right. \\ \left. -GM_0^2\frac{\omega_{p,2}^2 v_{T,1}^2}{(\omega_{p,1}^2+\omega_{G,2}^2)v_0^2}\right]\log(k_0 v_0/\omega_{p,1}), \tag{73}$$

where we have assumed $k_0^2 v_0^2 \gg \omega_{p,1}^2$, $\omega_{G,2}^2$ and $k_0^2 v_{T,1}^2 \ll \omega_{p,1}^2$, $\omega_{G,2}^2$. As can be seen, the energy losses in a bicomponent medium cannot turn to zero for any relation between $\omega_{p,1}$, $\omega_{G,1,2}$ and $k^2 v_{T,1,2}^2$. Indeed, the dispersion relation has at least two real solutions. However, the particle will not be decelerated if its charge and mass are related so as to nullify the square brackets in Equation (73).

Generally, both charged and neutral particles moving through a multicomponent self-gravitational plasma are capable of exciting waves similar to the longitudinal waves of conventional plasmas.

## 3.2 Longitudinal Waves in Dusty Plasmas

Digressing for a time from the effects of self-gravitation, let us consider the propagation of electric waves through just a dusty plasma. Of particular interest are longitudinal waves on which we will concentrate in this Section. The presence of dust grains in a plasma, even if they are fixed (i.e. $m_d \to \infty$), can affect the properties of ion-acoustic waves, and the special "dusty" effects are much more varied with an account of the dust grain dynamics, continuity of their size spectrum and variations of charge.

### 3.2.1 Eigenwave Spectrum. Dust-Sound Waves

The longitudinal waves in a dusty plasma are governed by the characteristic (modal) equation that can be obtained from Equation (33) by setting $\omega_{G,\alpha} = 0$. Assuming the simplest possible model of the medium, namely a three-component mixture of electrons, $e$, ions, $i$, and macroscopic grains of a single sort, $d$, we have

$$1 = \frac{\omega_{p,e}^2}{\omega^2 - k^2 v_{T,e}^2} + \frac{\omega_{p,i}^2}{\omega^2 - k^2 v_{T,i}^2} + \frac{\omega_{p,d}^2}{\omega^2 - k^2 v_{T,d}^2}. \tag{74}$$

To analyze the wave modes that can be supported by this plasma medium, let us multiply Equation (74) by $\omega^2$ and bring it to the form $f(v_{ph}^2) = \omega^2$, with

$$f(v_{ph}^2) = \frac{\omega_{p,e}^2}{1 - v_{T,e}^2/v_{ph}^2} + \frac{\omega_{p,i}^2}{1 - v_{T,i}^2/v_{ph}^2} + \frac{\omega_{p,d}^2}{1 - v_{T,d}^2/v_{ph}^2}, \tag{75}$$

where $v_{ph} = \omega/k$. The function is shown diagrammatically in Figure 7. Solutions of the modal Equation (74) are given by intersections of $f(v_{ph}^2)$ with $\omega^2 = \text{const}$. As can be seen from the Figure, the equation has a set of "high frequency" (i.e. $\omega^2 \gg \sum_{(\alpha)} \omega_{p,\alpha}^2$) solutions, characterized by $v_{ph} \simeq v_{T,\alpha}$ (with $\alpha = e, i, d$).

In the low frequency domain ($\omega^2 \ll \sum_{(\alpha)} \omega_{p,\alpha}^2$), the roots of Equation (74) can be estimated approximately by setting $\omega^2 = 0$. Additional simplifications are possible if the intersection points of $f(v_{ph}^2)$ with the abscissa are sufficiently separated from one another. Then the root lying between $v_{T,e}$ and $v_{T,i}$ can be found with the use of the inequality

$$v_{T,i} \ll v_{ph,i} \ll v_{T,e}, \tag{76}$$

while the solution between $v_{T,i}$ and $v_{T,d}$ obeys

$$v_{T,d} \ll v_{ph,d} \ll v_{T,i}. \tag{77}$$

In the first case, the modal equation becomes

$$-\frac{\omega_{p,e}^2 v_{ph,i}^2}{v_{T,e}^2} + \omega_{p,i}^2 + \omega_{p,d}^2 \simeq 0. \tag{78}$$

In view of $\omega_{p,i}^2 \gg \omega_{p,d}^2$ the latter term can be neglected (which strictly corresponds to the case of fixed dust grains, $m_d \to \infty$), to yield the familiar

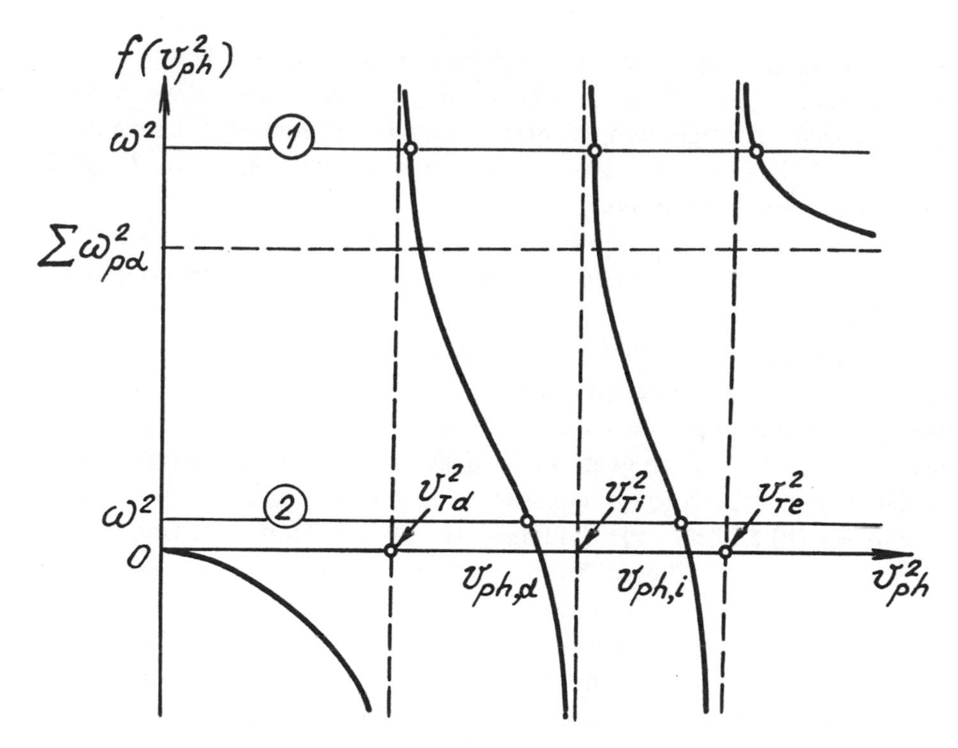


Figure 7: Graphic Analysis of the Dispersion Relation (Equation (74)): Level 1 Is for the High Frequency Wave and Level 2 for the Low Frequency Wave.

phase velocity of the ion-acoustic wave in a "pure" electron-ion plasma,

$$v_{ph,i} \simeq \left(\frac{Q_i^2 n_{0,i} T_e}{e^2 n_{0,e} m_i}\right)^{1/2} . \tag{79}$$

The ion-acoustic waves propagate with a low attenuation if $v_{ph,i} \gg v_{T,i}$, which is equivalent to the condition $T_e \gg T_i$. In the isothermal plasma, $T_e \simeq T_i$, the mode is severely damped.

The matter is quite different in the dusty plasma where the ion-acoustic mode can exist even with $T_e \simeq T_i$ (Shukla, 1992; Rosenberg, 1993). This result can be easily understood if we recall the quasineutrality condition (with allowance for the charge on dust grains and $Q_i = -e$),

$$e(n_{0,i} - n_{0,e}) + Q n_{0,d} = 0. \tag{80}$$

Re-writing Equation (79) as

$$v_{ph,i} \simeq (\delta T_e / m_i)^{1/2}, \tag{81}$$

where $\delta = n_{0,i}/n_{0,e}$, we observe that $\delta$ increases to become $\delta > 1$ if the dust grains are charged negatively and hence $n_{0,e}$ reduces. The phase velocity $v_{ph,i}$ increases accordingly and the attenuation is diminished. The condition $T_e \gg T_i$ is replaced by $T_e \gg T_i/\delta$, which can well hold with $T_e \simeq T_i$, provided $\delta$ is sufficiently large.

If the dust grains carry a positive charge, $\delta$ is lower than 1, so that the ion-acoustic wave can only exist in a highly non-isothermal plasma (Rosenberg, 1993).

The dusty plasma demonstrates another property that can affect the ion-acoustic mode. The background charge formed by the dust grains is not uniform but rather represents a discrete spatial structure of characteristic scale $d \sim (n_{0,d})^{-1/3}$. Since every grain is surrounded by an ionic cloud (in the case of $Q < 0$), it proves necessary to account for the spatial nonuniformity of $n_{0,i}(\mathbf{r})$ and $n_{0,e}(\mathbf{r})$. Evidently, the propagation conditions for the ion-acoustic wave are different from the "pure" plasma. The wave will be reflected from enhancements in $n_{0,i}(\mathbf{r})$, which effect may give rise to the so-called wave traps (de Angelis *et al.*, 1988). The effect is particularly bright in the one-dimensional case when it is allowable to speak of plane-parallel resonance cavities, with the wave amplitude increasing inside.

The writers used the effect to explain the burst in the intensity of electrostatic noise detected in the *Vega* experiment, with the peak near $\omega \simeq (1 \text{ to } 3) \times 10^3\,\mathrm{s}^{-1}$. The burst was observed when the spacecraft passed through a region with enhanced dust density (Waisberg *et al.*, 1987), well in agreement with the concept of resonantly increased ion-acoustic waves.

In the vicinity of its other root, namely $v_{ph,d}$ obeying the inequality Equation (77), the modal Equation (74) can be written as

$$-\frac{\omega_{p,i}^2 v_{ph,d}^2}{v_{T,i}^2} + \omega_{p,d}^2 \simeq 0 \text{ in view of } v_{ph,d} \ll v_{T,e}.$$

The solution is

$$v_{ph,d} \simeq \left(\frac{Q_d^2 n_{0,d} T_i}{Q_i^2 n_{0,i} m_d}\right)^{1/2}, \tag{82}$$

which is similar to the phase velocity of the ion-acoustic wave. By analogy, the mode has been named the dust-acoustic wave (Rao *et al.*, 1990).

The "conventional" ion-acoustic wave clearly demonstrates the cooperative properties of the plasma where it propagates. The major contribution to the pressure is given by electrons (*cf.* the temperature $T_e$ in Equation (79)), while inertia of the oscillations is provided by the heavier

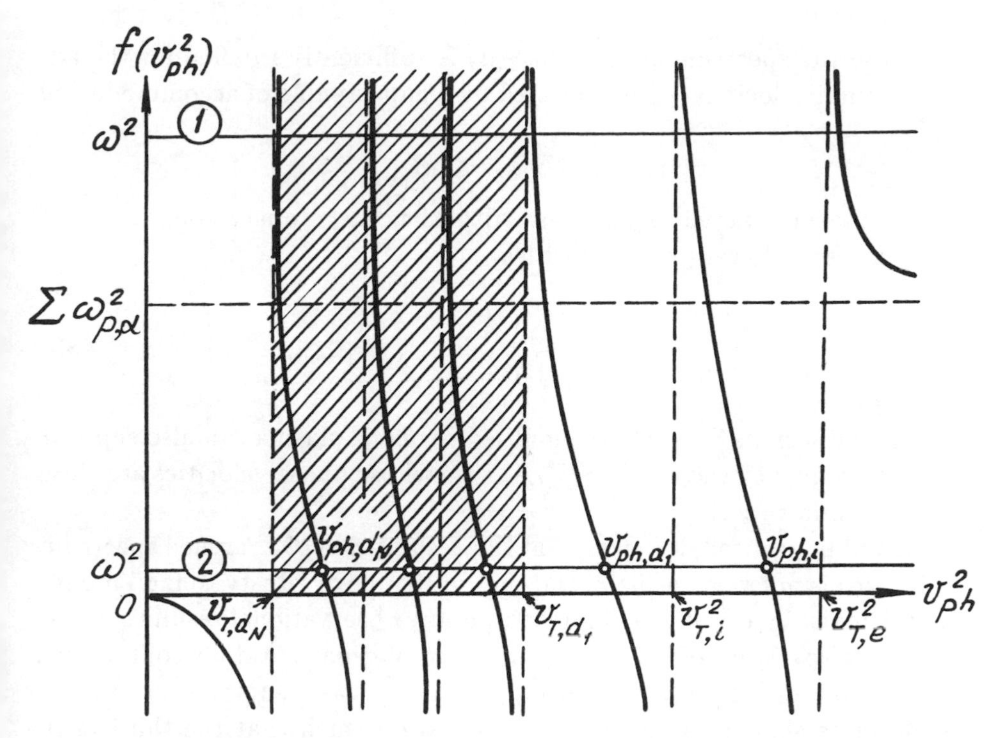


Figure 8: Graphic Analysis of the Dispersion Relation for a Multicomponent Dusty Plasma ($a_1 < a_2 < \cdots < a_N$). In the Case of a Continuous Spectrum of Grain Sizes, $a_{min} \leq a \leq a_{max}$, the Dust-Acoustic Waves with Parameters Lying in the Hatched Area Are Attenuated (see Figure 9).

particles which are ions (the mass $m_i$ in Equation (79)). In the case of its "dusty" analog, the lighter component is represented by ions (*cf.* the temperature $T_i$ in Equation (82)), while dust grains play the part of the heavier particles ($m_d$ in Equation (82)). Equation (82) can be readily extended to the case of a multicomponent dusty medium with grains of different sizes $a_j$. The last term of Equation (75) should be replaced by the sum $\sum_{j=1}^{N} \omega_{p,d_j}^2 (1 - v_{T,d_j}^2 / v_{ph,d}^2)^{-1}$ over dust grains of all sizes. The corresponding thermal velocities $v_{T,d_j} = (T_{d_j}/m_{d_j})^{1/2}$ being all different, the function $f(v_{ph}^2)$ would have $N$ vertical "dusty" asymptotes at the points $v_{T,d_j}$ ($j = 1, 2, \ldots N$). In Figure 8, the grains are ordered in accordance with their size, i.e. $a_1 < a_2 < \ldots a_N$. The highest phase velocity among the "dusty" acoustic waves is $v_{ph,d_1}$ given by Equation (82) with $n_{0,d} = n_{0,d_1}$, $Q_d = Q_{d_1}$ and $m_d = m_{d_1}$.

Suppose the spectrum of grain sizes $a_j$ is sufficiently rarefied for the adjacent thermal velocities to be separated by a gap capable of accommodating $v_{ph,j}$, i.e.

$$v_{T,d_{j+1}} \ll v_{ph,j} \ll v_{T,d_j}.$$

Then the plasma can support, along with the "fast" dusty sound, $N-1$ acoustic waves of lower phase velocities,

$$v_{ph,j} \simeq \left( \frac{Q^2_{d_{j+1}} n_{0,d_{j+1}} T_{d_j}}{Q^2_{d_j} n_{0,d_j} m_{d_{j+1}}} \right)^{1/2} \qquad (j = 1, 2, \ldots N). \tag{83}$$

As can be seen in Figure 8, the multicomponent plasma can also support $N+2$ high-frequency waves ($\omega^2 > \sum_{(\alpha)} \omega^2_{p,\alpha}$) whose phase velocities are close to $v_{T,e}$; $v_{T,i}$ and $v_{T,d_j}$.

It should be realized, however, that the assumption of a rarefied spectrum of grain sizes is not very realistic, and the multitude of "dusty sound" modes can hardly exist in any object. Moreover, many observational results suggest that the size spectrum of dust grains in space plasmas is rather continuous, occupying some range $a_{min} \leq a \leq a_{max}$. In this case, summation over all sorts of grains should be replaced by integration, such that the third term of Equation (74) will become

$$\int_{a_{min}}^{a_{max}} da\, \tilde{\omega}^2_{p,d}(a)[\omega^2 - k^2 v^2_{T,d}(a)]^{-1}, \tag{84}$$

with $\tilde{\omega}^2_{p,d} = 4\pi Q^2_d \tilde{n}_d(a)/m_d$ and $\tilde{n}_d(a)$ denoting the size-spectral density of $n_d$ (note the dimension of $\tilde{n}_d(a)$ to be $[\tilde{n}(a)] = \mathrm{cm}^{-4}$). The grain potential is often determined by the plasma temperature alone, being independent of size. Then the integral can be written as

$$J = \frac{3\psi^2}{\rho\omega^2} \int_{a_{min}}^{a_{max}} \frac{\tilde{n}_d(a)\, da}{a(1 - v^2_{T,d}(a)/v^2_{ph})},$$

with $v^2_{T,d}(a) = 3T/(4\pi\rho a^3)$, and $\rho$ denoting the mass density of the grain material. The integral can be further simplified by introducing the variable $z = a^3$,

$$J = \frac{\psi^2}{\rho\omega^2} \int_{z_{min}}^{z_{max}} \frac{w(z)\, dz}{z - z_T}, \tag{85}$$

where $w(z) \equiv \tilde{n}_d[a(z)] \equiv \tilde{n}_d(z^{1/3})$ and $z_T = 3T/(4\pi\rho v^2_{ph})$. If $z_T$ lies without the interval $(z_{min}, z_{max})$, then the integral is a real magnitude and the dusty plasma eigenwaves remain purely propagational despite the continuous size spectrum of the dust grains. The matter is different with $z_{min} \leq z_T \leq z_{max}$, when the integrand has a first order pole at $z = z_T$ (provided that $w(z_T) \neq 0$). Applying the Landau rule, one can obtain

$$J = \frac{\psi^2}{\rho\omega^2}\left[P\int_{z_{min}}^{z_{max}} \frac{w(z)\,dz}{z - z_T} + i\pi w(z_T)\right], \tag{86}$$

where $P$ denotes the principal value, and the modal equation receives the imaginary term $\mathrm{Im}J = i\pi w(z_T)\psi^2/\rho\omega^2$. The eigenwave thus acquires attenuation. The attenuated waves will be discussed below, while at this point we analyze the modal equation for propagational waves in a dusty plasma with a specific spectrum of sizes, $\tilde{n}_d(a)$. According to literature data, $\tilde{n}_d(a)$ often is a power-law function $\sim a^{-\beta}$ (in particular, it is believed to be such in Saturn's rings), with the exponent $\beta$ lying between $\sim 0.9$ and 4.5 (e.g. Burns *et al.*, 1984; Mendis and Rosenberg, 1994). Until the object to be considered has been specified, we can choose values of $\beta$ from the above range so as to simplify the calculations. For this purpose, $\beta = 3$ seems rather convenient. Along with the total number density $n_{0,d}$ of the dust grains, let us introduce another parameter, $a_0$ of length dimension, normalizing the distribution function $\tilde{n}_d(a)$ as $\int_{a_{min}}^{a_{max}} \tilde{n}_d(a)\,da = n_{0,d}$, with $\tilde{n}_d(a) = n_{0,d}a_0^2a^{-3}$. Then

$$a_0^2 = 2a_{max}^2a_{min}^2/(a_{max}^2 - a_{min}^2). \tag{87}$$

For a fairly wide range of sizes ($a_{max} \gg a_{min}$), $a_0 \simeq \sqrt{2}a_{min}$, while a narrow spectrum yields $a_0 \simeq \bar{a}^2/\sqrt{\bar{a}}\Delta a$, where $\bar{a} = 1/2(a_{max} - a_{min})$ is the mean size and $\Delta a = a_{max} - a_{min}$. Evaluating the integral of Equation (85) for this distribution function, we can obtain the modal equation of form

$$\begin{aligned}\omega^2 = f(v_{ph}^2) \quad &= \quad \frac{\omega_{p,e}^2}{1 - v_{T,e}^2/v_{ph}^2} + \frac{\omega_{p,i}^2}{1 - v_{T,i}^2/v_{ph}^2} + \frac{\Omega_{p,d}^2 v_{ph}^2}{3v_{T,d}^2(a_0)}\times \\ &\times \log\frac{1 - v_{T,d}^2(a_{max})/v_{ph}^2}{1 - v_{T,d}^2(a_{min})/v_{ph}^2},\end{aligned} \tag{88}$$

with $\Omega_{p,d}^2 = 4\pi n_{0,d}Q^2(a_0)/m_d(a_0)$. (Remember this equation to be valid with $v_{ph} < v_{T,d}(a_{max})$ or $v_{ph} > v_{T,d}(a_{min})$. The schematic of $f(v_{ph}^2)$ is shown in Figure 9. It is the same as in Figure 8, except the hatched region corresponding to damped dusty sound waves for a continuous size spectrum of the dust grains, where $f(v_{ph}^2)$ is complex. Therefore, Figure 9 shows $\mathrm{Re}\,f(v_{ph}^2)$ and $\mathrm{Im}\,f(v_{ph}^2)$. The discrete set of undamped waves with phase

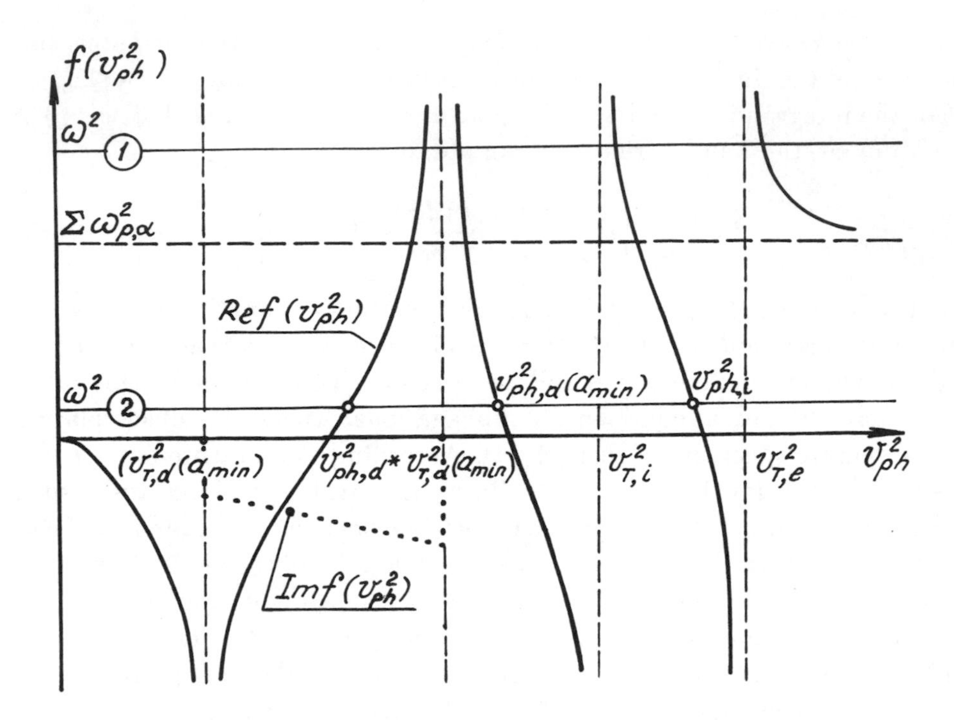


Figure 9: Graphic Analysis of the Dispersion Relation for the Case of a Continuous Spectrum of Grain Sizes. Note the Area where Dust-Acoustic Waves Are Attenuated (*cf.* Figure 8).

velocities $v_{ph,j}$ is thus replaced by a single fast wave whose phase velocity is given by an equation similar to Equation (82). With $v_T(a_{min}) \ll v_{ph} \ll v_{T,i}$ it is

$$v_{ph,d} \simeq \left[ \frac{T_i}{3m_d(a_{min})} \cdot \frac{Q^2(a_0)n_{0,d}}{Q_i^2 n_{0,i}} \right]^{1/2} .$$

Once again, the value is determined by the ion temperature and mass of the lightest dust grains; however the formula involves values characteristic of the whole mass spectrum (i.e. $n_{0,d}$ and $a_0$). There is also a slow wave with the phase velocity $v_{ph,d*}$ lying between $v_{T,d}(a_{max})$ and $v_{T,d}(a_{min})$; however it is a damped wave as $\mathrm{Im}\, f(v_{ph}^2) \neq 0$ over this range of velocities. Besides, the slope of $f(v_{ph}^2)$ is different, which is evidence for *anomalous dispersion.* For our further purposes, we shall need order-of-value estimates of the dust sound phase velocity. Such estimates are listed in Table 1 for plasma shells of different objects in space. The calculations after Equation (82) have been

Table 1:

| Medium | Scale size of grains, microns | Phase velocity of the "dusty sound", m/s |
|---|---|---|
| Interstellar and interplanetary dust | $10^{-2} \div 10^{-1}$ | $10^2 \div 10$ |
| Halo of Jupiter's ring | $10^{-1} \div 1$ | $10 \div 1$ |
| Narrow Saturn rings (E and G) and Jupiter's ring | $10^{-2} \div 1$ | $10^2 \div 1$ |
| Comet tails | $10^{-1} \div 10$ | $10 \div 0.1$ |

done for $T_i \sim 10^2$ K and $Q_i \sim= e$. The electric potential of macroscopic particles was assumed $|\psi| \sim 10$ V and the mass density of the dust material $\rho \sim 1\,\mathrm{g/cm^3}$. The electron component was neglected for simplicity.

### 3.2.2 Wave Damping in a Plasma with Fixed Dust Grains. Nonlinear Landau Damping

As has been shown in the preceding Subsection, even fixed dust grains can influence the waves propagating through a dusty plasma. Under certain conditions they may bring forth a sizable reduction in the attenuation rate of ion-acoustic waves. Meanwhile, an opposite effect is also possible, namely a characteristic attenuation of waves owing to the presence of dust grains, although the latter do not participate in the wave motion. The effect is not the collisional damping which is allowed for theoretically by including friction in the equations of motion of the plasma particles. Whereas in a broader sense, the attenuation in question results from collisions with the dust grains experienced by waves rather than the plasma electrons and ions. In other words, it is an effect associated with scattering of waves in a medium containing refractive index inhomogeneities. In the presence of randomly distributed charged macroparticles, the electron and ion densities $n_{0,e}(\mathbf{r})$ and $n_{0,i}(\mathbf{r})$ become random functions; hence the waves propagate through a medium with the refractive index $\mu(\mathbf{r}) = \bar{\mu} + \delta\mu(\mathbf{r})$, where $\bar{\mu}$ is the average value and $\delta\mu$ the fluctuating part. The electric component $E$ of the wave can be similarly represented as $E = \overline{E} + \delta E$, with the fluctuation $\delta E$ being the scattered field and $\overline{E}$ the field averaged over the ensemble of fluctuations $\delta\mu$. (In virtue of the ergodicity theorem, the ensemble average is equal to the spatially averaged value). In a homogeneous medium the field components are $\delta E = 0$ and $\overline{E} = E_0 \exp(ikx - i\omega t)$. If we assumed the average amplitude $E_0$ to remain constant in the random medium as well, the results would contradict the energy conservation law. Indeed, the average energy flux ($\sim \overline{|E^2|}$) should remain constant if true losses (absorption) are absent. Since $\overline{|E^2|} = |\overline{E}|^2 + \overline{|\delta E^2|}$, the appearance of $\delta E$ and growth of the value with distance along the propagation path should be accompanied by a decrease of $|\overline{E}|$. Thus, the functional dependence of $\overline{E}$ is $\overline{E} = E_0(x)\exp(ikx - i\omega t)$, with $E_0(x)$ being a decreasing function that we intend to find.

The propagation of waves through random media and, in particular, the lossless damping of the average field have by now been given a comprehensive analysis. Perhaps, it is sufficient to quote one or two pioneering papers on the subject (Lifshitz and Rosenzweig, 1946; Kaner and Bass, 1959) and the famous texts (Ishimaru, 1978; Rytov *et al.*, 1978).

For a statistically uniform and isotropic medium, the refractive index correlation function $B(\rho) = \overline{\delta\mu(\mathbf{r}_1)\delta\mu(\mathbf{r}_2)}/\sigma_\mu^2$ can be assumed to depend solely on $\rho = |\mathbf{r}_1 - \mathbf{r}_2|$, with the variance $\sigma_\mu^2 \equiv \overline{\delta\mu^2} = \text{const}$. An equivalent statistical description of $\delta\mu(\mathbf{r})$ is given by the spectral function $\Phi_\mu(\kappa) = (2\pi)^{-3}\int B(\rho)\exp(i\kappa\rho)\, d^3\rho$. Generally, $B(\rho)$ drops off quite rapidly, so that it is allowable to introduce the characteristic spatial scale $l$ of the fluctuations, or the correlation length.

We will limit the analysis to the so-called short-wavelength approximation, when $kl \gg 1$. It is the case where the scattered waves deviate only slightly from the initial propagation direction $\mathbf{k}$ (the characteristic scatter angle being $(kl)^{-1}$), so that the effect of the inhomogeneities essentially reduces to random phase fluctuations $S(x)$ of the wave,

$$E = E_0 \exp(ikx - i\omega t + iS(x)). \tag{89}$$

If the number of inhomogeneities along the propagation path of length $x$ is sufficiently large, $N = x/l \gg 1$, then the random phase $S(x)$ will be distributed normally with any distribution law of $\delta\mu$, and hence

$$\overline{E} = E_0 \exp(ikx - i\omega t)\overline{\exp(iS)} = E_0 \exp(ikx - i\omega t)\exp(-\sigma_s^2/2), \tag{90}$$

where $\sigma_s^2$ is the phase fluctuation variance. Since the random phase increments $S_i$ owing to individual inhomogeneities are independent, the variance

$\sigma_s^2$ is an additive value, $\sigma_s^2 = \sum_{i=1}^N \sigma_{s,i}^2 = N\sigma_{s,1}^2$ ($\sigma_{s,i}^2 = \sigma_{s,1}^2$ for all $i$'s as the medium is statistically uniform). Within the quasioptical approximation $kl \gg 1$, we have $\sigma_{s,1}^2 \simeq k^2 l^2 \sigma_\mu^2$ and

$$\sigma_s^2 \simeq k^2 l \sigma_\mu^2 x. \tag{91}$$

By substituting this into Equation (90) we obtain

$$\begin{aligned} \overline{E} &\simeq E_0 \exp(ikx - i\omega t)\exp(-k^2 l \sigma_\mu^2 x/2) = \\ &= E_0 \exp(-i\omega t)\exp[ix(k + i\sigma_\mu^2 k^2 l/2)]. \end{aligned} \tag{92}$$

The attenuation due to wave scattering can be described in terms of an imaginary addendum to the wavenumber, $k \to k + i\kappa$. The spatial damping also can be converted to attenuation in time by introducing a complex phase velocity $v_{ph} = v + iu$, either as $v_{ph} = \omega/(k + i\kappa)$ or $v_{ph} = (\omega + i\nu)/k$, whence $\kappa = -k^2 u/\omega$, $\nu = uk$ or $\nu = -\omega\kappa/k$, provided $\kappa \ll k$.

The crude estimates of $\kappa$ we have given are in agreement with more rigorous results (e.g., Kaner and Bass, 1959); however our task here is to relate the refractive index variance $\sigma_\mu^2$ with parameters of the dusty plasma. In the short-wavelength approximation, it proves sufficient to consider only the electronic contribution to the dielectric constant, assuming $\mu \simeq (1 - \omega_p^2/\omega^2)^{1/2}$, with $\omega_p^2 = 4\pi e^2 (n_{0,e} + \delta n_{0,e})/m_e$. Since fluctuations of the background electron density are associated with the transfer of some electrons to dust grains, it is plausible to assume $\overline{(e\delta n_{0,e})^2} \simeq (n_{0,d}Q)^2$. Hence,

$$\sigma_\mu^2 \simeq \frac{1}{4} \cdot \frac{\omega_{p,0}^2}{\omega^2} \left( \frac{Q n_{0,d}}{e n_{0,e}} \right)^2, \tag{93}$$

where $\omega_{p,0}^2 = 4\pi e^2 n_{0,e}/m_e$ and $\gamma = Q n_{0,d}/e n_{0,e}$ is the principal small parameter related to the presence of charged dust grains.

Until now, we have considered the wave attenuation owing to the transfer of energy from the coherent component $|\bar{E}|^2$ to the fluctuations, $\overline{|\delta E|^2}$. In fact, there is another channel of energy transfer, namely from the wave to plasma particles, or Landau damping. At a first glance, the effect should be negligible for waves of sufficiently high frequency, such that $\omega \gg k v_{T,e}$; however this is true only for homogeneous media. If the electron density is nonuniformly distributed in space, then the Landau damping can occur even for "high-frequency" waves that satisfy the above inequality (de Angelis *et al.*, 1989). We mean the so-called nonlinear Landau damping, when the necessary resonance condition $\omega = \mathbf{k}\mathbf{v}$ results from interaction of two waves.

One would be the electromagnetic wave under consideration (wave vector $\mathbf{k}$), while the other the wave $\mathbf{K}$, associated with electron density fluctuations, $\delta n_{0,e}(\mathbf{r})$. The result of their interaction would be the scattered, or "beat" wave, with the wave vector

$$\mathbf{k}_s = \mathbf{k} - \mathbf{K}. \tag{94}$$

Strictly speaking, this equation which is the momentum conservation law for the three waves should be complemented by a similar relation for their frequencies (i.e. conservation of energy),

$$\omega_s = \omega - \Omega, \tag{95}$$

where $\Omega$ refers to the fluctuations, $\delta n_{0,e}(\mathbf{r}, t)$. Since $\delta n_{0,e}$ is independent of $t$, the frequency $\Omega$ is zero. If $|\mathbf{k} - \mathbf{K}| \to k$, the resonance condition $\omega = \mathbf{v}(\mathbf{k} - \mathbf{K})$ can be satisfied even with $\omega > \mathbf{kv}$. Still the nonlinear Landau damping is negligibly weak within the approximation we have considered until now. The condition $kl \gg 1$ implies $K \sim l^{-1} \ll k$, and the magnitude $k - K \leq |\mathbf{k} - \mathbf{K}| \leq k + K$ remains close to $k$. The situation may be different for $kl \leq 1$.

To analyze the Landau damping, the modal equation should be derived in the kinetic rather than hydrodynamic theory, which in fact was done in the paper by de Angelis *et al.* (1989). The modal equation $\varepsilon(\omega, \mathbf{k}) = 0$ was analyzed numerically for a Gaussian correlation function of the electron density fluctuations, $B(\rho) = \exp(-\rho^2/l^2)$, and Maxwellian velocity distribution (in this case, the wave attenuation is determined by $f_{e,0}(v)|_{v=\omega/|\mathbf{k}+\mathbf{K}|} = n_{0,e}\pi^{-3/2}v_{T,e}\exp(-\omega^2/v_{T,e}^2|\mathbf{k} + \mathbf{K}|^2))$. The results were shown graphically, as $\nu = \nu(k)$ curves, where $\nu$ is the imaginary part of the wave frequency, the real part being $\omega_r \simeq \omega_p$. For a moderately rarefied plasma, $l = 3\lambda_D$, the $\nu = \nu(k)$ dependence can be represented approximately as

$$\nu(k) \simeq -2kl\omega_p\gamma^2 \tag{96}$$

over the wavenumber range $k \leq 2l^{-1}$. This result may be compared with the rate of attenuation owing to angular scattering that we have estimated before. To do so, let us express Equation (92) in terms of $\nu = -\omega\kappa/k = -\omega\sigma_\mu^2 kl/2$ and substitute $\sigma_\mu^2$ from Equation (93). With account of $\omega \simeq \omega_p$, the result is $\nu \simeq -kl\omega_p\gamma^2/8$. Apart from a numerical factor, both attenuation mechanisms have resulted in the same rate value. This might seem somewhat odd as the derivation of the scatter-produced attenuation never involved the electron temperature, while the Landau damping definitely should be temperature dependent and vanish at $T_e = 0$. In fact the two formulas are being compared for $l = 3\lambda_D = 3(T_e/e^2 n_{e,0})^{1/2}$, i.e. for a certain nonzero temperature of electrons. Meanwhile, the analytic expressions for $f_{e,0}(v)$ given in de Angelis *et al.* (1989) clearly show that $f_{e,0}(v)|_{v=\omega/|\mathbf{k}+\mathbf{K}|} \to 0$ with $T_e \to 0$, i.e. the nonlinear Landau damping really vanishes.

### 3.2.3 The Landau Damping due to the Continuous Mass Spectrum of Dust Grains

At this point, we turn once again to Equation (85) to consider attenuation of the dusty sound. The appearance of a damping effect of the same kind as the Landau damping has been rather unexpected within the hydrodynamic approach as the Landau damping is an essentially kinetic effect associated with specific features of the velocity distribution function of electrons. From a formal point of view, the explanation is quite simple, since the velocity distribution $f_{0,e}(v)$ of dust grains is represented by $\tilde{n}_d(a)$ (or $w(z)$) which is the grain size distribution. While in the classical kinetic theory an imaginary term in the dispersion relation arises from

$$-\int_{\infty}^{\infty} \frac{\partial f_0(v)/\partial v}{v - \omega/k} dv,$$

the integral that is responsible in the case of a dusty plasma is

$$\int_{a_{min}}^{a_{max}} \frac{w(z)\, dz}{z - z_T},$$

and the resemblance is only too obvious. The Langmuir wave is damped if $\partial f_0/\partial v(v = \omega/k) < 0$, while the corresponding requirement to $w(z)$ can be formulated as $w(z = z_T) > 0$. Note this condition to be always met, no matter what the specific form of $w(z)$, since the latter is a positive function by definition.

The formal analogy is in fact based on a solid physical foundation. First, let us recall the meaning of the condition $\partial f_0/\partial v(v = v_{ph}) < 0$. Among the particles at resonance with the wave (i.e. $v \simeq v_{ph}$) the number of those moving at a lower speed, $v < v_{ph}$, is greater than that of faster particles, $v > v_{ph}$. The slow particles are accelerated by the wave, giving them some of its energy, while the faster ones are slowed down as they transfer their energy to the wave. With $\partial f_0/\partial v < 0$ the wave amplitude decreases. Similar considerations do not apply directly to dust grains with a continuous distribution function $w(z)$. Indeed, it is essential to comprehend why the grains of different sizes interact differently with the wave field. The grain size itself is not reflected in the equations of motion as all the grains are regarded as points (the radius is involved indirectly through the parameters $Q_d$ and $m_d$).

In fact, introduction of a continuous spectrum of sizes is equivalent to a transition from hydrodynamics to a kinetic description with a distribution function $F_{0,d}(v)$ characterized by $\partial F_{0,d}/\partial v < 0$. Note that the dispersion relation Equation (74) for dust grains of a single sort can be obtained in the kinetic theory with the "water bag" distribution function $f_{0,\alpha}(v)$ with $v = \mathbf{vE}/E$ (Goertz, 1989): $f_{0,\alpha}(v) = 1/(2v_{T,\alpha})$ for $|v| \leq v_{T,\alpha}$ and $f_{0,\alpha}(v) = 0$ for $|v| > v_{T,\alpha}$, $\alpha = e, i, d$. If a continuous spectrum of grain sizes is introduced, then $f_{0,d}(v)$ is replaced by $f_{0,d}(v,a)$, while $F_{0,d}(v)$ can be obtained through integration over the grain sizes. In the case under consideration, all the $f_{0,d}(v,a)$ are "water bags", however their walls $v_{T,d}(a)$ are located at different places depending on the grain size. The general form of $F_{0,d}(v)$ can be understood from the diagram of Figure 10, even without specifying $\tilde{n}_d(a)$. Shown in the left corner are distribution functions $f_{0,d}(v,a)$ for $a_1 < a_2 < a_3$. The sum of the three functions, $F_{0,d} = \sum_{i=1}^{3} f_{0,d}(v,a_i)$ is represented in the right corner. All the functions are added over the range $0 < v \leq v_{T,d_3}$, while for $v_{T,d_3} < v \leq v_{T,d_2}$ only $f_{0,d}(v,a_2)$ and $f_{0,d}(v,a_3)$ are. Farther out, $f_{0,d}(v,a_1)$ remains alone and finally $F_{0,d}(v)$ becomes zero for $|v| > v_{T,d_1}$ (i.e. the thermal velocity of the smallest grains). The discontinuities of $F_{0,d}(v)$ are smoothed over for the continuous spectrum, such that the function decreases monotonically through the range $v_{T,d}(a_{\mathrm{max}}) \leq v \leq v_{T,d}(a_{\mathrm{min}})$.

The velocity distribution $F_{0,d}(v)$ of dust grains can be easily derived from their size distribution. E.g., for the power-law spectrum of Subsection (3.2.1), $\tilde{n}_d(a) = n_{0,d}a_0^{\beta-1}a^{-\beta}$ (with $\beta = 3$ for $a_{min} \leq a \leq a_{max}$), the formulas are

$$
\begin{aligned}
F_{0,d}(v) &= 2\gamma_0(a_{min}^{-1/2} - a_{max}^{-1/2}) && (0 \leq v \leq v_{T,d}(a_{max}))\\
F_{0,d}(v) &= 2\gamma_0[a_{min}^{-1/2} - v^{1/3}(4\pi\rho/3T)^{1/6}] && (v_{T,d}(a_{max}) < v \leq v_{T,d}(a_{min}))\\
F_{0,d}(v) &= 0, && (v > v_{T,d}(a_{min})),
\end{aligned}
$$

where $\gamma_0 = n_{0,d}a_0^2(\pi\rho/3T)^{1/6}$. The derivative of this function, $\partial F_{0,d}/\partial v$, is negative in the range $v_{T,d}(a_{max}) < v < v_{T,d}(a_{min})$, giving rise to the Landau damping. Outside this region, $\partial F_{0,d}/\partial v = 0$ and the damping is absent.

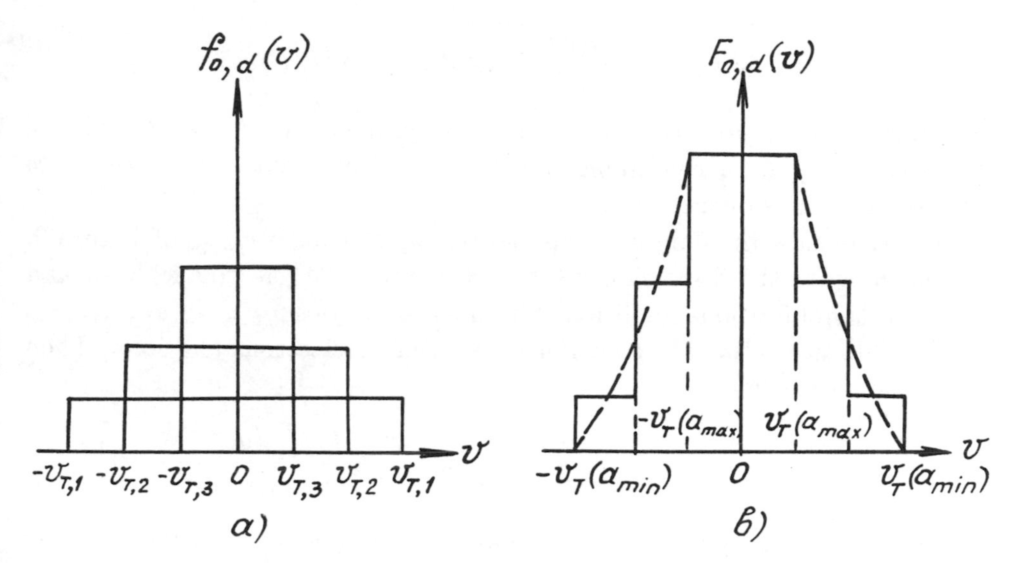


Figure 10: a) "Water-Bag" Velocity Distribution Functions $f_{0,d}(v)$ for Dust Grains of Different Sizes $a_1 < a_2 < a_3$ and b) the Resultant Distribution Function $F_{0,d}(v)$.

It might seem that a knowledge of $F_{0,d}(v)$ should permit abstracting from the size spectrum of dust grains and simply make use of the familiar kinetic theory expressions for the attenuation rate. In fact this is not so. Let us represent the supposed dispersion relation as

$$\varepsilon(\omega, k) = 1 - \frac{\omega_{p,d}^2}{k} \int \frac{\partial F_{0,d}/\partial v}{\omega - kv} dv = 0.$$

The question that arises naturally is for which values of $a$ should the dust-plasma frequency $\omega_{p,d}$ be taken in this equation. To answer it correctly, the kinetic equations should be written for all possible sizes $a$ and then integration over $a$ performed in the final results, including the dispersion relation. That would result in effective averaging of both $f_{0,d}(a, v)$ and $\tilde{\omega}_{p,d}^2(a)$. In fact, this procedure has been performed in this Section, however in the hydrodynamic approximation.

We have discussed this point in some detail to stress the fact that dust-acoustic waves experience the Landau damping in a plasma with a continuous spectrum of grain sizes even when the damping is absent for each particular sort (size) of dust grains. In case the collisionless damping occurs with particles of any size $a$ (e.g., with a Maxwellian distribution function)

$$f_0(a, v) = \left(\frac{m(a)}{2\pi T}\right)^{1/2} \exp(-m(a)v^2/2T),$$

the question does not arise at all. Naturally, dust-acoustic waves will be damped in a plasma containing dust grains of different sizes if they are damped by grains of each size.

Consider now the damping experienced by the wave $v_{ph,d*}$ of Figure 9. Let us re-write the dispersion relation Equation (74) as $\varepsilon(\omega, \mathbf{k}) = 0$ and make the simplifications conditioned by the inequalities $v_{ph,d*} \ll v_{T,i} \ll v_{T,e}$. These allow neglecting the contributions owing to electrons and ions. Then the equation becomes

$$\varepsilon(\omega, k) = \varepsilon_r + i\varepsilon_i \simeq 1 - \frac{\Omega^2_{p,d}}{3k^2 v^2_{T,d}(a_0)} \left[\log \frac{\omega^2 - k^2 v^2_{T,d}(a_{max})}{k^2 v^2_{T,d}(a_{min}) - \omega^2} - \pi i\right].$$

Assuming the relative attenuation rate to be small (the necessary conditions will be formulated below), we can write $\omega = \omega_r + i\nu$, $\nu \ll \omega_r$. The real frequency $\omega_r$ will be found as the solution of $\varepsilon_r(\omega, k) = 0$, while $\nu$ is given by the equation

$$\nu \simeq -\varepsilon_i(\omega_{r,k})/(\partial\varepsilon_r(\omega_{r,k})/\partial\omega_r).$$

The explicit relations are

$$\omega_r \simeq k(1 + b)^{-1/2} \left[b v^2_{T,d}(a_{min}) + v^2_{T,d}(a_{max})\right]^{1/2}$$

and

$$\nu \simeq \frac{k\pi b}{2}[v^2_{T,d}(a_{min} - v^2_{T,d}(a_{max})](1 + b)^{-1/2}[b v^2_{T,d}(a_{min}) + v^2_{T,d}(a_{max})]^{-1/2},$$

with $b = \exp(3k^2\lambda^2_{D,d})$ and $\lambda_{D,d} = v_{T,d}(a_0)/\Omega_{p,d}$. The latter value is an effective Debye length for the dust particles, dependent on the entire spectrum of grain sizes.

The expressions for $\omega_r$ and $\nu$ can be further simplified, because normally $v^2_{T,d}(a_{min}) \gg v^2_{T,d}(a_{max})$ (in view of $a_{max} \gg a_{min}$), and $b \geq 1$ by definition. Accordingly,

$$\omega_r \simeq k v_{T,d}(a_{min})\sqrt{\frac{b}{1+b}}$$

and

$$\nu \simeq k v_{T,d}(a_{min}) \frac{\pi}{2(1+b)} \sqrt{\frac{\beta}{1+b}}.$$

As can be seen, the phase velocity of damped dust-acoustic waves is close to the thermal velocity of the smallest dust grains

$$\left( v_{ph} = \frac{\omega_r}{k} \simeq v_{T,d}(a_{min}) \sqrt{\frac{b}{1+b}} \right),$$

and the relative attenuation rate, $\nu/\omega_r \simeq \pi/[2(1+b)]$ is small for $b \gg 1$. These are the conditions of validity of the above given formulas for $\omega_r$ and $\nu$.

### 3.2.4 Attenuation due to Grain Charge Variations

Still another mechanism of dusty sound attenuation has been considered recently by Melandsø*et al.* (1993a; 1993b). The effect arises from variations of the grain charge in the course of wave propagation. The charge $Q(t)$ is varied by the currents flowing to the dust grain from the ambient plasma, the source of the necessary energy being the propagating wave. In contrast to the mechanisms discussed in the preceding Subsections, this is a dissipative (lossy) attenuation, associated with inelastic collisions of electrons and ions with the dust grains.

It is important that variations of the grain charge are characterized by some delay with respect to changes in potential, and the corresponding phase shift other than $\pi/2$ between the current and voltage involved. Consider the process, following the analogy with a capacitor being charged through a resistor. The steady-state grain charge is $Q = C(\psi - \Phi)$, where $\Phi$ is the plasma potential. The latter varies in the course of wave propagation and the relation between the variables $\Phi(t)$ and $Q(t)$ is given by

$$\Phi + R\frac{dQ}{dt} + \frac{Q}{C} = 0. \tag{97}$$

Here $C \simeq a$ is the grain capacitance and $R$ the resistance in the charging circuit. An order of magnitude estimate of $R$ is $R \simeq \psi/J_e$ but a better accuracy is not necessary as Equation (97) can be re-written as

$$\Phi = -\frac{1}{a}\left(Q + t_0\frac{dQ}{dt_0}\right), \tag{98}$$

where the relaxation time $t_0 = RC$ has been estimated in Subsection 1.1.1 as $t_0 \simeq (T_e m_e)^{1/2}/(e^2 n_0 a)$. Equation (98) for the new variable $Q(t)$ should be added to the equation set for $n_\alpha$, $v_\alpha$ and $\Phi$ ($\alpha = e, i, d$, with all the dust grains being of the same size). To linearize the set, we assume $Q(t) = Q_0 + q(t)$ ($q \ll Q_0$) and

$$\frac{dQ}{dt} = \frac{\partial Q}{\partial t} + v_d \frac{\partial Q}{\partial x} \simeq \frac{\partial q}{\partial t}.$$

Taking, as before, the harmonic time and space dependence $\exp[-i(\omega t - kx)]$ for all the variables, we arrive at the algebraic set

$$v_\alpha = k\left(\frac{Q_\alpha}{m_\alpha}\Phi + \frac{v_{T,\alpha}^2}{n_{0,\alpha}} n_\alpha\right) \qquad (Q_e = e;\ Q_i = -e \text{ and } Q_d = Q_0)$$

$$\omega n_\alpha + k n_{0,\alpha} v_\alpha = 0$$

$$k^2 \Phi = 4\pi(n_{0,d} q + n_d Q_0) + 4\pi e(n_i - n_e)$$

$$\Phi = -\frac{q}{a}(1 - i\omega t_0). \tag{99}$$

The modal equation that can be obtained in a standard way is

$$1 = -\frac{4\pi n_{0,d} a}{k^2(1 - i\omega t_0)} + \frac{\omega_{p,d}^2}{\omega^2 - k^2 v_{T,d}^2} + \sum_{\alpha=e,i} \frac{\omega_{p,\alpha}^2}{\omega^2 - k^2 v_{T,\alpha}^2}. \tag{100}$$

Allowance for variations in $Q(t)$ has resulted in a new term which assumes complex values with $t_0 \neq 0$. To analyze the low frequency dusty sound wave, we can assume as before $v_{T,d} \ll v_{ph,d} \ll v_{T,i}$ (or $kv_{T,d} \ll \omega \ll kv_{T,i}$), which allows simplifying Equation (100),

$$1 \simeq -\frac{4\pi n_{d,0} a}{k^2(1 - i\omega t_0)} + \frac{\omega_{p,d}^2}{\omega^2} - \frac{\omega_{p,i}^2}{k^2 v_{T,i}^2}. \tag{101}$$

Setting $\omega = \omega_r + i\nu$ with $\nu \ll \omega_r$, we can find approximate solutions for $\omega_r$ and $\nu$ if the first term in the right-hand side of Equation (101) is also treated as a small correction. $\omega_r$ follows from the zero-order approximation in $a$,

$$\omega_r \simeq \frac{\omega_{p,d} \lambda_D k}{(1 + k^2 \lambda_D^2)^{1/2}}, \tag{102}$$

where $\lambda_D = v_{T,i}/\omega_{p,i}$. The phase velocity $v_{ph} = \omega_r/k$ of this wave in the long-wavelength limit $k^2 \lambda_D^2 \ll 1$ is $v_{ph} \simeq T_i/m_d$ which corresponds to the dusty sound. A nonzero imaginary correction $\nu$ appears in the first order in $a$ and $\omega t_0$, *viz.*

$$\nu \simeq -\frac{2\pi a n_{0,d} t_0 \omega_r^2 \lambda_D^2}{(1+k^2\lambda_D^2)(1+\omega_r^2 t_0^2)} = -\frac{2\pi t_0 \omega_r^2}{1+k^2\lambda_D^2} P, \tag{103}$$

where $P = a n_{0,d} T/(e^2 n_{0,e})$ is a dimensionless parameter often encountered in the theory of dusty plasmas (Havnes *et al.*, 1984; Melandsø *et al.*, 1993). The negative sign of $\nu$ shows that the dusty sound wave is indeed attenuated. However the formalism in which the value has been derived, employing some effective integrated parameters like the circuit resistance $R$, seems too schematic. The result is not very convincing, at least with regard to numerical factors.

The correct approach should be based on Equation (1) of Chapter 1 relating $dQ/dt$ with the currents flowing to the grain from the ambient plasma (Melandsø *et al.*, 1993), *viz.*

$$\begin{aligned} J_i &= 4\pi a^2 e v_{T,i} n_{0,i} \exp(-e\Phi/T)(1-eU/T) \\ J_e &= -4\pi a^2 e v_{T,e} n_{0,e} \exp[(\Phi+U)e/T]. \end{aligned} \tag{104}$$

By denoting deviations from the equilibrium values $Q_0$, $\Phi_0$ and $U_0$ by the same letters without subscripts, we can linearize the grain charging equation which becomes

$$\frac{\partial Q}{\partial t} = \frac{\partial}{\partial \Phi}\left(\sum_{\alpha=e,i} J_\alpha\right)\Phi + \frac{\partial}{\partial U}\left(\sum_{\alpha=e,i} J_\alpha\right) U = -a\Omega_\Phi \Phi - \Omega_U Q. \tag{105}$$

Thus, we have introduced the frequencies $\Omega_\Phi = -1/a[\partial/\partial\Phi(\sum_{\alpha=e,i} J_\alpha)]$ and $\Omega_U = -1/a[\partial/\partial U(\sum_{\alpha=e,i} J_\alpha)]$, where the derivatives are taken at $\Phi = \Phi_0$ and $U = U_0$. For the harmonically varying values, Equation (105) takes the form

$$\Phi = -\frac{Q}{a}\left(\frac{\Omega_U}{\Omega_\Phi} - \frac{i\omega}{\Omega_\Phi}\right), \tag{106}$$

which will be used now instead of Equation (99). In fact, the two equations are almost coincident as the frequencies $\Omega_U$ and $\Omega_\Phi$ are values of the same order (to be more precise, $\Omega_\Phi \simeq 1.53\,\Omega_U$ – Melandsø *et al.* (1993a)). By their physical meaning, both frequencies are the reciprocal relaxation time,

$$\Omega_\Phi \simeq \Omega_U \simeq 1/t_0, \tag{107}$$

whence the equivalence of Equation (106) and the last Equation (99) becomes quite evident. Note the definition of $t_0$ as $\Omega_\Phi^{-1}$ to be better justified physically than the estimate of Equation (3) of Chapter 1.

The equation set used by Melandsø *et al.* (1993a) is somewhat simpler than Equation (99). Having in mind the analysis of low-frequency waves only, from the very beginning the writers have neglected the inertia of electrons and ions, assuming them Boltzmann-distributed in a field of potential $\Phi$. This allows considering the dynamics of dust grains alone, leaving in the set Equation (99) just the equations with $\alpha = d$ and representing Poisson's equation as

$$k^2\Phi + \Phi/\lambda_D^2 = 4\pi(n_d Q_0 + n_{0,d} q).$$

The dispersion relation follows in the usual way, and the phase velocity of low-frequency waves happens to be complex. For a low density of dust grains ($P \ll 1$) and low-attenuated waves ($\omega = \omega_r + i\nu$, $\nu \ll \omega_r$), the $\omega_r$ of Melandsø *et al.* (1993a) is the same as Equation (102), while

$$\nu \simeq -\omega_r^2 \Omega_\Phi P / [4(1 + k^2\lambda_D^2)(\Omega_U^2 + \omega_r^2)]. \tag{108}$$

Apart from a numerical factor, this coincides with the above given estimate, provided that the condition $\omega_r^2/\Omega_U^2 \ll 1$ (i.e. $\omega^2\tau^2 \ll 1$ ) holds.

For the general case of arbitrary dust grain density ($P$ is not small), the phase velocity and attenuation of waves can be evaluated numerically. Such estimates may be used to analyze the propagation of dusty sound waves under various realistic conditions, e.g. in planetary rings. In the above cited paper, Melandsø and coauthors took the parameters $T = 50\,\mathrm{eV}$; $n_{0,e} = n_{0,i} = 10\,\mathrm{cm}^{-3}$ and the grain radius $a = 1\,\mu$. Three values of the parameter $P$ were used, $P = 10^{-5}$ (Saturn's E-ring); $P = 10^{-3}$ (Saturn's G-ring) and $P = 1$ (Jupiter's ring). Since the density variations that are observed occur over much shorter spatial scales than the ring circumference, the initial value of the wavenumber $k$ taken for the calculations was $k = 10^{-6}\,\mathrm{m}^{-1}$. With an account of $\lambda_D = 12\,\mathrm{m}$, such wavenumbers correspond to oscillation periods ranging between 10 days and a few years. The whole system can hardly be assumed to retain stationarity over time periods as long as this.

On the other hand, it seems plausible to assume that waves can only propagate in an essentially steady-state way if their frequency is considerably higher than, e.g. the planetary orbital frequency $\Omega_p$ or the Kepler frequency, $\Omega_K$. By taking $\omega_r \geq 10^{-4}\,\mathrm{s}^{-1}$, we obtain wavelengths less than about 1 km for the $E$-ring of Saturn or $\Lambda \leq 150\,\mathrm{km}$ for the Jupiter ring. The attenuation rates are low for most cases, except waves of short period in the Jupiter ring where $\nu/\omega_r \sim 0.1$. The conclusion that can be made is that dust-acoustic waves can exist only in compact regions of the rings.

The dust-acoustic waves, with allowance for variations in the grain charge $Q(t)$, have also been studied in a more general formulation proceeding from a kinetic equation (Melandsø *et al.*, 1993b). The approach allowed a simultaneous account of the Landau damping and the grain drift at a velocity $\mathbf{w}$ relative the plasma. In a planetary ring, the drift motion can result from the difference of angular velocities of dust grains and the rigidly corotating plasma (see Chapter 2). As follows from numerical estimates, $w$ is always lower than the electron thermal velocity $v_{T,e}$, while it can be of the same order as the ion velocity, $v_{T,i}$ (in the outer rings of Saturn, $w > v_{T,i}$). This may prove important in evaluating the ionic component $J_i$ of the grain charging current.

Similar to the hydrodynamic analysis, allowance for charge variations brings forth an extra term in the modal equation $D(\omega, \mathbf{k}) = 0$ that proves to be complex,

$$D(\omega, \mathbf{k}) = 1 + \sum_{\alpha=e,i,d} \frac{\omega_{p,\alpha}^2}{n_{0,\alpha} k^2} \int \frac{\mathbf{k}\nabla_v f_{0,\alpha}}{\omega - \mathbf{kv}} d^3\mathbf{v} + i\frac{4\pi a \Omega_\Phi}{k^2} \int \frac{f_{0,d}}{\omega - \mathbf{kv} + i\Omega_U} d^3\mathbf{v}. \tag{109}$$

The equation was analyzed for Maxwellian velocity distributions of particles of all sorts. The phase velocities and attenuation rates of dust-acoustic waves were obtained either analytically (in the case of small $P$) or numerically. The kinetic analysis gave more support to the earlier conclusions of the same writers that dust-acoustic waves in planetary rings are possible only at the frequencies that are a few times higher than $\Omega_K$ for the appropriate radial distance in the ring.

Since the distribution functions $f_{0,\alpha}$ allow for the grain drift in the plasma, the modal equation with the $D(\omega, \mathbf{k})$ of Equation (109) also describes instabilities of the dust grain flow. A rough estimate of the dust-sound growth rate (without account for attenuation) can be written for $\mathbf{k} \parallel \mathbf{w}$ and $w \gg \omega/k$ as $\nu/\omega_r \simeq (\sqrt{\pi}/4) w/v_{T,i}$.

In this Section, we have discussed separately two physical mechanisms responsible for the attenuation of dust-acoustic waves, namely one associated with the continuous size spectrum of the dust grains and the other arising from variations of the grain charge. In principle, there is no difficulty in analyzing the combined action of the two mechanisms.

## 3.3 Electromagnetic Wave Scattering in the Dusty Plasma

The scattering of electromagnetic waves by charged dust grains has been discussed on two occasions in this book. First, in Subsection 2.3.1. we considered the radiation pressure effect and found out that the magnitude of the grain charge was practically of no importance. The major role belonged to the grain geometry and electrical properties of the material. However, that was only true for dust in a vacuum, while in a plasma environment the grain charge is a parameter of utmost importance. As will be shown below, the radiation pressure acting on a charged grain in a plasma may be much greater than in a vacuum.

Wave scattering was mentioned for a second time in Subsection 3.2.2, treating the nonlinear Landau damping of dust-acoustic waves. In that case, we were only interested in the behavior of some average field rather than the scattered waves, the excitation of which actually gave rise to the special kind of attenuation of the primary propagating wave.

In contrast to this, the scattered waves are the principal object of investigation in this Subsection. As was shown before, the dusty plasma can support different wave modes: According to the graphical analysis of the modal Equation (74), there are five longitudinal modes with different phase velocities (Figure 7), of which three lie in the high-frequency and two in the low-frequency domain. These should be complemented by yet another high-frequency mode which is a transverse wave, so that the total number of normal eigenwaves supported by the dusty plasma is six. Any two of these can interact with each other, provided their dispersion laws $\omega_k = \omega(\mathbf{k})$ are compatible with the conservation laws

$$\omega_{k'} + \omega_{k''} = \omega(\mathbf{k})$$

$$\mathbf{k}' + \mathbf{k}'' = \mathbf{k}.$$

Similar processes are described as *scattering* of wave $\omega_{\mathbf{k}'} = \omega(\mathbf{k}')$ by wave $\omega_{k''} = \omega(\mathbf{k}'')$ if $\omega_{k'} = \omega(\mathbf{k}')$ and $\omega_k = \omega(\mathbf{k})$ belong to the same modal type. If they are different, the effect is referred to as *mode conversion*. The six normal modes of the dusty plasma formally could be grouped into $6^3 = 216$ different schemes of wave scattering (or conversion) involving the conservation laws Equations (94) and (95) (also known as the resonance conditions). In fact only some of these can be implemented, namely such where the resultant $\omega_k$ and $\mathbf{k}$ are real-valued, with allowance for both the resonance conditions and dispersion laws (Bass and Blanc, 1962).

### 3.3.1 Wave Scattering in a Plasma with Fixed Dust Grains

The waves propagating through a uniform medium whose electric properties are characterized by the dielectric constant $\varepsilon = \text{const}$ are not scattered. As soon as fluctuations in the medium, including the dusty plasma, are taken into account, the medium can no longer be regarded as uniform. The dielectric constant fluctuations $\delta\varepsilon(\mathbf{r}, t)$ produce scattering of the waves. Let the scale time of variations in $\delta\varepsilon$ be $\tau_\varepsilon$. Then the wave interaction in the medium will give rise to waves of frequencies differing from the incident frequency $\omega$ by $\Omega \simeq \tau_\varepsilon^{-1}$. In the case of high-frequency waves, $\omega \gg \Omega$, this small difference often may be neglected, thus allowing for spatial variations of $\delta\varepsilon(\mathbf{r})$ only. If the plasma inhomogeneities are produced by charged dust grains, these latter may be considered as fixed. Allowance for their vibrations under the action of the incident wave's electric field would give rise to scattered waves of the same frequency, however of vanishingly low intensity (see Subsection 2.2.1). Indeed, the scattering cross-section of an isolated particle of mass $m_\alpha$ and charge $Q_\alpha$ is $\sigma_\alpha = (8\pi/3)(Q_\alpha^2 m_\alpha^{-1} c^{-2})^2$ (Equation (22) of Chapter 2). Therefore, even ions which are particles of a greatly lower mass than dust grains, are characterized by much smaller scattering cross-sections than electrons ($\sigma_i \leq 10^{-6}\sigma_e$). Thus, electrons are the major (in fact, the only) source of scattered radiation as long as individual particles are considered. In a plasma, both ions and dust grains may have a substantial effect on wave scattering by electrons as the heavier particles produce nonuniform electric fields and disturbances of the electron density. Let us consider the effects due to a fixed charged dust grain, following Tsytovich *et al.* (1989). The spherical domain of scale size $\lambda_D$ around the charged grain contains a compensating electric space charge where the electron density $n_e$ is different from the unperturbed value. Denoting the magnitude of the grain charge in terms of $e$ as $Z_d = |Q|/e$, the number of electrons in the Debye sphere will be $Z \sim Z_d$. With $\lambda \gg \lambda_D$, where $\lambda$ is the incident wavelength, the waves scattered by all of these electrons are in phase (coherent scattering). The scattering cross-section area characterizing this ensemble of electrons can be written as $\sigma \sim Z^2\sigma_e \sim Z_d^2\sigma_e$. In space plasmas, magnitudes as high as $Z_d \simeq 10^3 \div 10^4$ are quite common, hence $\sigma$ greatly exceeds $\sigma_e$. All the electrons from the Debye sphere being connected to the central grain by Coulombian forces, the radiation pressure on the grain sharply increases too.

To estimate the total cross-section area per unit volume of the plasma, let us recall that scattering by the electrons lying outside the Debye sphere is incoherent; hence the corresponding cross-section areas are simply added. The number density of such free electrons is $n_{0,e} - n_{o,d}Z_d$, therefore

$$\sigma_{tot} \simeq \sigma_e[(n_{0,e} - n_{0,d}Z_d) + n_{0,d}Z_d^2].$$

(The dust concentration is sufficiently low for each grain to be an independent scatterer). In view of $Z_d \gg 1$, $\sigma_{tot} \simeq \sigma_e n_{0,e}[1 + (n_{0,d}/n_{0,e})Z_d^2]$, while the cross-section area in the absence of dust grains is $\sigma_{tot} = \sigma_e n_{0,e}$. Because of charged dust grains in the plasma (fixed particles!), the cross-section area has increased by a factor of $1 + (n_{0,d}/n_{0,e})Z_d^2$ which may prove substantial if $Z_d \gg (n_{0,e}/n_{0,d})^{1/2}$ (de Angelis *et al.*, 1992). This is a quite realistic condition, even with a low concentration of dust grains, $n_{0,d}/n_{0,e} \sim 10^{-6}$, and a few volts' electric potentials of micron-size grains. As has been mentioned above, $Z_d$ is only a rough estimate of the number of electrons in the Debye sphere. Speaking of wave scattering, it proves possible to introduce an effective value $Z = Z_{eff}$ and distinguish between low-potential, or linear ($|\psi|e/T \ll 1$) and nonlinear ($|\psi|e/T \gg 1$) Debye screening of the charged grain. In the former case, the screening charge is formed by equal deviations $\delta n_e$ and $\delta n_i$ of the electron and ion number densities from their equilibrium values. Since electrons alone are involved in the process of wave scattering, $Z_{eff} = Z_d/2$. These considerations apply to waves of sufficient length only, $k\lambda_D \ll 1$. In the case of $k\lambda_D \sim 1$, allowance for the structure of the screening electron cloud is necessary. The equations for the scattering cross-section area would involve Fourier components of the screening charge density, $Z_{\mathbf{k},eff} = (2\pi)^{-3}\int(n_e - n_i)\exp(-i\mathbf{kr})\,d^3\mathbf{r}$. The appropriate analysis was performed by Tsytovich *et al.* (1989) who considered the transition scattering of longitudinal waves by charged grains in the plasma and obtained

$$Z_{\mathbf{k},eff} = Z_d/(1 + T_e/T_i + k^2\lambda_D^2).$$

The cross-section area increases with $\lambda$, getting saturated at $\lambda \gg \lambda_D$ where $\sigma \sim \sigma_e Z_d^2$. With $\lambda \ll \lambda_D$, the area drops down as $\sigma \sim \sigma_e Z_d^2(\lambda/\lambda_D)^4$. The value is independent of the charge sign sgn$Q$ in the case of low potential screening, whereas in the nonlinear case ($|\psi|e \gg T$) the sign may prove very important. The effect of nonlinear screening upon wave scattering has been considered both for longitudinal (Tsytovich *et al.*, 1989) and transverse waves (Bingham *et al.*, 1991). The restriction upon grain size, namely $a \ll \lambda_D$, is always satisfied in dusty space plasmas. The other parameter of importance is the radiation wavelength to the Debye length ratio. The analysis is particularly difficult for $a \ll \lambda \ll \lambda_D$, when the cross-section area is highly sensitive to the character of the electron density distribution $n_e(\mathbf{r})$ in the screening layer, at small distances from the grain surface, $a \leq r \ll \lambda_D$. It is the range of distances where the distribution is drastically different for positively or negatively charged grains.

The scattering in a medium of low dust density is considered to occur as from individual grains, however when estimating the cross-section area of a plasma volume containing many scatterers, the question of their correlation arises (de Angelis *et al.*, 1991). The simplest model assuming randomly distributed independent grains may prove insufficient if the energy $Q^2/d$ of electrostatic interaction between the grains is comparable with the energy $\varepsilon_{th}$ of their thermal motion (Goertz, 1989; de Angelis *et al.*, 1992). The amount of order in the grain disposition can be characterized by the parameter $\Gamma = Q^2/(d\varepsilon_{th})$. Special regular structures known as Coulombian lattices may appear with sufficiently high values of the parameter. These are characterized by sizable correlation of widely separated grains, resulting in diffractional peaks of the scattered radiation intensity along specific directions. Goertz (1989) suggested very high values of the parameter, $\Gamma \gg 1$, for dusty plasmas in space; however the current knowledge of basic parameters of the medium is too uncertain for positive statements.

All the scattering effects heretofore discussed have related to electrically charged dust grains. In principle, electromagnetic effects are also possible with $Q = 0$, provided the dust grains are magnetized (de Angelis *et al.*, 1991). The magnetic field in an individual magnetic domain near a surface of a micron-size ferromagnetic grain may be as high as $B \geq 10^4$ G and the appropriate magnetic moment $\mu_d \simeq Ba^3$ would be rather high. Obviously, the magnetic field is not screened whatsoever in the plasma and the nearby electrons oscillating in the field of an incident electromagnetic wave are acted upon by the d.c. magnetic field. The resulting a.c. currents generate scattered waves. The magnetized dust grains occurring in planetary magnetospheres are oriented along the "external" magnetic field of the planet. This preferred orientation should manifest itself in the coherence properties of the scattered radiation.

In the dusty plasma, wave scattering effects may accompany many phenomena of astrophysical significance (Tsytovich *et al.*, 1989). The increased radiation pressure on dust grains may be the cause of their drift resulting in electromagnetic instabilities.

The difference in the magnitude of pressure exerted on dust grains and microparticles sweeps the grains out of the plasma. This is one of the possible effects controlling the structure of planetary rings.

The increased cross-section area of electromagnetic scattering owing to coherent effects was reported in experiments on radar sounding of the summer time mesopause. According to Havnes *et al.* (1990), the observed echo intensities could be accounted for by a dust component of density $n_d \sim 10\,\text{cm}^{-3}$, with the grain size $a \sim 5 \times 10^{-2}\,\mu$. In a later paper (Havnes *et al.*, 1992) the writers allowed for dust grain motion through turbulent vortices of the neutral atmosphere, and for the gravity force.

### 3.3.2 Scattering by Dust-Acoustic Waves

Consider now what new effects may result from abandoning the assumption of statically distributed dust grains. Speaking of the motion of heavy particles, we mean primarily their oscillations associated with the dust-acoustic waves, a component of the thermal fluctuation spectrum in the dusty plasma. While their contribution to the total scattering cross-section is negligible, they may essentially influence the scattered spectrum.

As is known, allowance for the motion of ions in a "pure" plasma (which are also heavy-weight particles not involved directly in wave scattering) results in the appearance of Doppler-shifted lines in the scattered spectrum, $\omega_s = \omega \pm K v_{ph,i}$, where $\mathbf{K} = \mathbf{k}_s - \mathbf{k}$ and $v_{ph,i} = (T_e/m_i)^{1/2}$ is the ion-sound phase velocity. The "sharpness" of these peaks is determined by the attenuation rate of the ion sound. They are quite distinct in a non-isothermal plasma, $T_e \gg T_i$, thus making investigation of the ionic composition possible. Along with evaluation of the scattering cross-section area, spectral measurements underly the *incoherent scatter* technique, a widely used method of ionospheric diagnostics using the radiation of ground-based radars at $\omega \gg \omega_{p,max}$ (where $\omega_{p,max}$ is the plasma frequency near the F-layer maximum). The ion-acoustic mode is supported by the dusty plasma too, although the phase velocity is somewhat different (see Equation (81)). More importantly, the attenuation is low at $T_e n_{0,i} \gg T_i n_{0,e}$, which condition can be satisfied even at $T_e \sim T_i$ if $Q < 0$ and the number of electrons concentrated on dust grains is high.

The natural question is, what parameters of the dusty plasma could be recovered from observations of the electromagnetic wave scattering by the dusty sound. The satellite spectral lines would be shifted by

$$\pm\Omega = K v_{ph,d},$$

where the phase velocity $v_{ph,d}$ is given by Equation (82), i.e.

$$\Omega = K[Q^2 n_{0,d} T_i / (Q_i^2 n_{0,i} m_d)]^{1/2}. \tag{110}$$

Numerical estimates can be obtained from the data of Table 1. If the magnitude of $K$ were known, then ratios of some dust grain parameters could be estimated from the measured frequency shift $\Omega$. As follows from the momentum conservation law (Equation (94)),

$$K = 2k \sin \theta/2, \tag{111}$$

$\theta$ being the angle between $\mathbf{k}_s$ and $\mathbf{k}$ (the scattering angle). The difference between $k = |\mathbf{k}|$ and $k_s = |\mathbf{k}_s|$ is neglected in Equation (111) in view of $\Omega \ll \omega$ and $\omega_s \simeq \omega$, however $\theta$ may vary over a wide range from $\theta \simeq \pi$ (backscatter in the case of radar sounding) to $\theta \ll 1$ (forward scattering). Accordingly, $K \simeq 2k$ for the first case and $K \simeq k\theta \ll k$ for the second. Small-angle (forward) scattering is of special interest because of its possible applications for sensing dust grain parameters in far-away objects like the interstellar medium.

Today's concept of the medium composition is that of a rarefied plasma containing dust particles of a few microns or fractions of a micron in size. Listed in Table 2 (Voschinnikov, 1986) are the physical parameters of the interstellar dust grains that can be determined from observational data. As can be seen from the Table, all the basic parameters, except the electric charge, can be estimated more or less reliably (at least within some model).

As for the grain charge, it has been evaluated until now only through model calculations (the principal charging mechanisms are described in Chapter 1). For this reason, every new experiment suggesting evaluation of the grain charge would be most important. One such experiment might be observation of the radiation scattered by eigenwaves of the medium. The general outlay is shown in Figure 11. A sufficiently powerful source of radiation of frequency $\omega$ is at point $S$, while the scattered wave $(\mathbf{k}_s, \omega_s)$ is received by a radio telescope at point $O$. To identify the weak scattered signal and record the spectral peaks at $\omega \pm \Omega$, it would be desirable that the scattering angle $\theta$ and frequency shift $\Omega$ were as large as possible. According to Equations (110) and (111), that would require a high observation frequency $\omega$ and $k = \omega/c$. Meanwhile, there is a constraint on the dusty sound wavelength $\Lambda = 2\pi/K$ which should be much greater than the Debye length. Since $\theta \ll 1$ and $k \simeq K\theta$, the condition $K\lambda_D \ll 1$ becomes

$$k\theta\lambda_D \ll 1.$$

On the other hand, the scattered signal should be detectable in space if the scattering angle were greater than the angular width of the reception

Table 2: Dust grain parameters that can be recovered from observational data.

| Measurement of / Parameter | 1 | 2 | 3 | 4 | 5 | 6 | 7 | 8 |
|---|---|---|---|---|---|---|---|---|
| Chemical composition | 0 | 0 | 0 | 0 | + | (0) | − | + |
| Size | + | + | 0 | 0 | 0 | − | 0 | − |
| Geometry | (0) | + | 0 | − | 0 | − | − | − |
| Temperature | − | (0) | − | + | 0 | 0 | − | 0 |
| Electric charge | − | − | − | − | − | − | − | − |

1 Absorption in the interstellar medium
2 Polarization
3 Wave scattering
4 Infra-red continuum
5 Infra-red emission bands
6 Interstellar bands
7 Diffuse interstellar bands
8 Abundance of chemical elements

Notation:

| | |
|---|---|
| + | parameter can be determined reliably |
| 0 | parameter is determined unreliably |
| (0) | parameter has but a small effect on the measurements |
| − | parameter cannot be determined from observational data. |

pattern, $\theta > (kD)^{-1}$, $D$ being the scale size of the antenna. Thus we arrive at the double inequality

$$(kD)^{-1} < \theta \ll (k\lambda_D)^{-1}$$

and a constraint on the size of the antenna,

$$D \gg \lambda_D.$$

The Debye length in the interstellar medium is $\lambda_D \sim 10^2$ cm; hence the antenna should be a few tens of meters in size. This should be regarded now as a difficult but real technical requirement, and identification of the scattered signal in space seems possible. However the problem of spectral

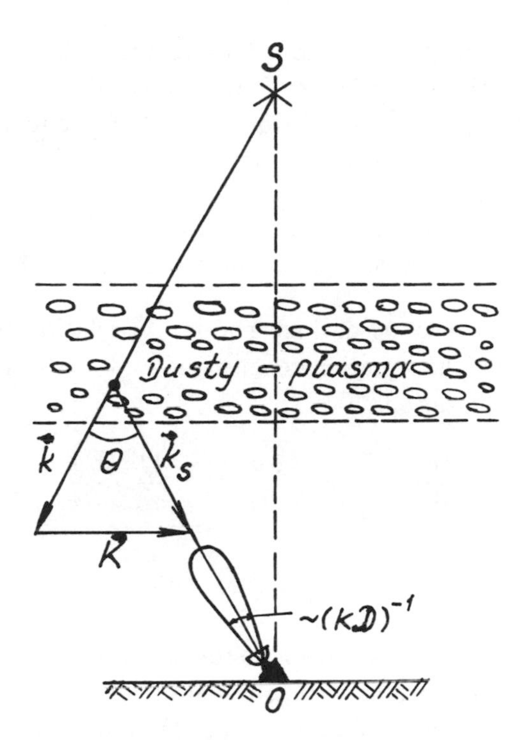


Figure 11: Radiation from a Source $S$ Scattered in the Dusty Interstellar Plasma.

measurements is more complex. Because of $K \ll \lambda_D^{-1} \sim 10^{-2}\,\mathrm{cm}^{-1}$ and $v_{ph,d} \sim 10^3\,\mathrm{cm}\cdot\mathrm{s}^{-1}$ (*cf.* the data of Table 1 for $a \simeq 10^{-2}\,\mu$), the satellite peaks cannot be separated from the central frequency by more than $\Omega \sim 10^2\,\mathrm{s}^{-1}$. This little shift is hardly detectable, especially because of the lack of a quasimonochromatic source of radiation characterized by a spectrum width $\Delta\omega \ll \Omega$. Even space masers with bandwidths of $10^5 \div 10^6\,\mathrm{s}^{-1}$ (Avrett, 1976) seem too broadbanded to be straightforwardly applicable to the observations described here. Yet a broadband source still could be usefully employed if the scattered signal carrying the imprint of the dusty sound wave were compared with the direct wave as a reference.

## 3.4 Nonlinear Waves

### 3.4.1 Dust-Acoustic Solitons

Once again, consider wave motions in the plasma medium using the problem formulation of Subsection 3.2.4. The motions will be assumed slow enough to allow the electron and ion density distributions to adiabatically follow variations of the potential $\Phi$. Since we consider dust-acoustic waves with $v_{ph,d} \simeq (T_i/m_d)^{1/2}$, the condition $v_{ph} \ll v_{T,i}$ and hence $v_{ph} \ll v_{T,e}$ is certainly met. The microparticle densities $n_e$ and $n_i$ are given by the Boltzmann distributions

$$n_e = n_{0,e} \exp(e\Phi/T_e) \text{ and } n_i = n_{0,i} \exp(-e\Phi/T_i). \tag{112}$$

As for the number density $n_d$ of dust grains, its relation to the potential $\Phi$ should be found from the equations of motion and continuity which will not be subjected to linearization,

$$\frac{\partial v_d}{\partial t} + v_d \left( \frac{\partial v_d}{\partial x} \right) = \frac{Q}{m_d} \cdot \frac{\partial \Phi}{\partial x} \tag{113}$$

and

$$\frac{\partial n_d}{\partial t} + \frac{\partial}{\partial x}(n_d v_d) = 0.$$

(For the sake of simplicity, we are considering a one-dimensional problem for a single sort of dust grains with a constant charge $Q < 0$ and constant mass, $m_d$). The equation set Equations (113) involving three variables, $n_d$, $v_d$ and $\Phi$, is complete with the Poisson equation which is also nonlinear,

$$\frac{\partial^2 \Phi}{\partial x^2} = -4\pi e[n_i(\Phi) - n_e(\Phi) - Q n_d/e]. \tag{114}$$

This set of partial differential equations permits of solutions depending on the combination $\xi = x - ut$, where $u$ is a constant value, rather than upon $x$ and $t$. Such solutions represent waves propagating at a velocity $u$ without variations of their waveform.

By introducing the new variable, $\xi$, instead of $x$ and $t$, we have transformed Equations (112)-(114) into a set of ordinary equations that can be solved with respect to $n_d$,

$$n_d = n_{0,d} \frac{u}{u - v_d} = n_{0,d} u [u^2 + 2Q\Phi/m_d]^{-1/2}. \tag{115}$$

(The boundary conditions have been $\Phi = 0$, $v_d = 0$ and $n_d = n_{0,d}$ for $\xi \to \infty$). An equation for $\Phi$ is obtained from Equation (115) by multiplying it by $d\Phi/d\xi$,

$$\frac{1}{2} \left( \frac{d\Phi}{d\xi} \right)^2 + V_s(\Phi, u) = 0. \tag{116}$$

This is known as the "energy integral", and $V_s(\Phi, u)$ is the so-called Sagdeev potential,

$$V_s(\Phi,u) = 4\pi\Big\{n_{0,i}T_i\Big[1-\exp(-e\Phi/T_i)\Big]+$$

$$+n_{0,e}T_e\Big[1-\exp(e\Phi/T_e)\Big]+n_{0,d}m_d u^2\times$$

$$\times\Big[1-(1+2Q\Phi/(u^2 m_d))^{1/2}\Big]\Big\}.$$

Given Equation (116), determination of the nonlinear dust-acoustic waveform $\Phi(\xi)$ has reduced to taking a quadrature. The equation was obtained and analyzed in the very first paper on dust-acoustic waves (Rao *et al.*, 1990; see also Verheest, 1992; 1993a; 1993b). Of special interest are the solutions dying away, with their derivatives, both at $\xi \to \infty$ and $\xi \to -\infty$. They are known as solitary waves, or solitons. There is vast literature on solitons in different media, plasmas in particular. In connection with dust-acoustic solitons, we have mentioned the review paper by Verheest (1993a) where more references can be found.

To consider waves of a finite but not extremely high amplitude, the potential $V_s(\Phi, u)$ can be expanded in powers of $\Phi$. The lowest-order term is $\Phi^2$, while the linear power is absent by virtue of the medium quasineutrality. Retaining this first term alone, we obtain

$$\left(\frac{d\Phi}{d\xi}\right)^2 = A\Phi^2, \qquad \text{or} \qquad \frac{d\Phi}{d\xi} = \pm\sqrt{A}\Phi \tag{117}$$

with

$$A = \frac{\omega_{p,i}^2}{v_{T,i}^2}\left(1+\frac{n_{0,e}T_i}{n_{0,i}T_e}\right) - \frac{\omega_{p,d}^2}{u^2}.$$

The second term in the brackets can be neglected as small, since $n_{0,e} < n_{0,i}$ owing to the negatively charged dust grains (even $n_{0,e} \ll n_{0,i}$ is plausible), while $T_e$ normally is greater than $T_i$. Then $A$ can be written in the compact form

$$A = \mathrm{M}^2 - 1/(\lambda_D^2\mathrm{M}^2), \tag{118}$$

where $\mathrm{M} = u/v_{ph,d}$ is the Mach number; $\lambda_D = v_{T,i}/\omega_{p,i}$ the ionic Debye length, and $v_{ph,d}$ the phase velocity of dust-acoustic waves as introduced before. The two solutions of Equation (117) are

$$\Phi_\pm = \Phi_0\exp(\pm\sqrt{A}\xi). \tag{119}$$

Since Equation (116) has reduced to a linear equation, it might seem that no parameters of the soliton could be determined within this approximation. In fact, this is not so, as the linear approximation may be regarded as an asymptotic form of the exact solution, since $\Phi$ is very small at the soliton "tails" $\xi \to \pm\infty$. First, it proves possible to immediately establish the condition imposed on the soliton velocity $u$. It follows from the demand that $A > 0$ (otherwise, Equation (119) would not represent a dying away function but rather oscillate at $\xi \to \pm\infty$). Therefore, $\mathrm{M} > 1$ or $u > v_{ph,d}$, i.e. the soliton should be a supersonic disturbance. The scale width $\Delta$ of the soliton can be easily estimated too as it is natural to expect the characteristic length of variations of $\Phi$ to remain the same for all $\xi$'s. Then $\Delta \sim 1/\sqrt{A} \simeq \lambda_D \mathrm{M}/\sqrt{\mathrm{M}^2 - 1}$, and the width decreases with growing M. For solitons of moderate velocity we can set $\mathrm{M}^2 - 1 = \mu^2$ ($\mu^2 \ll 1$) to obtain

$$\Delta \sim \lambda_D/\mu. \tag{120}$$

In the further analysis, $\mu$ will be regarded as a second small parameter, along with $e\Phi/T \sim Q\Phi/T \ll 1$.

Now we attempt to estimate the soliton amplitude, i.e. the maximum value $\Phi_m$ to be reached, judging by the symmetric behavior of the asymptotic solutions Equation (119), near $\xi = 0$. The true solutions are not linearly independent near $\xi = 0$ (Figure 12). While $\Phi_m$ can be roughly estimated as $\Phi_m \sim \Phi_\pm(0) = \Phi_0$, its real value can only be found through retaining the next term in the expansion of $V_s(\Phi, u)$,

$$\left(\frac{d\Phi}{d\xi}\right)^2 \simeq A\Phi^2 + B\Phi^3. \tag{121}$$

Here

$$B = 4\pi \left[\frac{Q^3 n_{0,d}}{2u^4 m_d^2} - \frac{e^3}{6}\left(\frac{n_{0,i}}{T_i^2} - \frac{n_{0,e}}{T_e^2}\right)\right]$$

or, with the same simplifying assumptions as before,

$$B = \left(\frac{3}{\mu^4} - 1\right) / (24\pi e n_{0,i} \lambda_D^4).$$

Since $\Phi = \Phi_m$, where $d\Phi/dx = 0$, the soliton amplitude is given by the equation $V_s(\Phi_m, u) = 0$, *viz.*

$$\Phi_m = -A/B = 24\pi e n_{0,i} \lambda_D^2 \mathrm{M}^2(\mathrm{M}^2 - 1)/(\mathrm{M}^4 - 3) \simeq -12\pi\mu^2 e n_{0,i} \lambda_D^2. \tag{122}$$

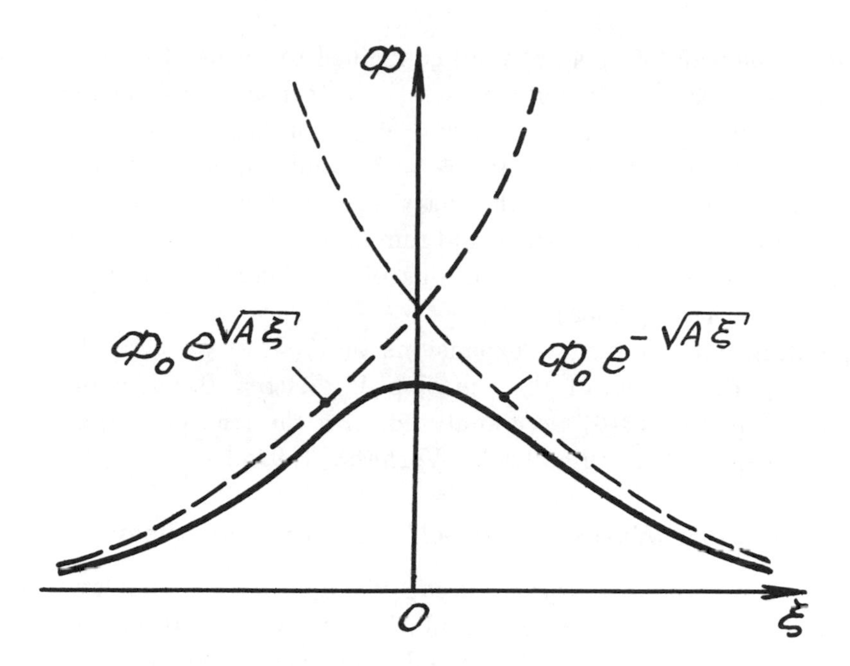


Figure 12: Formation of a Dust-Acoustic Soliton.

Note the dust-acoustic soliton to be a negative disturbance (at least within the present approximation) (Rao *et al.*, 1990; Verheest, 1993a), in contrast to the ion-acoustic soliton which is a positive pulse of the potential $\Phi$. (The maximum value $\Phi_m$ of the ion-acoustic soliton is given by the same Equation (121), with $en_{0,i}$ replaced by $-en_{0,e}$ and $\lambda_D^2$ by $v_{T,e}^2/\omega_{p,e}^2$). The total charge $Q_{s,tot}$ carried by the solitary wave is $Q_{s,tot} \simeq \Phi_m\Delta \simeq -12\mu\pi en_{0,i}\lambda_D^3 \simeq -12\mu\pi Q n_{0,d}\lambda_D^3$. To the order of magnitude, the total charge in the soliton should be equal to the charge of the dust-and-ions plasma component contained in the Debye sphere (with account of the small parameter $\mu$).Both the soliton amplitude, $\Phi_m$ and the total charge $Q_{s,tot}$ increase with $\mu$.

The semi-qualitative estimates of $\Delta$ and $\Phi_m$ given in this Section can be checked against the values that follow from the exact solution of Equation (116) obtainable for the two-term representation Equation (121) of $V_s(\Phi, u)$, *viz.*

$$\Phi = -(A/B)\mathrm{sech}^2(\sqrt{A}\xi/2). \qquad (123)$$

Naturally enough, the asymptotic forms of Equation (123) at $\xi \to \pm\infty$

coincide with Equation (119), and the maximum value is $\Phi_m = \Phi(\xi = 0) = -A/B$.

If the expansion of $V_s(\Phi, u)$ were continued to terms of order $\Phi^4$, then the equation $V_s(\Phi, u) = 0$ could possess, with certain combinations of the parameters, another double root at $\Phi = \Phi_m$. The derivative $d\Phi/d\xi$ would be equal to zero both at $\Phi = 0$ and $\Phi = \Phi_m$ and the potential would vary between $\Phi = 0$ and $\Phi = \Phi_m$ with $\xi$ varying from $\xi = -\infty$ to $\xi = \infty$. The characteristic scale of variations would remain $\Delta\xi \sim \sqrt{A}$, with the potential changing sharply near $\xi = 0$. Solutions of this kind (double layers) were analyzed in Verheest (1993a).

Operating with power-law expansions of $V_s(\Phi, u)$ we take the risk of obtaining spurious roots of the equation $V_s(\Phi, u) = 0$. The problem is removed if Equation (116) were analyzed with the true potential $V_s(\Phi, u)$ which only can be done numerically (Verheest, 1993a,b).

### 3.4.2 Nonlinear Waves in the Self-Gravitational Plasma

The analysis of nonlinear waves is greatly complicated if self-gravitation is taken into account. The mere formulation of basic equations encounters difficulties. Strictly speaking, they should have been discussed in the linear theory when introducing small deviations of all variables from their equilibrium values (see Equation (10)). It was implied then that all the variables vanished if the particle number densities were not perturbed (i.e. $n'_\alpha = 0$). In fact, this assertion is not true for the gravitation potential $\psi_G$ as the mass densities do not satisfy any condition analogous to electric quasineutrality, $\sum n_{0,\alpha} Q_\alpha = 0$.

The equilibrium potential is not zero, $\psi_{0,G} \neq 0$; however it can be assumed that the oscillatory motion of particles is not affected by this gravitational d.c. background jointly produced by all the participating masses. Accordingly, the values $n_i$ and $n_d$ involved in the Poisson equation for $\psi_G$,

$$\frac{\partial^2 \psi_G}{\partial x^2} = -4\pi G(n_d m_d + m_i n_i), \tag{124}$$

represent *deviations* from the equilibrium values rather than total number densities of ions and dust grains. (It is assumed once again that all the plasma electrons have been absorbed by dust grains, the plasma consisting just of ions and grains of a single sort). The Boltzmann distribution of ions then takes the form

$$n_i = n_{0,i}\{\exp[(-e\psi_E + m_i\psi_G)/T] - 1\}, \tag{125}$$

where the electric potential has been denoted $\psi_E$ instead of $\Phi$ to emphasize the symmetry of electric and gravitation forces.

Equations (124) and (125) should be complemented by the equations of motion and continuity, and Poisson's equation for $\psi_E$,

$$\begin{aligned}
&\frac{\partial v_d}{\partial t} + v_d \frac{\partial v_d}{\partial x} = \frac{Q}{m_d} \cdot \frac{\partial \psi_E}{\partial x} + \frac{\partial \psi_G}{\partial x} \\
&\frac{\partial n_d}{\partial t} + \frac{\partial}{\partial x}(n_{0,d} + n_d) v_d = 0 \\
&\frac{\partial^2 \psi_E}{\partial x^2} = 4\pi (Q n_d - e n_i).
\end{aligned} \tag{126}$$

The complete set Equations (124-126) of partial differential equations can be reduced, upon introduction of the new coordinate $\xi = x - ut$, to two ordinary nonlinear equations with respect to $\psi_E$ and $\psi_G$,

$$\begin{aligned}
\frac{\partial^2 \psi_E}{\partial \xi^2} &= W_E(\psi_E, \psi_G, u) \\
\frac{\partial^2 \psi_G}{\partial \xi^2} &= W_G(\psi_E, \psi_G, u),
\end{aligned} \tag{127}$$

with

$$\begin{aligned}
W_E &= 4\pi \left\{ Q n_{0,d} \left[ 1 + \frac{2}{u^2} \left( \frac{Q}{m_d} \psi_E + \psi_G \right) \right]^{-1/2} - \right. \\
&\left. - e n_{0,i} \exp(-e\psi_E/T + m_i \psi_G/T) \right\}
\end{aligned}$$

and

$$\begin{aligned}
W_G &= -4\pi G \left\{ m_d n_{0,d} \left[ \left[ 1 + \frac{2}{u^2} \left( \frac{Q}{m_d} \psi_E + \psi_G \right) \right]^{-1/2} - 1 \right] + \right. \\
&\left. + m_i n_{0,i} \exp(-e\psi_E/T + m_i \psi_G/T) - 1 \right\}.
\end{aligned}$$

No first integral of the set has been found so far, hence it has not been investigated analytically. Expanding $W_E$ and $W_G$ in powers of $\psi_E$ and $\psi_G$ does not yield the desired tractable result either, because of cross-products. Within the linear approximation, the set Equations (127) yields the familiar dispersion relation Equation (28) in which "electronic" terms are absent ($\omega_{p,e} = 0$), $\omega$ has been set zero in the "ionic" terms (owing to the use of Boltzmann's distribution) and the thermal velocity $v_{T,d}$ of dust grains is zero in the dust-dependent terms.

By neglecting self-gravitation effects (i.e. setting $G = 0$ and $\psi_G = 0$), Equation (127) can be reduced to the above discussed Equation (116). On the other hand, electrostatic interaction of particles can be excluded in a similar way, by setting $Q = 0$, $e = 0$ and $\psi_E = 0$. That would leave a nonlinear equation representing wave processes in a neutral self-gravitating medium (with the same simplifying assumptions as before, namely of the Boltzmann-distributed light-weight component and zero thermal velocity for the heavier particles). The "energy integral" for that case is

$$\frac{1}{2}\left(\frac{\partial \psi_G}{d\xi}\right)^2 + V_G(\psi_G, u) = 0, \tag{128}$$

with

$$\begin{aligned} V_G(\psi_G, u) &= 4\pi G\Big\{ m_d n_{0,d} u^2 \left[\left(1 + \frac{2}{u^2}\psi_G\right)^{1/2} - 1\right] + \\ &\quad + m_i n_{0,i} v_{T,i}^2 \left[\exp(m_i \psi_G / T) - 1\right] - \psi_G (m_d n_{0,d} + m_i n_{0,i})\Big\}. \end{aligned}$$

This equation can be analyzed in the same pattern as Equation (116). Expanding $V_G(\psi, u)$ in powers of $\psi_G$ to the third-order term, we have

$$\left(\frac{d\psi_G}{d\xi}\right)^2 \simeq A_G \psi_G^2 + B_G \psi_G^3, \tag{129}$$

where

$$A_G = -\frac{\omega_{G,i}^2}{v_{T,i}^2} + \frac{\omega_{G,d}^2}{u^2}$$

and

$$B_G = -\frac{\omega_{G,i}^2}{v_{T,i}^4} - \frac{\omega_{G,d}^2}{u^4}.$$

The solution is

$$\psi_G = -(A_G/B_G)\mathrm{sech}^2(\sqrt{A_G}\xi/2), \tag{130}$$

which represents a self-gravitational soliton, provided that $A_G > 0$. This infers the condition to be met by the soliton velocity $u$,

$$u < v_{T,i}\frac{\omega_{G,d}}{\omega_{G,i}} = \left(\frac{T n_{0,d} m_d}{n_{0,i} m_i^2}\right)^{1/2}. \tag{131}$$

The right-hand side of this inequality is the velocity of the very low frequency sound, which can be seen from the linear dispersion relation Equation (28). With the same simplifications as before, nullification of the second factor in Equation (28) (the coupling term $K_T$ turns to zero with $Q_\alpha = 0$) yields

$$1 - \frac{\omega_{G,i}^2}{k^2 v_{T,i}^2} + \frac{\omega_{G,d}^2}{\omega^2} = 0, \tag{132}$$

whence the phase velocity of the sound is

$$v_s = v_{T,i} \left( \frac{\omega^2 + \omega_{G,d}^2}{\omega_{G,i}^2} \right)^{1/2} = \left[ T_i(\omega^2 + \omega_{G,d}^2)/(m_i \omega_{G,d}^2) \right]^{1/2}. \tag{133}$$

This sound wave deserves several comments. First, the familiar rule for composing the phase velocity of "hybrid" acoustic waves (like the ion-acoustic or dust-acoustic mode), namely substituting the temperature of the lighter component and the mass of the heavier into the standard formula for sound velocity, does not apply here. While the temperature $T$ does relate to the lighter particles, the "effective mass" is $m_i^2 n_{0,i}/m_d n_{0,d}$, which depends upon the masses and number densities of both components. The reason is that the characteristic frequencies $\omega_{p,i}$ and $\omega_{p,d}$ of the media controlled by electric interaction are *inversely proportional* to $m_i^{1/2}$ and $m_d^{1/2}$, whereas the similar values $\omega_{G,i}$ and $\omega_{G,d}$ of a gravitation controlled medium show a *direct* proportionality to $m_i^{1/2}$ and $m_d^{1/2}$.

Oddly enough, Equation (131) does not involve the gravitation constant $G$, while the lighter and heavier particles are coupled only through gravitation. With $G \to 0$, the dependence on $m_d$ admittedly should vanish. In fact it is seen from Equation (133) that $G$ cannot be set zero in Equation (131), lest the condition $\omega^2 \ll \omega_{G,d}^2$ used in the derivation should be violated. If necessary, the limiting transition can be made in the the dispersion relation Equation (132) with allowance for the inertia of light-weight particles. The result would be the correct relation for the sound velocity, $v_s = \sqrt{T_i/m_i}$, demonstrating loss of coupling between the lighter and heavier particles at $G \to 0$.

Since Equation (131) does not account for the inertia of lighter particles, care should be taken to satisfy the condition $\omega^2 \ll k^2 v_{T,i}^2$, or $v_s \ll v_{T,i}$. This leads to the following condition on the dust parameters

$$\frac{n_{0,d} m_d}{n_{0,i} m_i} \ll 1. \tag{134}$$

By way of example, consider oxygen atoms, $m_i \simeq 2.7\times10^{-23}$ g and loose dust grains of $\rho \simeq 0.5\,\mathrm{g/cm^3}$, $m_d \sim 2\times10^{-18}a^3$ ($a$ is in microns). With parameters like these, Equation (134) will be satisfied if $n_{0,d}/n_{0,i} \ll 10^{-5}/a_\mu^3$, which seems quite realistic for submicron dust grains.

Another restriction to be remembered is associated with the replacement of $V_G(\psi_G, u)$ by its power-law expansion Equation (129). This is allowed if the dimensionless potential $\psi_G/v_{T,i}^2$ is small, i.e.

$$m_i\psi_G/T \ll 1. \tag{135}$$

The maximum value of $\psi_G$ can be estimated as

$$\psi_{G,m} = |A_G/B_G| \simeq \frac{3}{4}v_{T,i}^2\left(\frac{v_s^2}{u^2} - 1\right),$$

hence the condition for the validity of Equation (130) reduces to $u \leq v_s$ which is in agreement with Equation (131). Thus, the most restrictive condition is Equation (134).

## 3.5 Waves in Magnetized Plasmas

Like in the isotropic plasma, charged dust grains in a magnetized plasma can influence wave processes in the medium even if they do not move themselves. Allowance for the dynamic properties of the grains in the presence of a magnetic field and self-gravitation greatly complicates the analysis. In particular, the modal equation acquires three new series of characteristic frequencies, i.e. $\omega_{p,d}$; $\omega_{B,d}$ and $\omega_{G,d}$ or the grain-Langmuir, grain-gyro and the Jeans frequency.

### 3.5.1 Magnetized Plasma with Static Dust Grain Configurations

The waves supported by a plasma medium can "feel" the presence of fixed charged dust grains for two possible reasons. First, part of the plasma electrons are excluded from plasma vibrations being absorbed by the dust grains. Second, the charged grains produce a nonuniform distribution of the electric potential, against which background wavelike disturbances can develop. Both mechanisms have been discussed regarding the isotropic plasma in Subsections 3.2.1 and 3.2.2. The same problem formulation is retained here (Salimullah *et al.*, 1991).

Consider the propagation of low-frequency electrostatic waves through a plasma placed in a d.c. magnetic field, $\mathbf{B}_0$, and d.c. electric field $\mathbf{E}_0$. The magnetic field is uniform, while the electric, $\mathbf{E}_0(\mathbf{r}) = -\nabla\Phi(\mathbf{r})$ is produced by an ensemble of randomly distributed charged dust grains. We will confine the analysis to the case of rarely distributed dust grains ($d \gg \lambda_D$, where $d$ is the average separation between the grains) and reasonably low mean potential $\Phi_0 \ll T/e$, where $\Phi_0$ is given by Equation (29) of Chapter 1 and $T$ denotes the ion and electron temperature. Then the linearized kinetic equations governing the oscillatory corrections $f_\alpha(\mathbf{r},\mathbf{v},t)$ to the electron ($\alpha = e$) and ion ($\alpha = i$) distribution functions (the Vlasov equations) can be written as

$$\frac{\partial f_\alpha}{\partial t} + \mathbf{v}\cdot\nabla f_\alpha + \frac{Q_\alpha}{m_\alpha}\cdot\frac{(\mathbf{v}\times\mathbf{B}_0)}{c}\cdot\nabla_{\mathbf{v}} f_\alpha + \frac{Q_\alpha}{m_\alpha}\mathbf{E}(\mathbf{r},t)\cdot\nabla_{\mathbf{v}} f_0(\mathbf{r},\mathbf{v}) - \frac{Q_\alpha}{m_\alpha}\cdot\nabla\Phi_0(\mathbf{r})\cdot\nabla_{\mathbf{v}} f_\alpha = 0, \tag{136}$$

with $\mathbf{E}(\mathbf{r},t)$ denoting the electric vector of the propagating wave. The equilibrium distribution function $f_{0,\alpha}$ can be represented as the product

$$f_{0,\alpha}(\mathbf{r},\mathbf{v}) = C_\alpha(\mathbf{r})F_{0,\alpha}(\mathbf{v}), \tag{137}$$

with

$$C_\alpha(\mathbf{r}) = (1-\mu_\alpha-\mu_\alpha^2)\left[1-\mu_\alpha\sigma(\mathbf{r})+\frac{1}{2}\mu_\alpha^2\sigma^2(\mathbf{r})\right],$$

$$F_{0,\alpha}(\mathbf{v}) = \frac{n_{0,\alpha}}{\pi^{3/2}v_{T,\alpha,\perp}^2 v_{T,\alpha,\parallel}}\exp\left(-\frac{v_\perp^2}{v_{T,\alpha,\perp}^2}-\frac{v_\parallel^2}{v_{T,\alpha,\parallel}^2}\right)$$

and $\mu_\alpha = Q_\alpha\Phi_0(1/T_\perp + 1/T_\parallel)$ (de Angelis *et al.*, 1989; Krall and Trivelpiece, 1973). The magnetic field manifests itself through the anisotropy in temperature (the subscripts $\parallel$ and $\perp$ denote directions along and across $\mathbf{B}_0$, respectively). $\mu_\alpha$ are small parameters to allow for the electric field of charged dust grains. Equation (136) can be solved if the last term is treated as a perturbation. The random function

$$\sigma(\mathbf{r}) = (1/T_\perp + 1/T_\parallel)Q_\alpha\Phi(\mathbf{r})/\mu_\alpha$$

takes into account fluctuations $\Phi(\mathbf{r})$ in the electrostatic potential. It is represented by its statistical characteristics, i.e. either the autocorrelation function or power spectrum. In the final results, the latter will be assumed Gaussian,

$$S(\gamma) = \frac{1}{\pi^{3/2}\gamma_0^3}\exp(-\gamma^2/\gamma_0^2),$$

where $\gamma_0^{-1}$ characterizes the spatial scale of the random distribution of dust grains ($\gamma_0^{-1} \sim r_0$). Equations (136) are used to obtain Fourier components $f_\alpha(\mathbf{k},\omega)$ of the distribution function, and then the corresponding number densities $n_\alpha(\mathbf{k},\omega)$ of the plasma particles and the dielectric constant $\varepsilon(\mathbf{k},\omega)$. Details of the derivation can be found, e.g. in Salimullah *et al.* (1991). We will only quote the results of analyzing the modal equation $\varepsilon(\mathbf{k},\omega) = 0$ in particular cases. These can be classified according to the frequency range considered and the preferred propagation direction.

Ion-acoustic frequencies ($kv_{T,i} \ll \omega \ll kv_{T,e}$)

The dispersion relation for the real frequency in the case of parallel propagation ($k_\parallel = k$, $k_\perp = 0$) is

$$\omega^2 \simeq k^2 c_s^2 \delta/(1 + k^2\lambda_{D,e}^2 - 2\mu_e + \mu_e^2/2), \tag{138}$$

with $c_s^2 = T_e/m_i$ and $\lambda_{D,e}^2 = T_e/4\pi n_{0,e}e^2$. Comparison of this equation with the similar formula for the conventional plasma reveals the effects owing to the electric field of the dust grains. It contains extra terms $\sim \mu_e$ and $\mu_e^2$ in the denominator and the $\delta = n_{0,i}/n_{0,e}$ factor in the numerator.

Equation (138) is an approximate formula representing the "principal value" of the $\omega = f(k)$ dependence from which $B_0$ has dropped out. The effect of $B_0$ on the real frequency $\omega$ and the imaginary part (i.e. attenuation rate of the wave) is described by formulas of better precision.

In the case of transverse propagation (i.e. $k_\parallel = 0$ and $k_\perp = k$) and with $\omega_{B,e} \gg kv_{T,e}$ the $\omega = f(k)$ relation takes the form

$$\omega^2 \simeq (\sqrt{\pi}/4)\mu_i^2(k\gamma_0)v_{T,i,\perp}^2. \tag{139}$$

In contrast to the previous relation, this mode has no analogs in the "pure" plasma and vanishes at $\mu_i = 0$ or $\gamma = 0$.

Very low frequencies ($\omega \ll kv_{T,i}$)

If the dust grain dynamics and thermal motion were taken into account, then the frequency range we are discussing here could be referred to as that of dust-acoustic waves. At these frequencies, the presence of a magnetic field gives rise (in the case of parallel propagation, $k_\parallel = k$ and $k_\perp = 0$) to a new mode,

$$\omega \simeq \mu_i \gamma_0 v_{T,i}/2. \tag{140}$$

By comparing this equation with the relation between $\omega_p$ and $\lambda_D$ in the "pure" plasma ($\omega_p \simeq v_T/\lambda_D$), one could say that Equation (140) represents ionic vibrations in a nonuniform dusty structure whose spatial scale is the average dust grain separation, $d \sim \gamma_0^{-1}$, rather than the Debye length.

The dispersion relations for transversely propagating waves ($k_{\|} = 0$ and $k_{\perp} = k$) are, depending on the relative magnitude of $\omega_{B,\alpha}$ and $kv_{T,\alpha}$,

$$\omega \simeq \omega_{B,i}\left[1+\omega_{p,i}^2\frac{4-\mu_i^2(\gamma_0/k)\sqrt{\pi}}{4\omega_{B,i}^2}\right]^{1/2} \qquad (\omega_{B,\alpha} \gg kv_{T,\alpha}) \tag{141}$$

or

$$\omega \simeq \omega_{B,i}\left[1+\omega_{B,i}\frac{1-2\mu_i+\mu_i^2(1+\gamma_0/k)/2}{k^3 v_{T,i}\lambda_{D,i}^2\sqrt{2\pi}}\right]^{1/2} \qquad (\omega_{B,\alpha} \ll kv_{T,\alpha}). \tag{142}$$

The waves of Equations (141) and (142) can be regarded as dust-modified analogs of ionic Bernstein modes (Krall and Trivelpiece, 1973). Like in the conventional plasma, these waves propagate without attenuation (at least in the approximation assumed in this Subsection).

### 3.5.2 Magnetized Dusty Self-Gravitational Plasma: Dielectric Constant Tensor

In this Subsection, the dynamics of dust grains and their contribution to cooperative processes via self-consistent electric and gravitation fields will be essential. Restricting the analysis by the simplest formulation, let us consider linearized equations of motion for particles of species $\alpha$,

$$\frac{\partial \mathbf{v}_\alpha}{\partial t} = \frac{Q_\alpha}{m_\alpha}(\mathbf{E}+\mathbf{v}_\alpha\times\mathbf{B}_0/c) - \nabla\psi_G. \tag{143}$$

Once again, $\psi_G$ is the gravitation potential given by Equation (124). If all the values vary harmonically with time and coordinates as $\exp[-i(\omega t - \mathbf{kr})]$, Equation (143) becomes an algebraic set,

$$\mathbf{v}_\alpha = \frac{i}{\omega}\left[\frac{Q_\alpha}{m_\alpha}(\mathbf{E}+\mathbf{v}_\alpha\times\mathbf{B}_0/c)\right] - \frac{\mathbf{k}}{\omega^2 k^2}\sum_\alpha \omega_{G,\alpha}^2(\mathbf{k}\mathbf{v}_\alpha). \tag{144}$$

In the frame of reference where $\mathbf{k}$ lies within the plane $xOz$ making an angle $\theta$ with the magnetic field $\mathbf{B}_0$, projections of $\mathbf{v}_\alpha$ onto the axes $x$, $y$ and $z$ can be easily found with allowance for quasineutrality. Substituting these into the current density equation, $\mathbf{j} = \sum_\alpha Q_{o,\alpha} n_{0,\alpha} \mathbf{v}_\alpha$, one specifies components of the conductivity tensor, $\sigma_{ij}$ and the dielectric constant $\varepsilon_{ij}$ (Bliokh and Yaroshenko, 1987), *viz.*

$$\begin{aligned}
\varepsilon_{xx} &= 1 - \sum_\alpha \frac{\omega_{p,\alpha}^2}{\omega^2 - \omega_{B,\alpha}^2} + \sin^2\theta \frac{\omega^2 K_B^2}{\Omega^2} \\
\varepsilon_{xy} &= -\varepsilon_{yx} = -i\left[\sum_\alpha \frac{\omega_{p,\alpha}^2 \omega_{B,\alpha}}{\omega(\omega^2 - \omega_{B,\alpha}^2)} - \omega \sin^2\theta \frac{\tilde{\Omega} K_B}{\Omega^2}\right] \\
\varepsilon_{yy} &= 1 - \sum_\alpha \frac{\omega_{p,\alpha}^2}{\omega^2 - \omega_{B,\alpha}^2} + \sin^2\theta \frac{\tilde{\Omega}^2}{\Omega^2} \\
\varepsilon_{zz} &= 1 - \sum_\alpha \frac{\omega_{p,\alpha}^2}{\omega^2}
\end{aligned} \tag{145}$$

and $\varepsilon_{xz} = \varepsilon_{zx} = \varepsilon_{yz} = \varepsilon_{zy} = 0$, where

$$K_B = \sum_\alpha \frac{\omega_{p,\alpha}\omega_{G,\alpha}}{\omega^2 - \omega_{B,\alpha}^2}, \qquad \tilde{\Omega} = \sum_\alpha \frac{\omega_{p,\alpha}\omega_{G,\alpha}\omega_{B,\alpha}}{\omega^2 - \omega_{B,\alpha}^2},$$

and

$$\Omega^2 = \omega^2 + \sum_\alpha \omega_{G,\alpha}^2 \left(\cos^2\theta + \frac{\omega^2 \sin^2\theta}{\omega^2 - \omega_{B,\alpha}^2}\right).$$

If self-gravitation were "switched off" (i.e. $G = 0$), then the tensor components of $\varepsilon_{ij}$ would go over to the respective values for the conventional magnetoactive plasma. On the other hand, Equation (145) would coincide at $B_0 = 0$ with Equations (58) and (59) if $v_{T,\alpha}$ were set to zero in the latter equations (the present formulas relate to the cold plasma) and the coupling factor $K_T$ were also nullified.

Generally, the waves supported by the magnetized plasma cannot be separated into transverse and longitudinal. Therefore, it proves necessary to analyze the modal equation of the general form,

$$\det|k^2\delta_{ij} - k_i k_j - \frac{\omega^2}{c^2}\varepsilon_{ij}(\omega, \mathbf{k})| = 0.$$

Given the specific structure of $\varepsilon_{ij}$ in the frame of reference that has been adopted, this becomes

$$\left(k^2 - \frac{\omega^2}{c^2}\varepsilon_{yy}\right)\left[\frac{\omega^2}{c^2}\varepsilon_{xx}\varepsilon_{zz} - k^2(\varepsilon_{zz}\cos^2\theta + \varepsilon_{xx}\sin^2\theta)\right] -$$

$$-\frac{\omega^2}{c^2}\varepsilon_{xy}\epsilon_{yx}\left(k^2\sin^2\theta-\frac{\omega^2}{c^2}\varepsilon_{zz}\right)=0. \tag{146}$$

Once again, the analysis can be somewhat simplified for the particular cases of parallel ($\theta = 0$) and transverse ($\theta = \pi/2$) propagation.

### 3.5.3 Parallel Propagation

With $\theta = 0$ and $\Omega^2 \neq 0$, Equation (146) becomes a product of two factors corresponding to two classes of solutions well known in the theory of the common plasma, *viz.*

$$\varepsilon_{zz} = 1-\sum_\alpha \omega_{p,\alpha}^2/\omega^2 = 0 \tag{147}$$

and

$$k^2 = \frac{\omega^2}{c^2}(\varepsilon_{xx} \pm \sqrt{\varepsilon_{xy}\varepsilon_{yx}}). \tag{148}$$

The first of these describes longitudinal plasma vibrations, uneffected either by the magnetic field or even self-gravitation. This might seem unexpected and certainly deserves special discussion. In fact, the dielectric constant tensor $\varepsilon_{ij}$ describes the response of the medium (an ensemble of charged particles!) to *electric* fields. As for self-gravitation, it can produce cooperative motion of particles *in the absence* of an electric field. Such motions are not allowed for by $\varepsilon_{ij}$, although the combined effects of the electric and gravitation fields are reflected by $K_B$, $\Omega$ and $\tilde{\Omega}$. Note that Equation (147) has been derived under the additional condition $\Omega^2 \neq 0$. The demand that $\Omega^2 = 0$ with $\theta = 0$ would correspond to the Jeans instability characterized by $\omega^2 = -\sum_\alpha \omega_{G,\alpha}^2$. However, setting simultaneously $\theta = 0$ and $\Omega^2 = 0$ in Equations (145) renders $\varepsilon_{xx}$ and $\varepsilon_{yy}$ uncertain. The uncertainty can be avoided by appealing to the initial set of basic equations (including Equation (143)) and considering the case of $\theta = 0$, or $\mathbf{v}_\alpha \parallel \mathbf{B}_0$. Then the electric would be a potential field, $\mathbf{E} = -\nabla\psi_E$, and the equation set for $\psi_G$ and $\psi_E$ would take the form

$$\begin{aligned}\psi_G &= -\frac{4\pi G}{\omega^2}\sum_\alpha n_{0,\alpha}m_\alpha\left(\frac{Q_\alpha}{m_\alpha}\psi_E+\psi_G\right)\\ \psi_E &= \frac{4\pi}{\omega^2}\sum_\alpha n_{0,\alpha}Q_\alpha\left(\frac{Q_\alpha}{m_\alpha}\psi_E+\psi_G\right).\end{aligned} \tag{149}$$

(This is quite similar to Equation (27) of Chapter 1, except $v_T = 0$, and hence $K_T = 0$). Owing to quasineutrality of the medium ($\sum_\alpha n_{0,\alpha}Q_\alpha = 0$), the two equations are independent, *viz.*

$$\psi_G(1 + \frac{1}{\omega^2}\sum_\alpha \omega_{G,\alpha}^2) = 0 \qquad \text{and} \qquad \psi_E(1 - \frac{1}{\omega^2}\sum_\alpha \omega_{p,\alpha}^2) = 0.$$

Accordingly, two modes of longitudinal wavelike disturbances are possible,

$$a) \qquad \psi_E \neq 0, \qquad \psi_G = 0 \qquad \text{and} \qquad \omega^2 = \sum_\alpha \omega_{p,\alpha}^2$$

the Langmuir vibrations
and

$$b) \qquad \psi_E = 0, \qquad \psi_G \neq 0 \qquad \text{and} \qquad \omega^2 = -\sum_\alpha \omega_{G,\alpha}^2$$

the Jeans instability involving no electric field. It would be of interest to analyze exactly why the oscillatory motion of heavier particles in case a) fails to disturb the gravitation field and, contrary to this, the exponential growth of the perturbed number density of charged particles in case b) does not excite an electric field. The perturbed number densities of charged particles determined for the two cases from the equations of motion and continuity are

$$n_\alpha = -\frac{k^2}{\omega^2} n_{0,\alpha} \frac{Q_\alpha}{m_\alpha} \psi_E \qquad \text{case a)}$$

and

$$n_\alpha = -\frac{k^2}{\omega^2} n_{0,\alpha} \psi_G \qquad \text{case b).}$$

In case a), the number density fluctuations are proportional to the particle charge $Q_\alpha$, which value is positive for one portion of the particles and negative for the other. As a result, the fluctuating number density $n_\alpha$ varies with $\alpha$ in *counterphase.* The net density of the source term $\sum_\alpha m_\alpha n_\alpha$ in Poisson's equation for the gravitation potential proves to be zero,

$$\sum_\alpha m_\alpha n_\alpha = -\frac{k^2 \psi_E}{\omega_\alpha^2} \sum_\alpha n_{0,\alpha} Q_\alpha = 0.$$

In case b), all the number density fluctuations are *in phase*, hence the source term for the electric potential,

$$\sum_\alpha n_\alpha Q_\alpha = -\frac{k^2}{\omega^2} \psi_G \sum_\alpha n_{0,\alpha} Q_\alpha$$

is also zero. This explains the absence of electric perturbations in a Jeans instability. The essential condition for the separation of electrostatic and gravitation perturbations is the field-aligned motion of particles along $\mathbf{B}_0$.

Now we turn to transverse waves. With $\theta = 0$, the dispersion relation Equation (146) differs from the standard one solely by the greater number of terms, *viz.*

$$\frac{k^2c^2}{\omega^2} = 1 - \sum_\alpha \frac{\omega_{p,\alpha}^2}{\omega(\omega \pm \omega_{B,\alpha})}. \tag{150}$$

Let us assume briefly that all the dust grains are of the same species and their charge $Q_d$ is negative, $Q_d < 0$. Denoting the refractive index as $\eta_{R,L} = kc/\omega$ (with the subscripts $R$ and $L$ referring to the right and left circular polarized modes, respectively), we can bring Equation (150) to the form

$$\eta_{R,L}^2(\omega) = 1 - \frac{\omega_{p,e}^2}{\omega(\omega \mp \omega_{B,e})} - \frac{\omega_{p,i}^2}{\omega(\omega \pm \omega_{B,i})} - \frac{\omega_{p,d}^2}{\omega(\omega \mp \omega_{B,d})}, \tag{151}$$

where all the $\omega_{p,\alpha}$s are positive. The dispersion laws $\eta_{R,L}^2 = f(\omega)$ correspond to different modes. The Alfvén mode with the phase velocity

$$v_{ph} = v_A(1 + v_A^2/c^2)^{-1/2} \simeq v_A$$

exists in the low frequency domain $\omega \ll \omega_{B,d}$. The Alfvén velocity is $v_A = B_0/\sqrt{4\pi\rho_{tot}}$, where $\rho_{tot} = \sum_\alpha n_{0,\alpha}m_\alpha$ is the total mass density of the medium, i.e. the sum of the plasma density $\rho_p = n_{0,e}m_e + n_{0,i}m_i$ and that of the dust, $\rho_d = n_{0,d}m_d$. Normally $\rho_d \gg \rho_p$, hence the Alfvén velocity is greatly lower in a dusty plasma (Verheest, 1993). The $\eta_{R,L}^2 = f(\omega)$ curves have three vertical asymptotes corresponding to three cyclotron resonances. Two of these appear in the right circular polarized mode, namely the electron cyclotron resonance at $\omega = \omega_{B,e}$ and the dust-cyclotron resonance at $\omega = \omega_{B,d}$. The latter one is unknown in the common plasma.

The frequency range $\omega_{B,i} \ll \omega \ll \omega_{B,e}$ supports, like in the common plasma, the whistler mode with the dispersion law

$$\eta_R^2 \simeq 1 + \frac{\omega_{p,e}^2}{\omega\omega_{B,e}} \simeq \frac{\omega_{p,e}^2}{\omega\omega_{B,e}}.$$

A similar wave appears, owing to the presence of the dust component, at $\omega \ll \omega_{B,d}$. Its dispersion law is

$$\eta_R^2 \simeq \frac{\omega_{p,d}^2}{\omega\omega_{B,d}}.$$

The third resonance is observed in the left circular polarized mode near the ion gyro frequency $\omega = \omega_{B,i}$. Another whistler mode exists here at frequencies below the resonance, $\omega_{B,i} \gg \omega \gg \omega_{B,d}$ with

$$\eta_L^2 \simeq \frac{\omega_{p,i}^2}{\omega \omega_{B,i}} \qquad \text{or} \qquad \omega \simeq k^2 c^2 \omega_{B,i}/\omega_{p,i}^2 = k^2 c^2 \omega_{B,e}/\omega_{p,e}^2.$$

Mendis and Rosenberg (1994) suggested a better approximation to the dispersion law, *viz.*

$$\omega \simeq k^2 c^2 \omega_{B,e}/\omega_{p,e}^2(\delta - 1)$$

with $\delta = n_{0,i}/n_{0,e}$, and indicated the boundaries of the range of existence, $(\delta - 1)\omega_{B,i}/\delta \gg \omega \gg \omega_{B,d}$.

Now we can consider the dusty plasma with dust grains of different sorts. If the spectrum of grain sizes (masses) is discrete, then the right circular polarized mode is characterized by a multitude of dust-cyclotron resonances at the frequencies $\omega_{B,d}^{(j)}$ determined by the grain charges $Q_d^{(j)} = \psi_0 a_j$ and masses $m_d^{(j)} = (4/3)\pi \rho a_j^3$. The number of vertical asymptotes of the dispersion curve $\eta_R^2 = f(\omega)$ in the range $\omega_{B,d,\text{min}} \leq \omega \leq \omega_{B,d,\text{max}}$ is equal to the number of dust grain species, the situation being quite similar to the dust-acoustic waves in the isotropic plasma (cf. Figure 8).

However, a discrete mass spectrum can be hardly expected in real dusty plasmas; hence we will consider the case of a continuous spectrum of sizes in more detail. The "dusty" part of the sum in Equation (151) will be replaced by an integral, $I_d$, to be analyzed for the right circular polarized mode only. (The left circular polarized mode is somewhat modified too, but no extra resonances appear if all the grains are charged negatively). The integral,

$$I_d = \int_{a_1}^{a_2} \frac{\tilde{\omega}_{p,d}^2(a)\, da}{\omega[\omega - \omega_{B,d}(a)]}, \tag{152}$$

can be brought to the form

$$I_d = \frac{3\psi_0^2}{\rho_0 \omega^2} \int_{a_1}^{a_2} \frac{\tilde{n}_d(a) a\, da}{a^2 - (3B_0\psi_0)/(4\pi c \rho_0 \omega)},$$

or

$$I_d = \frac{3\psi_0^2}{2\rho_0 \omega^2} \int_{z_1}^{z_2} \frac{\tilde{n}_d[a(z)]\, dz}{z - z_B},$$

where $z = a^2$ and $z_b = 3B_0\psi_0/(4\pi c \rho_0 \omega)$. In specifying the size spectrum $\tilde{n}_d(a)$, we will be guided by the only criterion of derivation simplicity. At this stage we are interested in revealing the effects of spectrum continuity, without any relevance to specific objects. From this point of view, the simplest choice is to consider a uniform distribution of grain sizes over some interval $\Delta a = a_2 - a_1$. Such a spectrum is $\tilde{n}_d(a) = n_0/\Delta a$, where $n_0$ is the total volume density of grains. The formulas will also involve the mean size $a_0 = (a_1 + a_2)/2$, e.g. $z_B = \omega_{B,d}(a_0)a_0^2/\omega$. The integral $I_d$ becomes

$$I_d = \frac{\omega_{p,d}^2(a_0)a_0}{2\omega^2 \Delta a} \int_{z_1}^{z_2} \frac{dz}{z - z_B}, \tag{153}$$

and the result depends on whether $z_B$ lies within or without the integration interval $(z_1, z_2)$. In fact, it is the question whether the frequency range considered lies within or without the domain of dust-cyclotron resonances as $z_B$ is a frequency-dependent value. For $z_B < a_1^2$ or $z_B > a_2^2$,

$$I_d = \frac{\omega_{p,d}^2}{2\omega^2} \log\left(\frac{a^2 - z_B}{a_1^2 - z_0}\right) \cdot \frac{a_0}{\Delta a}, \tag{154}$$

while in the other case the principal value of the integral and the contribution from the pole at $z = z_B$ (contoured, according to Landau's rule, from below) is

$$I_d = \frac{\omega_{p,d}^2}{2\omega^2} \left[\log\left(\frac{a^2 - z_B}{z_B - a_1^2}\right) \cdot \frac{a_0}{\Delta a} + \pi i\right], \qquad a_1^2 \leq z_B \leq a_2^2. \tag{155}$$

In the limiting case of a single sort of dust particles (i.e. $\Delta a/a_0 \to 0$), Equations (154) and (155) yield the dispersion relation that was derived before. In fact, it is only Equation (154) that remains valid, since the range of validity of Equation (155) reduces to zero. The limiting value at $\Delta a_0/a_0 \to 0$ of the log term in Equation (154) is $2a_0\Delta a/(a_0^2 - z_B)$, which yields

$$I_d \mid_{\Delta a/a_0 \to 0} \simeq \frac{\omega_{p,d}^2(a_0)}{\omega[\omega - \omega_{B,d}(a_0)]}$$

and the familiar dispersion relation

$$\eta_R^2 = 1 - I_d - \frac{\omega_{p,i}^2}{\omega(\omega + \omega_{B,i})} - \frac{\omega_{p,e}^2}{\omega(\omega - \omega_{B,e})}.$$

Generally, both $I_d$ and $\eta_R^2$ are complex values. Schematic frequency dependences of $\mathrm{Re}\,\eta_R^2$ and $\mathrm{Im}\,\eta_R^2$ are shown in Figure 13. Vertical asymptotes are present at the boundary frequencies $\omega_{B,d,\min}$ and $\omega_{B,d,\max}$ as before; however the singularities at these points are of logarithmic rather than power law

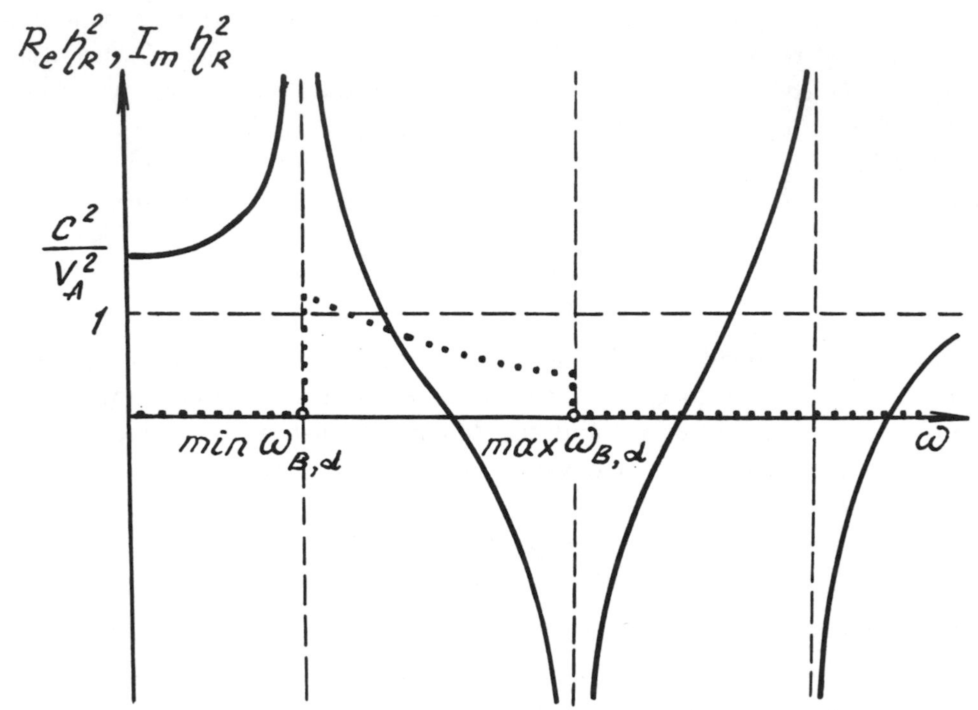


Figure 13: Frequency Dependences of $\mathrm{Re}\,\eta_R^2$ and $\mathrm{Im}\,\eta_R^2$.

character. In the vicinity of zero frequency, the logarithm can be simplified again (owing to $z_B \to \infty$), the result being practically the same dispersion law for the Alfvén mode as before. The logarithmic function alters the dispersion law only slightly near the electron cyclotron resonance as well. The major changes occur between $\omega_{B,d,\mathrm{min}}$ and $\omega_{B,d,\mathrm{max}}$ where *anomalous dispersion* is observed and the wave is attenuated owing to the dust-cyclotron absorption.

To simplify the analysis, we will estimate the imaginary part of the wavenumber $k + i\kappa$ only for the frequency $\omega_0$ where the argument of the log function equals 1. The frequency is determined from the condition $z_B = (a_1^2 + a_2^2)/2$. In the case of a narrow spectrum of grain sizes, $z_B \simeq a^2$ and $\omega_0 \simeq \omega_{B,d}(a_0)$. Assuming the absorption to be weak, we can write $\eta_R^2 \simeq (k^2 + 2ik\kappa)c^2/\omega^2$ and

$$k \simeq \frac{\omega}{c}\left(1 - \frac{\omega_{p,d}^2(a_0)}{\omega_{B,d}^2(a_0)} + \frac{c^2}{v_{Ao}^2}\right)^{1/2} \simeq \frac{\omega}{v_{Ao}}$$

$$\kappa \simeq \frac{\omega}{c}\left(\frac{\pi v_{Ao}}{4c}\cdot\frac{\omega_{p,d}^2(a_0)}{\omega_{B,d}^2(a_0)}\cdot\frac{a_0}{\Delta a}\right). \tag{156}$$

We have denoted the common Alfvén velocity $v_{Ao} = B_0/\sqrt{4\pi\rho}$ and assumed the number density of dust grains low enough to have $\omega_{p,d}^2(a_0)/\omega_{B,d}^2(a_0) \ll 1$, while $c^2/v_{Ao}^2 \gg 1$. The relative absorption rate,

$$\frac{\kappa}{k} \simeq \frac{\pi v_{Ao}^2\omega_{p,d}^2(a_0)}{4c^2\omega_{B,d}^2(a_0)}\cdot\frac{a_0}{\Delta a},$$

is independent of frequency and increases with lower scatter in the grain sizes. In real astrophysical objects the Alfvén waves are also subject to collisional damping (Pilipp *et al.*, 1987; 1990). Which of the attenuation mechanisms prevails, depends on the plasma parameters. Weakly attenuated Alfvén waves apparently can be excited only outside the dust-cyclotron absorption band.

### 3.5.4 Transverse Propagation

Setting $\theta = \pi/2$ in the general Equation (146) we arrive at the dispersion relation for the waves propagating across the magnetic field. Like in the common plasma, the equation is a product of two factors. By equating each of these to zero, one can derive the refractive indices for two modes of which one is a purely transverse wave known as the ordinary mode. It is characterized by $E_z \neq 0$ and $E_x = E_y = 0$, and the refractive index is

$$\eta_0^2 \equiv \frac{k^2c^2}{\omega^2} = \varepsilon_{zz} = 1 - \sum_\alpha \frac{\omega_{p,\alpha}^2}{\omega^2}.$$

This wave is not affected either by the magnetic of gravitation field, hence it is of little interest in this context.

The other or extraordinary mode ($E_z = 0$; $E_x, E_y \neq 0$) possesses electric components both along ($E_x$) and across ($E_y$) the propagation direction. The refractive index is

$$\eta_E^2 = \frac{k^2c^2}{\omega^2} = \varepsilon_{yy} - \frac{\varepsilon_{yx}\varepsilon_{xy}}{\varepsilon_{xx}}. \tag{157}$$

The full forms of $\eta_0^2$ or $\eta_E^2$, with account of the specific expressions for $\varepsilon_{ij}$, Equation (145), are too complex. We will rather consider the simplest model of the medium, involving just two sorts of particles characterized by different signs of the electric charge. (A still simpler medium with particles of a single sort would not satisfy the quasineutrality condition.) For self-gravitation effects to remain significant, we assume the two-component medium to consist of dust grains of different sizes and free electrons, without ions. Even in this simplified model, tractable formulas for the refractive index follow only in some limiting cases. Let $m_1 \ll m_2$, then it seems natural to assume

$$\omega_{p,1}^2 \gg \omega_{p,2}^2; \qquad \omega_{G,1}^2 \ll \omega_{G,2}^2 \qquad \text{and} \qquad \omega_{B,1}^2 \gg \omega_{B,2}^2.$$

Consider first the low frequency range $\omega \ll \omega_{B,2}$ where the common plasma supports the magneto-acoustic wave with the phase velocity

$$v_{ph} \simeq v_A = \frac{B_0}{\sqrt{4\pi(n_{0,1}m_1 + n_{0,2}m_2)}}.$$

Analysis of Equation (157) shows such a wave to exist in the self-gravitational plasma too, provided the condition is met $\omega_{G,2} < \omega_{B,2}$. The phase velocity is somewhat different, namely

$$v_{ph} \simeq \frac{v_A}{\sqrt{1 + \omega_{G,2}^2/\omega^2}},$$

i.e. the magnetic sound shows a frequency dispersion owing to self-gravitation.

Equation (157) reveals another property of the self-gravitational magnetized plasma. The off-diagonal components of $\varepsilon_{ij}$ can turn to zero at some frequency $\omega = \omega_0$. Without self-gravitation, this is only possible at $\omega_{B,\alpha} = 0$. Now, setting $\varepsilon_{xy} = \varepsilon_{yx} = 0$ with $\varepsilon_{xx} \neq 0$ and $\varepsilon \neq 0$ in Equation (157), we obtain

$$\eta_E = \sqrt{\varepsilon_{yy}(\omega_0)}. \tag{158}$$

This refractive index corresponds to a wave for which the magnetized plasma *is not a gyrotropic medium.* The anisotropy vanishes at

$$\omega_0 \simeq (\omega_{G,2}^2 + \omega_{B,2}^2 + \omega_{p,2}^2\omega_{B,2}\omega_{B,1}/\omega_{p,1}^2)^{1/2}.$$

Owing to the quasineutrality condition, $n_{0,1}Q_1 = -n_{0,2}Q_2$, the last term can be transformed as

$$\frac{\omega_{p,2}^2\omega_{B,2}\omega_{B,1}}{\omega_{p,1}^2} = \frac{n_{0,2}Q^3B_0^2}{m_2^2 n_{0,1}Q_1c^2} = -\frac{Q_2^2B_0^2}{m_2^2c^2} = -\omega_{B,2}^2,$$

and $\omega_0^2 \simeq \omega_{G,2}^2$. By substituting this into Equation (158) we find the phase velocity,

$$v_{ph} \simeq c\frac{\omega_{G,2}^2\omega_{p,2}}{\sqrt{2}\,\omega_{p,1}^2\omega_{B,1}}.$$

This mode is peculiar to the self-gravitational magnetized medium and obviously has no common plasma prototype. Hybrid resonances occurring at the frequencies where $\eta_E^2 \to \infty$ are also influenced by gravitation. The resonance condition $\varepsilon_{xx}(\omega) = 0$ can be represented as

$$\varepsilon_p(\omega)\varepsilon_G(\omega) + K_B^2(\omega) = 0, \tag{159}$$

where

$$\varepsilon_p(\omega) = 1 - \sum_\alpha \frac{\omega_{p,\alpha}^2}{\omega^2 - \omega_{B,\alpha}^2} \qquad \text{and} \qquad \varepsilon_G(\omega) = 1 + \sum_\alpha \frac{\omega_{G,\alpha}^2}{\omega^2 - \omega_{B,\alpha}^2}$$

are the plasma and gravitation factors, respectively, and

$$K_B(\omega) = \sum_\alpha \frac{\omega_{p,\alpha}\omega_{G,\alpha}}{\omega^2 - \omega_{B,\alpha}^2}$$

is the coupling term. Equation (159) is very similar to the dispersion relation Equation (32); however $K_B(\omega)$ never becomes zero, in contrast to $K_T(\omega)$ which vanishes at $T = 0$. Neglecting the terms with $\omega_{G,1}$ and treating those with $\omega_{G,2}$ as corrections, one can use Equation (159) to obtain the lower and upper hybrid resonance frequencies, *viz.*

$$\begin{aligned} \omega_{LH}^2 &\simeq \omega_{B,1}\omega_{B,2} - \omega_{G,2}^2 \\ \omega_{UH} &\simeq \omega_{p,1}^2 + \omega_{B,1}^2 + \omega_{p,1}^2\omega_{p,2}^2/(\omega_{p,1}^2 + \omega_{B,1}^2). \end{aligned}$$

In the opposite limiting case of high $\omega_{G,2}$, the resonance conditions are greatly different. E.g., the lower hybrid resonance is absent at $\omega_{G,2}^2 \gg \omega_{B,2}^2$ (specifically, $\omega_{LH}^2 < 0$), while

$$\omega_{UH}^2 \simeq \omega_{p,1}^2 + \omega_{G,2}^2.$$

Bliokh and Yaroshenko (1987) analyzed the resonances for various relations between the frequency parameters $\omega_{G,\alpha}$; $\omega_{B,\alpha}$ and $\omega_{p,\alpha}$.

# References

Akhiezer, A.I., Akhiezer, I.A., Polovin, R.V. *et al.* 1974. Electrodynamics of Plasmas (Nauka Publ. Co: Moscow. English Translation: 1975, Pergamon Press).

de Angelis, U., Formisano, V. and Giordano, M. 1988, *J. Plasma Phys.* **40**, 399.

de Angelis, U., Bingham, R. and Tsytovich, V.N. 1989, *J. Plasma Phys.* **42**, 445.

de Angelis, U., Bingham, R., Havnes, O. and Tsytovich, V.N. 1991. In E. Sindoni and A.Y. Vong (Eds.) *Dusty Plasma in Earth Environment*, (SIF: Bologna), p. 175.

de Angelis, U., Forlani, A., Tsytovich, V.N. and Bingham, R. 1992, *J. Geophys. Res.* **97**, 6261.

Avrett, E.H.(Ed.) 1976. Frontiers of Astrophysics (Harvard University Press.).

Bass, F. and Blanc, A. 1962, *Soviet Phys.—JETP* **43**, 1479 (*in Russian*).

Bingham, R., de Angelis, U., Tsytovich, V.N. and Havnes, O. 1991, *Phys. Fluids*, **B3**, 811.

Bliokh, P.V. and Yaroshenko, V.V. 1985, *Sov. Astron.* **29**, 330 (*in Russian*).

Bliokh, P.V. and Yaroshenko, V.V. 1987. Self-Gravitational Plasmas (Institute of Radio Astronomy: Kharkov), Preprint No 8 (*in Russian*).

Bliokh, P.V. and Yaroshenko, V.V. 1988, *Sov. Radiophysics and Quantum Electronics* **31**, 778 (*in Russian*).

Bliokh, P.V. and Yaroshenko, V.V., 1989. In *Plasma Electronics* (Naukova Dumka: Kiev), p. 66 (*in Russian*).

Bliokh, P.V. and Yaroshenko, V.V. 1991. In H. Kikuchi (Ed.) *Environmental and Space Electromagnetics*, (Springer-Verlag), p. 341.

Burns, T.A., Showalter, M.R., and Morfill, G.E. 1984. In R. Greenberg and A. Brahic (Eds.) *Planetary Rings*, (Univ. of Ariz. Press: Tucson, AZ), p. 200.

Fridman, A.M. and Polyachenko, V.L. 1981, *Soviet Phys.—JETP* **81**, 13 (*in Russian*).

Fridman, A.M. and Polyachenko, V.L. 1983. Physics of Gravitating Systems (Springer-Verlag).

Goertz, C.K. 1989, *Rev. Geophys.* **27**, 271.

Hartquist, T.W., Havnes, O. and Morfill, G.E. 1992, *Fundamentals of Cosmic Physics* **15**, 107.

Havnes, O., Morfill, G.E. and Goertz, C.K. 1984, *J. Geophys. Res.* **89**, 10999.

Havnes, O., de Angelis, U., Bingham, R., Goertz, C.K., Morfill, G.E. and Tsytovich, V.N. 1990, *J. Atmos. Terr. Phys.* **52**, 637.

Havnes, O., Melandsø, F., La Hoz, C., Aslaksen, T.K. and Hartquist, T. 1992, *Phys. Scripta* **45**, 535.

Ishimaru, A. 1978. Wave Propagation and Scattering in Random Media (Academic Press).

Jeans, J. 1929. Astronomy and Cosmology (Cambrige University Press).

Kaner, E. and Bass, F. 1959, *Sov. Phys.—Doklady* **127**, 792 (*in Russian*).

Krall, N.A. and Trivelpiece, A.W. 1973. Principles of Plasma Physics (McGraw-Hill Inc).

Lifshitz, I.M. and Rosenzweig, L.N. 1946, *Soviet. Phys.— JETP* **16**, 967 (*in Russian*).

Melandsø, E., Aslaksen, T. and Havnes, O. 1993a, *Planet. Space Sci.* **41**, 321.

Melandsø, E., Aslaksen, T. and Havnes, O. 1993b, *J. Geophys. Res.* **98**, 13315.

Mendis, E.A. and Rosenberg, M. 1994, *Ann. Rev. Astron. Astrophys.* **32**, (*in press*).

Pilipp, W., Hartquist, T. and Havnes, O. 1990, *Mon. Not. Roy. Astron. Soc.* **243**, 685.

Pilipp, W., Hartquist, T., Havnes, O. and Morfill, G. 1987, *Astrophys. J.* **314**, 341.

Rao, N., Shukla, P. and Yu, M. 1990, *Plan. Space Sci.* **38**, 543.

Rosenberg, M. 1993, *Planet. Space Sci.* **41**, 329.

Rytov, S.M., Kravtsov, V.A., and Tatarski, V.I. 1978. Introduction to Statistical Radiophysics (Nauka: Moscow) (*in Russian*).

Salimullah, M., Hassan, M.H.A. and Sen, A. 1991. Low Frequency Electrostatic Modes in a Magnetized Dusty Plasma,IC/91/318 (International Centre for Theoretical Physics) p. 12.

Shukla, P.K. 1992, *Phys. Scripta* **45**, 504.

Tsytovich, V.N., de Angelis, U. and Bingham, R. 1989, *J. Plasma Phys.* **42**, 429.

Verheest, F. 1992, *Planet. Space Sci.* **40**, 1.

Verheest, F. 1993a. In R.W. Schrittweiser (Ed.) *Double Layers and Other Nonlinear Potential Structures in Plasmas*, (World Scientific Publ.: Singapore), p. 162.

Verheest, F. 1993b, *Phys. Scripta* **47**, 274.

Voschinnikov, N.V. 1986. Interstellar Dust (Nauka: Moscow) (*in Russian*).

Waisberg, O., Smirnov, V., and Omelchenko, A. 1987. In B. Battrick, E.J Rolfe and R. Renhard (Eds.) *Proceedings of International Symposium on Exploration of Halley's Comet*, **2**, p. 17.

Zeldovich, Y.B. and Novikov, I.D. 1975. Structure and Evolution of the Universe (Nauka: Moscow) (*in Russian*).

# 4 Unstable Disturbances in Flows of Dusty and Self-Gravitational Plasmas

Stationary magnetic, electric or gravitation fields are nearly always present in space, affecting the motion of different kinds of particles in different ways. Quite often, the result may be a relative motion of particles. Consider, for example, the vicinity of a magnetized planet. For heavy particles near the planet, gravity prevails independently of their electric charge, and hence such particles move through the gravitation field of the central body in accordance with Kepler's laws. Contrary to this, the motion of microparticles (i.e. electrons and ions) is governed by electromagnetic forces greatly exceeding the gravitation. In most cases, the microparticles are entrained by the planetary magnetic field and corotate with the planet. As for electrically charged dust grains, they "feel" both gravitation and electromagnetic forces. As a result, the macroscopic particles do not move around the planet at the Kepler velocity $V_K$ but rather at a somewhat different velocity $V_{0,\alpha}$. The difference $\delta V_\alpha = V_{0,\alpha} - V_K$ is determined by the charge/mass ratio, hence it may vary for particles with different $Q_\alpha/m_\alpha$ even at the same orbit. In particular, the linear velocity of a particle of species $\alpha$ moving along an equatorial-plane circular orbit of radius $R$ is

$$V_{0,\alpha} \simeq V_K \left[1 + \frac{\Omega_{B,\alpha}}{2\Omega_K^2}(\Omega_K - \Omega_p)\right]. \tag{1}$$

(*cf.* Equation (33) of Chapter 2). Here $V_K = \Omega_K R$ is the Kepler velocity; $\Omega_K = (GM_p/R^3)^{1/2}$ the Kepler frequency; $M_p$ and $\Omega_p$ the planet's mass and rotation frequency, respectively; $\Omega_{B,\alpha} = Q_\alpha B_0(R)/m_\alpha c$ the gyrofrequency of the particles of species $\alpha$; $B_0(R)$ the planetary magnetic field and $c$ the velocity of light.

As can be seen from Equation (1), of special interest is the radial distance $R = R_{co}$ known as the synchronous orbit radius, where $\Omega_K(R_{co}) = \Omega_p$. It is the distance where $V_{0,\alpha}$ is independent of $\alpha$, and hence where particles of different masses and charges move along the orbit all at the same velocity $V_K(R_{co}) = \Omega_p R_{co}$. The two parameters to control the difference of $V_{0,\alpha}$ at $R \neq R_{co}$ from the appropriate Kepler velocity are the radial distance from the planet and the sign of the particle's electric charge.

Now we are able to comprehend the pattern of mutually penetrating particle streams about the planet. At an orbit of radius $R$, electrons and ions corotate with the planet; dust grains characterized by different $Q_\alpha/m_\alpha$ ratios move at the velocities of Equation (1), while the heaviest (or electrically neutral) particles with $Q_\alpha/m_\alpha \to 0$ have the Kepler velocity $V_K$. A similar multistream system is formed by the particles moving along the adjacent orbit, the result being a complex distribution of spatially separated orbital streams. Apparently, similar multistream structures should exist not only in planetary rings but in other astrophysical objects characterized by the presence of a dust component.

As is well known, the scatter in unperturbed velocities is a prerequisite for the development of beam instabilities. This is true both for "pure plasma" beams and self-gravitating streams of neutral particles. In this Chapter, we shall analyze which (if any) are the special features of beam instabilities in the dusty and self-gravitational plasmas.

We shall also consider another instability mechanism characteristic of the dusty plasma. While moving around a planet, the majority of dust grains repeatedly pass through the solar terminator zone. The electric charge of particles is changed as a result of transition from the night to the day side of the planetary ring. The periodic changes in the particle charge may cause a parametric instability.

The wavelike disturbances that we discussed in the preceding Section were for the most part stable. From a formal point of view, this means that real values of the wavenumber $k$ corresponded to real solutions $\omega_j$ of the modal equation and each of these represented a wave $\sim \exp(i\mathbf{kr} - i\omega_j t)$. It was only in a few cases that the dispersion relation for the self-gravitational plasma yielded purely imaginary solutions $\omega_j = \pm i\nu_j$, such that the time dependence shown by the disturbance was $\sim \exp(\pm\nu_j t)$. One of these exponents represents a disturbance growing with time. This situation corresponds to the instable solutions (discussed in Chapter 3) that may be qualified as "Jeans type" instabilities. The present Chapter is devoted to unstable disturbances of a different kind that may be called beam instabilities. These are characterized by *complex* values of $\omega_j$ with real $\mathbf{k}$ or complex $\mathbf{k}_j$ with *real* $\omega$.

## 4.1 Instability of Mutually Penetrating Dusty Plasma Beams

First, we shall consider beams of just dusty plasma, neglecting the self-gravitation. The simplest model for analyzing the instability development would be a set of unbounded, mutually penetrating cold particle beams. Certainly, real beams have limited cross-sections (planetary rings being an obvious example). The scatter in particle velocities owing to their thermal motion and different charge-to-mass ratios might also be essential. These complicating effects will be allowed for below.

### 4.1.1 Two-Beam Instability of Unbounded Streams

Consider electrostatic wavelike disturbances in a set of unbounded particle beams of different species $\alpha$. Let all the particles move at constant velocities $\mathbf{V}_{0,\alpha}$ in the same direction that will be assumed for the direction of the vector $\mathbf{k}$. These one-dimensional disturbances obey the dispersion relation that differs from the familiar one (e.g. Krall and Trivelpiece, 1973) only by the greater number of terms corresponding to different particle species,

$$1 - \sum_{\alpha} \frac{\omega_{p,\alpha}^2}{(\omega - kV_{0,\alpha})^2} = 0. \tag{2}$$

A similar equation was analyzed graphically in Chapter 3 where it was brought to the form $\omega^2 = f(v_{ph})$ to find intersection points of $y = f(v_{ph})$ with $y = \omega^2$. Thus, it was presumed that the frequency $\omega$ could be only real, while the wavenumber could assume complex values corresponding to spatial attenuation of the wave. Strictly speaking, the algebraic Equation (2) with real coefficients can have complex solution solely in the form of complex-conjugate pairs $k = k_r \pm i\kappa$ representing damped and growing waves. If external sources of energy are absent, it is quite logical to assume that a disturbance arising at some point should die away at remote points. This is an extra boundary condition permitting the correct choice from the two complex-conjugate solutions $\mathbf{k}$ of the modal equation.

Meanwhile, we shall follow a different pattern in the analysis of Equation (2). By multiplying it by $k^2$, we bring it to the form

$$k^2 = f(v_{ph}) \equiv \sum_{\alpha} \frac{\omega_{p,\alpha}^2}{(v_{ph} - V_{0,\alpha})^2}$$

and seek for intersections of the curve $y = f(v_{ph})$ with $y = k^2$. In other words, we presume $k$ to be always real while suggesting complex-conjugate solutions for the wave frequency $\omega = \omega_r \pm i\nu$. Such complex roots appear if the number of real intersection points is less than the highest power of Equation (2).

In the case under analysis, there is no reason to discard the solutions with a positive imaginary part $\nu$ that grow with time. Indeed, the physical system is characterized by energy pumping from the exterior, namely the energy of the relative motion of beam particles. By way of example, consider the system involving two beams. Then Equation (2) is a fourth-degree equation with respect to $\omega$, and hence it should have four solutions $\omega_i = \omega_i(k)$. The $y = f(v_{ph})$ dependence is shown for the case in Figure 1. As can be seen, with large values of $k$, $k > k_0$ (short-wave disturbances) $y = f(v_p h)$ and $y = k^2$ intersect at four points. The corresponding four phase velocities $v_{ph,i}$ can be associated with four real solutions $\omega_i = kv_{ph,i}$ of the modal Equation (2). At small $k$s, i.e. $k < k_0$ (long-wave disturbances), $f(v_{ph})$ and $y = k^2$ intersect at only two points, hence there are just two real solutions, $v_{ph,1}$ and $v_{ph,4}$ (or $\omega_1$ and $\omega_4$). The other two roots of the dispersion relation, $v_{ph,2}$ and $v_{ph,3}$ (or $\omega_2$ and $\omega_3$) prove to be a complex-conjugate pair, of which one represents an exponentially growing wave that propagates along the particle stream.

In other words, the set of beams is unstable against long-wave disturbances. To establish the instability criterion, let us consider the problem analytically, which is easier to do for $\omega_{p,1} = \omega_{p,2} = \omega_p$. In view of the general quasineutrality (i.e. $n_{0,1}Q_1 = n_{0,2}Q_2$) this is equivalent to the condition that $|Q_1/Q_2| = m_1/m_2$ which can be easily met in a medium consisting of dust grains of different sizes and both signs of the electric charge. Another conceivable situation is when a beam of negatively charged dust grains $(Q_2, m_2)$ passes through an assembly of positive ions $(Q_1, m_1)$.For the ions of oxygen and dust grains of size $a \sim 10^{-2}\,\mu$ and mass density $\rho \sim 1\,\mathrm{g/cm^3}$ the ratio $Q_2/Q_1$ should be $\sim 10^5$ which can be regarded as a large but achievable magnitude.

Now the fourth-order Equation (2) would become biquadratic with respect to the Doppler-shifted frequency $\widetilde{\omega} = \omega - kv_0$, where $v_0 = (1/2)(V_{0,1} + V_{0,2})$. In the frame of reference moving at the speed $v_0$, the dispersion relation would take the form

$$1 = \frac{\omega_p^2}{(\widetilde{\omega} - k\Delta v)^2} + \frac{\omega_p^2}{(\widetilde{\omega} + k\Delta v)^2}, \tag{3}$$

with $\Delta v = (1/2)(V_{0,1} - V_{0,2})$. The four solutions of this are

$$\widetilde{\omega}_{1-4} = \pm[(\omega_p^2 + k^2\Delta v^2) \pm \omega_p(\omega_p^2 + 4k^2\Delta v^2)^{1/2}]^{1/2}. \tag{4}$$

Now it is easy to see that if the condition is met

$$k\Delta v < \omega_p\sqrt{2}, \tag{5}$$

then one of the roots is purely imaginary with $\mathrm{Im}\,\widetilde{\omega} > 0$. Specifically, for $k\Delta v \ll \omega_p$ it is

$$\mathrm{Im}\,\widetilde{\omega} = \nu \approx k\Delta v, \tag{6}$$

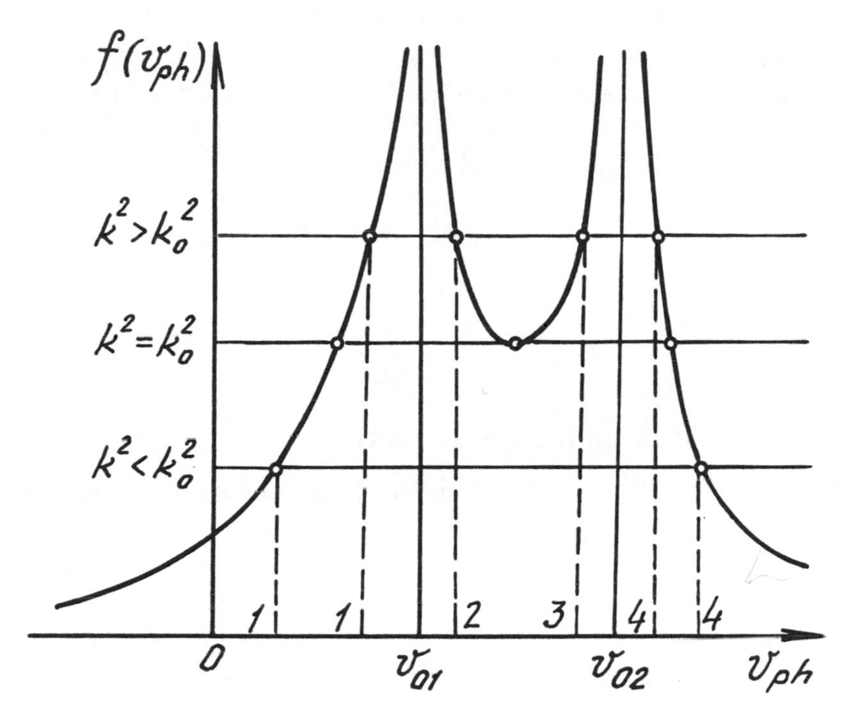


Figure 1: Graphic Analysis of Equation (2) for the Case of Two Particle Beams: Stable ($k^2 > k_0^2$) and Unstable ($k^2 < k_0^2$) Disturbances.

while the real part of the frequency is

$$\omega_r = \frac{1}{2}k(V_{0,1} + V_{0,2}). \tag{7}$$

Let us consider variations of the imaginary part, or the instability growth rate, as a function of the wavenumber $k$. As is evident from Equation (5), the instability does not develop for disturbances of sufficiently short wavelengths (large $k$s). By differentiating Equation (4) we find that the growth rate reaches its maximum value at

$$k = \frac{\sqrt{3}\omega_p}{2\Delta v} \tag{8}$$

and happens to be

$$\nu_{max} = \frac{\omega_p}{2}. \tag{9}$$

The growth rate characteristic of beam instabilities in the plasma is shown in Figure 2 as a function of $k$. Along with the case of equal plasma frequencies, other simplifying assumptions are known that can bring the dispersion relation to a tractable form. E.g., Krall and Trivelpiece (1973) considered the same two-beam model with $Q_1 = Q_2$ and $m_1 \gg m_2$, i.e. $\omega_{p,1} \ll \omega_{p,2}$ (the ion-electron plasma). Both the position of the maximum growth rate on the $k$-axis and the rate $\nu_{max}$ itself differed from the respective values of Equations (8) and (9) by numerical factors only, *viz.*

$$k \simeq \omega_{p,2}/2\Delta v \tag{10}$$

and

$$\nu_{max} \simeq \frac{\sqrt{3}}{2}\left(\frac{m_2}{2m_1}\right)^{1/3}\omega_{p,2}. \tag{11}$$

### 4.1.2 Beams of Bounded Cross-Sections

Excellent examples of multibeam systems of particles are given by planetary rings. Dust grains move in the rings at velocities that depend both on their charge-to-mass ratio $Q_\alpha/m_\alpha$ and the orbit radius. Direct application of Equation (2) to the analysis of wavelike disturbances in the rings is limited by some logic inconsistencies. If $V_{0,\alpha}$ were treated as the grain velocities at different orbits, then such beams could not be regarded as mutually penetrating. The greater is the velocity difference $\Delta v$, the farther separation between the beams.

Attempts of applying Equation (2) to the grains at a certain orbit, where the velocities $V_{0,\alpha}$ may differ because of different $Q_\alpha/m_\alpha$ ratios, do not face the difficulty of separated trajectories but suffer from another inconsistency. Equation (2) has been derived for unbounded particle beams of infinite cross-section, whereas specific orbits correspond to narrow, filamentary beams. A detailed discussion of these problems is left for the next Chapter, while here we shall just try to so "improve" the dispersion relation Equation (2) as to adapt it to beams of limited cross-section.

The dust particles move, as before, along parallel trajectories, i.e. the beam boundaries are fixed and known. Then the necessary modifications of the dispersion relation can be suggested from a simple analysis of dimensions. First, any changes will relate to $\omega_p^2$ alone, since the "kinematic" denominator $(\omega - kV_{0,\alpha})^2$, should not change as long as the trajectories remain unchanged.

Consider the limiting cases of particle beams bounded along one transverse coordinate (planar layer) or two coordinates (filament). In the first case, beam parameters are controlled by the *surface density* $\sigma_{0,\alpha}\,[\mathrm{cm}^{-2}]$ and in the second by the *linear density* $\gamma_{0,\alpha}\,[\mathrm{cm}^{-1}]$ of grains. Meanwhile, the equation for $\omega_p^2 \sim n_{0,\alpha}Q_\alpha^2/m_\alpha$ involves the *volume* density of particles,

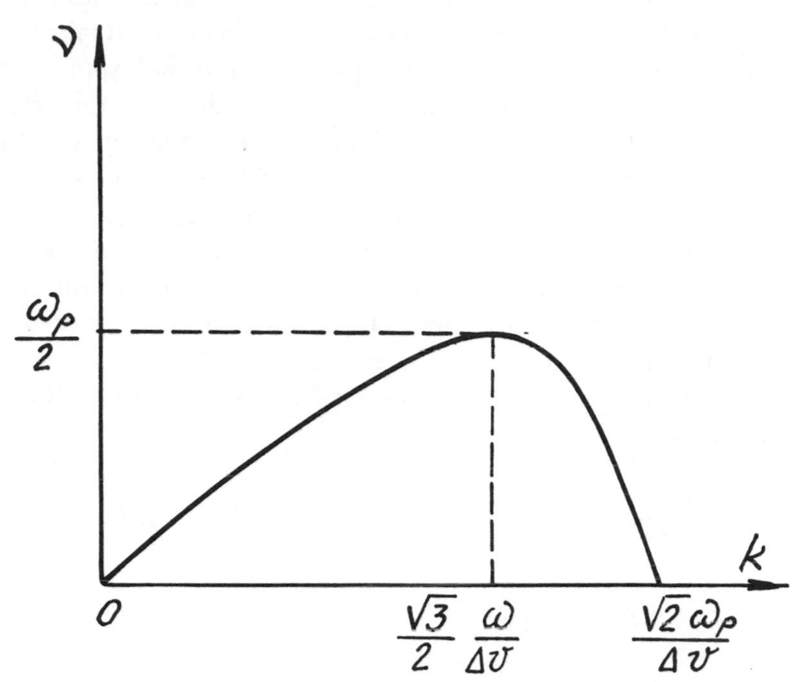


Figure 2: The Growth Rate of Unstable Disturbances in a Two-Beam System as a Function of the Wavenumber.

$n_{0,\alpha}$ [cm$^{-3}$], and hence $\sigma_{0,\alpha}$ or $\gamma_{0,\alpha}$ cannot be substituted in the equation in lieu of $n_{0,\alpha}$ in either of the cases of lower dimension. To correct the situation, $\sigma_{0,\alpha}$ should be multiplied by a value of the inverse length dimension, [cm$^{-1}$], while $\gamma_{0,\alpha}$ by a value of dimension [cm$^{-2}$]. The only parameter of length dimension involved in the model is the wavelength $\lambda \sim 1/k$. Therefore, the desired factors would be $|k|$ in the case of the planar layer and $k^2$ for the filament. (The absolute value $|k|$ appears because $\omega_p^2$ should be positive with either propagation direction). Thus, Equation (2) would retain its form for transversely bounded beams if $\omega_p^2$ were defined as

$$\omega_p^2 \sim |k|\sigma_{0,\alpha}Q_\alpha^2/m_\alpha \qquad \text{(planar beam)} \tag{12}$$

or

$$\omega_p^2 \sim k^2\gamma_{0,\alpha}Q_\alpha^2/m_\alpha \qquad \text{(filament)} \tag{13}$$

(apart from numerical factors that cannot be specified within the semiqualitative analysis of dimensions).

The infinitesimally thin planar layer and the filament actually are as ideal models as the uniform beam of infinite cross-section. Real beams are always characterized by some finite transverse size $d$ and *volume* density of particles, $n_{0,\alpha}$. The dimensionless parameter to describe the transverse size is $kd$. With $kd \gg 1$, the plasma frequency is given by the standard formula $\omega_p^2 \sim n_{0,\alpha}Q_\alpha^2/m_\alpha$. For a planar layer of thickness $d$, $kd \ll 1$, the plasma frequency $\omega_p^2$ is given by Equation (12) with $\sigma_{0,\alpha} = n_{0,\alpha}d$. In the case of a filamentary beam of circular cross-section and diameter $d$ (once again, $kd \ll 1$) Equation (13) with $\gamma_{0,\alpha} = \pi d^2 n_{0,\alpha}/4$ is valid. This definition of $\gamma_{0,\alpha}$ with finite values of $d$ permits avoiding the problem of a singular potential near an infinitesimally thin charged filament. Now the dispersion relation for waves in particle beams of transverse size $d$ takes the forms (Bliokh and Yaroshenko, 1985)

$$1 - \sum_\alpha \frac{\omega_{p,\alpha}^2}{(\omega - kV_{0,\alpha})^2} = 0 \qquad (\text{“thick” beam, } kd \gg 1) \tag{14}$$

$$1 - |k|d\sum_\alpha \frac{\omega_{p,\alpha}^2}{(\omega - kV_{0,\alpha})^2} \simeq 0 \qquad (\text{thin planar beam}) \tag{15}$$

or

$$1 - k^2d^2\sum_\alpha \frac{\omega_{p,\alpha}^2}{(\omega - kV_{0,\alpha})^2} \simeq 0 \qquad (\text{thin filament beam}), \tag{16}$$

where the plasma frequency is given by the common equation, $\omega_{p,\alpha}^2 = 4\pi n_{0,\alpha}Q_\alpha^2/m_\alpha$ in all the three cases. Precise numerical factors in front of the sums will be specified below.

In beams of finite cross-section, longitudinal waves can appear even in a *cold plasma*. For the simplest case of a single sort of particles and without drift motions, the longitudinal phase velocities can be estimated as $v_{ph} \sim \omega_p(d/k)^{1/2}$ (thin planar layer) or $v_{ph} \sim d\omega_p$ (filament). In an unbounded uniform plasma, longitudinal waves do not propagate under such conditions and only vibrations of frequency $\omega = \omega_p$ exist.

The simple considerations involving an analysis of dimensions can be illustrated by pictures of the electric field distribution across beams of different cross-sections (Figure 3). The electric fields arise from periodic charge density disturbances that accompany the longitudinal waves propagating through the beams (shown in the picture are the fields across segments of length $\lambda/2$). The density disturbances in an unbounded homogeneous beam appear as plane-parallel layers ($a$). These are replaced by parallel strips in

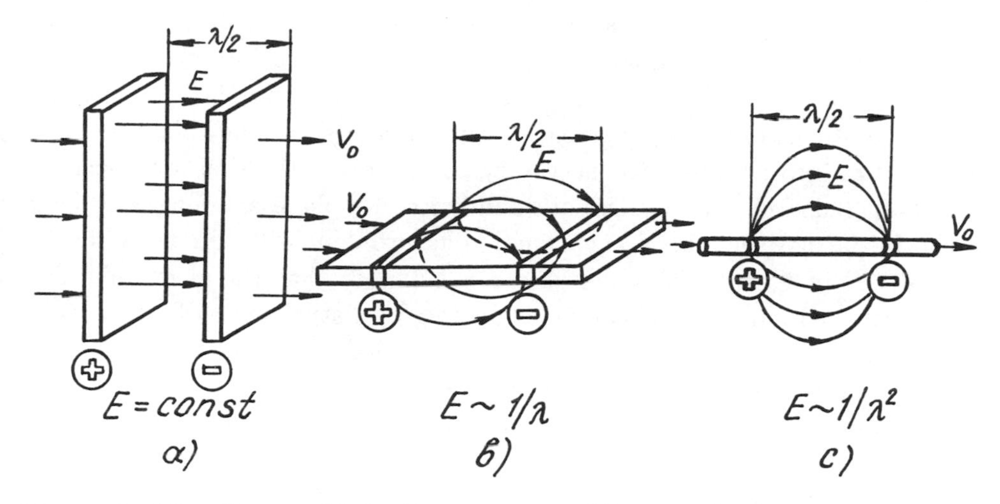


Figure 3: The Electric Field Components of Longitudinal Waves in Beams of Different Cross Section Geometries: a) Infinite Beam; b) Thin Layer and c) Filamentary Beam.

the planar layer ($b$) and by "points" in the filament ($c$). In case ($a$) the electric field strength $E$ is *independent* of $\lambda$ (accordingly, the plasma frequency does not contain $k$); in case ($b$) $E \sim 1/\lambda$ ($\omega_p^2$ involves the factor $|k|$), and in case ($c$) $E \sim 1/\lambda^2$ ($\omega_p^2$ involves $k^2$).

### 4.1.3 Instability of a Narrow Multicomponent Particle Beam with a Discrete Spectrum of Dust Grain Sizes

Consider a model planetary ring as a thin dusty plasma beam of circular cross-section of diameter $d$ orbiting within the equatorial plane at a distance $R$ from the planet center (for definiteness, we will assume $R > R_{co}$). Let the dust grains be negatively charged, while the particle stream as a whole is quasineutral. Then the dispersion relation for disturbances of a relatively short scale length $\lambda \ll R$ (this is the condition under which the beam curvature can be neglected) takes the form of Equation (16). With the proper numerical factors, it is

$$1 = \frac{k^2 d^2}{32} \sum_{\alpha} \frac{\omega_{p,\alpha}^2}{(\omega - kV_{0,\alpha})^2}. \tag{17}$$

Strictly speaking, the summation over $\alpha$ in Equation (17) should involve, along with all the dust grain species, the electrons and ions of the ambient plasma frozen in the planetary magnetic field. The motion of microparticles is synchronous with that of the planet (i.e. $V_{0,e,i} = \Omega_p R$), while the velocity of dust grains $V_{0,d}$ is given by Equation (1). Taking this difference in velocities into account, it is necessary to analyze if the ring may become unstable. This analysis requires allowance for the planetary magnetic field in the initial hydrodynamic equations. The problem will be considered in more detail in Chapter 5, while here we just note that the electronic and ionic terms in Equation (17) can be neglected for disturbances of a relatively low frequency, $\omega \ll \Omega_{B,e}; \Omega_{B,i}$. Then the simplified version of Equation (17) would relate to a plasma consisting of dust grains alone.

Suppose we are able to select particles of a specific size. Let the radii of particles of "adjacent" species differ by the same magnitude $\Delta a$, $a_\alpha = a_{\alpha-1} + \Delta a$ (the radius is assumed to increase with the species number $\alpha$). Then the unperturbed velocities $V_{0,\alpha}$ can be represented, with account of the above assumptions concerning the sign of the charge $Q_\alpha = \psi a_\alpha$ and distance from the synchronous orbit, as

$$V_{0,\alpha} = V_K(1 + S^2/a_\alpha^2), \tag{18}$$

where $S^2 = 3\psi B_0 \Omega_p(1 - \Omega_K/\Omega_p)/(8\pi\rho c \Omega_K^2) > 0$ for $R > R_{co}$. Thus, the higher velocities pertain to dust grains of the smallest size (i.e. $V_\alpha(a_{min})$), while the lower to the largest fragments with $V_0(a_{max}) \approx V_K$. The difference of "neighbor" velocities,

$$\Delta V_{\alpha,\alpha-1} \equiv V_{0,\alpha-1} - V_{0,\alpha} = \frac{2S^2 V_K \Delta a}{a_{\alpha-1}^3} \tag{19}$$

is the lower, the larger is the grain size. Once again, let us introduce $v_{ph} = \omega/k$ and rewrite Equation (17) as $f(v_{ph}) = 1$, with $f(v_{ph}) = (d^2/32) \times \sum_\alpha \omega_{p,\alpha}^2/(v_{ph} - V_{0,\alpha})^2$. This function is characterized by isolated second-order poles at $v_{ph} = V_{0,\alpha}$, *cf.* Figure 4. The minima of $f(v_{ph})$ lie between the poles. If at least one of the minima happens to lie above the level $f(v_{ph}) = 1$, then some of the roots of Equation (17) are complex, which suggests instability of the appropriate disturbances. Streams of larger particles are more liable to this instability, since the highest lying minimum of $f(v_{ph})$ corresponds to the larger in size dust grains (for which the velocity difference $\Delta V_{\alpha,\alpha-1}$ is minimal). Now we are able to find the $\Delta a = \Delta a_{min}$ for which the highest lying minimum is tangent to the straight line $f(v_{ph}) = 1$. The level of that minimum can be estimated without differentiating $f(v_{ph})$. To obtain the estimate, it is sufficient to take just two terms of the sum in Equation (17), namely those of arguments $a_{max}$ and $a_{max} - \Delta a$ corresponding to the closest lying poles of $f(v_{ph})$ to the left and right of the minimum. Since the minimum belongs to the narrow interval $\Delta V_{\alpha,\alpha-1} = V_0(a_{max} - \Delta a) - V_0(a_{max}) \simeq 2S^2 V_K \Delta a / a^3_{max}$ and $\omega_p^2(a_{max}) \simeq \omega_p^2(a_{max} - \Delta a)$, we have

$$f_{min} \simeq \frac{d^2}{32} \cdot \frac{2\omega_p^2(a_{max})}{(\Delta V_{\alpha,\alpha-1}/2)^2} \simeq \frac{d^2 a^6_{max} \omega_p^2(a_{max})}{16 S^4 V_K^2 (\Delta a)^2}.$$

The equation $f_{min} = 1$ yields

$$\Delta a_{min} \simeq \frac{d a^3_{max} \omega_p(a_{max})}{4 S^2 V_K}. \tag{20}$$

If dust grains are discretized finer than $\Delta a_{min}$, the particle beam may become unstable. Note the magnitude of $\Delta a_{min}$ to depend upon location of the ring with respect to the synchronous orbit, because $S^2 \sim \Omega_K(R) - \Omega_p$. Specifically, $\Delta a_{min}$ is the greatest near $R_{co}$, and development of the instability is hindered in that vicinity. All the effects controlled by the variance in velocities (Equation (1)) disappear at the synchronous orbit itself where the instability cannot develop.

In contrast to Equation (5) relating to infinite beams, the condition for the development of instability in particle beams of diameter $d$ is

$$\Delta V_{\alpha,\alpha-1} < \frac{d\omega_{p,\alpha}}{4\sqrt{2}}.$$

The instability growth rate is of the same order as in Equation (6), i.e. $\nu \sim k \Delta V_{\alpha,\alpha-1}$. It is clear now that, while larger grains are easily involved in the instability, the corresponding growth rate is very low. The maximum of $\nu$ is reached in streams of the smallest particles, i.e. dust grains of size $a_{min}$ and $(a_{min} + \Delta a)$.

Let us consider numerical estimates of the growth rate for Saturn's $F$-ring (i.e. $R \sim 1.4 \times 10^8\,\mathrm{m}$ and $d \sim 10^3\,\mathrm{m}$). The electric potential $|\psi|$ of micron-size particles ($a \sim 10^{-6}\,\mathrm{m}$) can be estimated as $\sim 10\,\mathrm{V}$ and $\Delta a \sim$

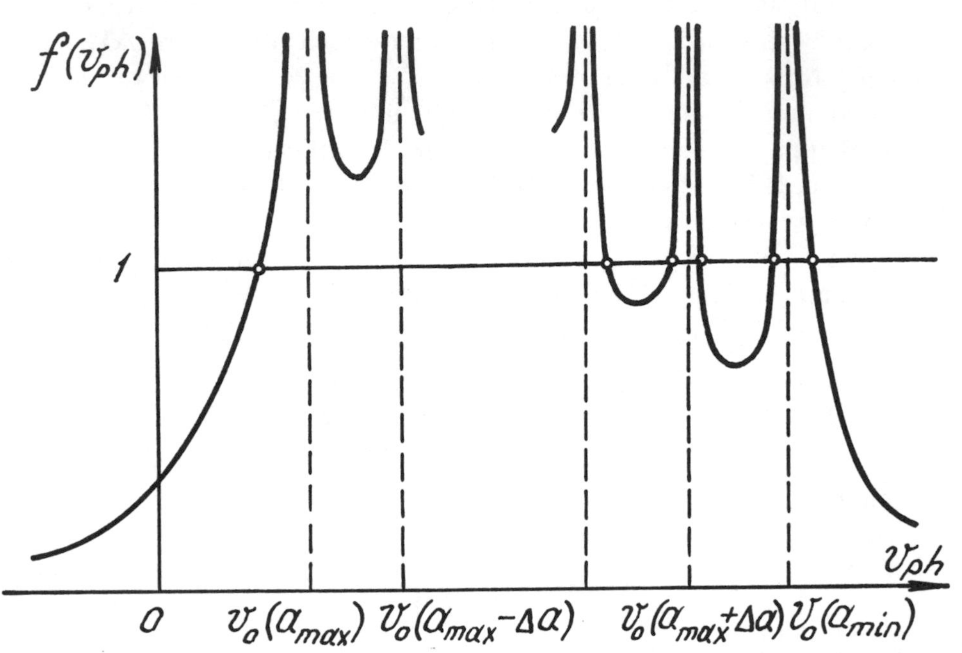


Figure 4: Graphic Analysis of Equation (17): The Instability Develops Owing to the Interaction of the Largest Grains Near $v_{ph} = V_0(a_{max})$.

$10^{-7}$ m, hence the velocity difference $\Delta V(a)$ is $\Delta V(a) = V_0(a) - V_0(a+\Delta a) \sim$ 10 m/s. Recalling that the scale size of the disturbances should obey the double inequality $R \gg \lambda \gg d$ we assume $\lambda \sim 10^5$ m. Then the estimate for $\nu$ is $\nu \sim k\Delta v(a) \sim 10^{-3}\,\mathrm{s}^{-1}$, which implies rise times varying between a few minutes and several hours.

### 4.1.4 Continuous Spectrum of Dust Grain Sizes. Effective Beam Temperature

The multistream instability cannot develop if the discrete ensemble of grain species is replaced by a continuum of sizes. This can be comprehended if we recall the results of Dawson (1960) who showed a plasma consisting of a great number of particle beams to be as unstable as a two-beam medium. However, the instability growth rate in such a plasma is

$$\nu \sim \Delta V \log \Delta V,$$

i.e. tends to zero with the difference $\Delta V$ of velocities in two adjacent streams. In the limiting case of a continuous velocity distribution, the plasma should be stable. This is also true for the model discussed in the preceding Section. Indeed, with $\Delta a \to 0$ the difference $\Delta V_{\alpha,\alpha-1}$ of Equation (19) tends to zero, and hence the instability does not develop.

For a continuum of particle velocities the summation over $\alpha$ in Equation (17) is replaced by integration over $a$, which integral can be represented, by analogy with the familiar kinetic theory equations, as

$$1 = \frac{k^2 d^2}{32} \int_{-\infty}^{\infty} \frac{F_0(V_0)\, dV_0}{(\omega - kV_0)^2}, \tag{21}$$

where

$$F_0(V_0) = \frac{3\psi^2 \tilde{n}_d[a(V_0)]}{\rho a} \left| \frac{da}{dV_0} \right| \simeq \frac{3\psi^2 \tilde{n}_d[a(V_0)] a^2(V_0)}{2\rho V_K S^2} \text{ for } V_{0,min} \leq V_0 \leq V_{0,max}$$

and $F_0(V_0) = 0$ for $V_0 < V_{0,min}$ or $V_0 > V_{0,max}$. The summation (integration) over $a$ in Equation (17) is from $a_{min}$ to $a_{max}$, while the corresponding integration limits over $V_0$ are $V_{0,max}$ and $V_{0,min}$. Once again, $\tilde{n}_d(a)$ denotes the spectrum of particle sizes. In principle, this equation suggests instabilities that are analogs to familiar kinetic effects. Let the size distribution be a power-law function (as in Chapter 3), $\tilde{n}_d(a) \sim a^{-\beta}$. According to literature data (e.g., Havnes *et al.*, 1990), the power exponent $\beta$ lies, in the case of planetary rings, between 0.9 and 4.5. Without specifying a concrete object, let us assume $\beta = 2$, for no better reason than to simplify the derivations. Then $F_0(V_0) = \text{const}$ for $a_{min} \leq a \leq a_{max}$, and the dispersion relation Equation (21) corresponds to wavelike disturbances in the ring. These are characterized by the mean velocity $V_0 = (1/2)[V_0(a_{max}) + V_0(a_{min})]$, and the scatter $\Delta V = (1/2)[V_0(a_{min}) - V_0(a_{max})]$. The scatter arises from the particle size distribution. Evaluating the integral, we obtain

$$1 \simeq \frac{k^2 d^2}{32} \cdot \frac{\omega_p^2(a_{min})}{(\omega - kV_0)^2 - k^2 \Delta V^2}, \tag{22}$$

with

$$\omega_p^2(a_{min}) = \frac{3\psi^2 n_{0,d}(a_{min})}{2\rho}, \qquad a_{min} \ll a_{max}.$$

Note $\Delta V$ to be involved in Equation (22) in much the same way as the thermal velocity in the dispersion relation for the conventional hot (i.e. $T_\alpha \neq 0$)

plasma (*cf.* Equation (74) of Chapter 3). Therefore, we are able to associate it with an effective "temperature"

$$T_{eff}\,[\mathrm{eV}] = \Delta V^2 m(a_{min}), \tag{23}$$

as if the longitudinal waves in a beam with a continuous spectrum of sizes were governed by particles of a *single* size $a_{min}$, and "thermal" scatter in velocities. In the case of Saturn's ring $F$ this "temperature" can reach extremely high values of $T_{eff} \sim 10^8 - 10^{10}\,\mathrm{K}$ because the mass of dust grains is much greater than that of microparticles. However, the value does not characterize in any degree the chaotic grain velocities since it has been related with the regular difference in their speeds.

For other power-law distributions in size than the quadratic, $F_0(V_0)$ acquires a nonzero derivative, $\partial F_0/\partial V_0 \neq 0$, which does not however lead to a kinetic instability. Indeed, the necessary condition for Equation (21) to possess growing solutions is the existence of at least one minimum of $F_0(V_0)$ over the integration range (Mikhailovsky, 1975), while $F_0(V_0)$ is a monotonous function for all power law distributions (Figure 5). If the distribution function $F_0(V_0)$ is characterized by a "hole", i.e. a reduction in the number of particles over some range of velocities (i.e. sizes), then according to the Penrose criterion, the distribution is unstable (Krall and Trivelpiece, 1973). Unfortunately, numerical estimates of the growth rates are not possible due to the lack of data on the particle distribution functions in planetary rings.

### 4.1.5 Three-Component Beam: Ion-Acoustic and Dust-Acoustic Instabilities

The instabilities discussed in the preceding Subsections can develop in cold streams of a dusty plasma. A finite temperature of the beam particles can be a stabilizing factor if the thermal velocities are greater than the drift velocity. Allowance for the particle temperature is therefore a necessity, however not for this reason alone. In case we are interested in low frequency disturbances like dust-acoustic or ion acoustic waves, then the very possibility of their existence is conditioned by finite values of the beam particle temperature. A complete analysis should be based on the kinetic theory; however some relations of interest can be derived within the hydrodynamic approximation. It is only necessary to allow for pressure in the equations of motion, the way we have done more than once before.

Once again, consider a thin beam of particles consisting of electrons, ions and dust grains. The two former components are characterized by

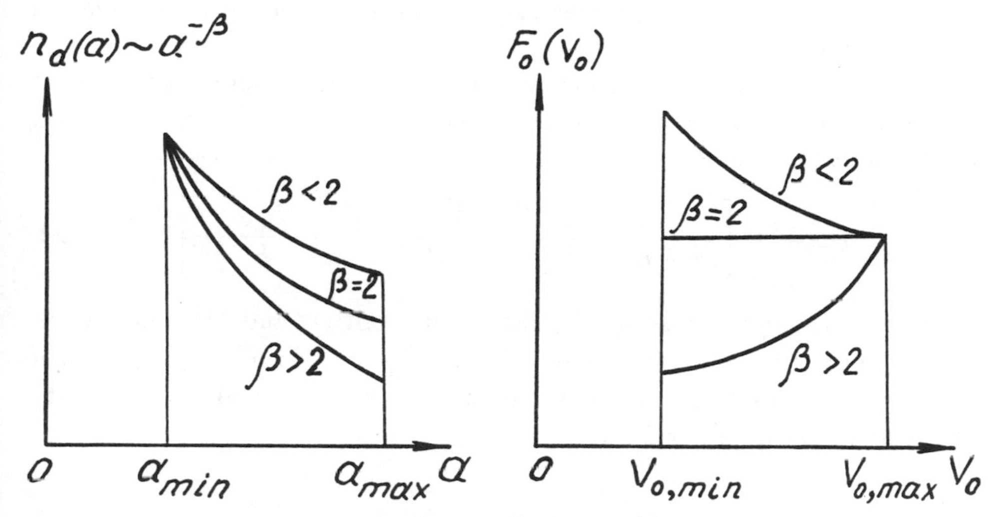


Figure 5: Power-Law Size Distributions, $\tilde{n}_d(a) \sim a^{-\beta}$, and the Corresponding Velocity Distribution Functions.

temperatures $T_e$ and $T_i$ or the corresponding velocities $v_{T,e}$ and $v_{T,i}$. The dust grains could be similarly described by $T_d$ and $v_{T,d}$ if particles of a single species were supposed. A model suggesting different sizes of the grains would be more realistic. In that case, the dispersion relation for the wavelike disturbances involves a dynamic scatter of particle velocities, $\Delta V$, that can be associated with a high effective temperature $T_{eff}$. Apparently, there is no sense in accounting for the normal temperature if $T_d \ll T_{eff}$, therefore $v_{T,d}$ will be replaced in what follows by $\Delta V$. To analyze the stability of a charged particle beam, it might be essential to know the relative values of $\Delta V$, $v_{T,i}$ and $v_{T,e}$, along with the drift velocities $V_{0,d}$, $V_{0,i}$ and $V_{0,e}$. While the relation of $v_{T,e}$ and $v_{T,i}$ is known (i.e. $v_{T,e} \gg v_{T,i}$), comparison of $\Delta V$ and $v_{T,i}$ should be based on numerical estimates.

Let $T_{eff} \sim 10^9$ K and $T_i \sim 10^5$ K, then

$$\left(\frac{\Delta V}{v_{T,i}}\right)^2 = \frac{T_{eff}\, m_i}{m_d T_i} \sim 10^4 \frac{m_i}{m_d}.$$

The ratio can vary in a wide range. E.g., in the case of hydrogen ions and micron-size dust grains, $a = 1\,\mu$, it is $m_i/m_d \simeq 10^{-12}$, while for oxygen and submicron grains ($a = 0.01\,\mu$) it becomes $m_i/m_d \simeq 10^{-5}$ (the mass density of the grain material is $\rho = 1\,\mathrm{g/cm^3}$ in both cases). Thus, the velocity ratio can be $(\Delta V/v_{T,i})^2 \sim 10^{-8} \div 10^{-1}$. In what follows, we will always assume $\Delta V \ll v_{T,i}$, although with greater $T_{eff}$ and lower $T_i$ the ratio may reach 1 or even exceed it.

Let us consider the low frequency waves in a three-component thin beam in the absence of drifts and for $\Delta V \ll v_{T,i} \ll v_{T,e}$. The corresponding dispersion relation is

$$1 \simeq k^2 d^2 \left[ \frac{\tilde{\omega}_{p,d}^2}{\omega^2 - k^2 \Delta V^2} + \frac{\tilde{\omega}_{p,i}^2}{\omega^2 - k^2 v_{T,i}^2} + \frac{\tilde{\omega}_{p,e}^2}{\omega^2 - k^2 v_{T,e}^2} \right], \tag{24}$$

where the argument of $\omega_{p,d}^2(a_{min})$ has been omitted and the plasma frequencies have been re-normalized, $\tilde{\omega}_{p,\alpha}^2 = \omega_{p,\alpha}^2/32$. To find the eigenwave spectrum, one should follow the same pattern as before and determine the intersection points of $y = 1$ and $y = f(v_{ph})$, where

$$f(v_{ph}) = d^2 \left[ \frac{\tilde{\omega}_{p,d}^2}{v_{ph}^2 - \Delta V^2} + \frac{\tilde{\omega}_{p,i}^2}{v_{ph}^2 - v_{T,i}^2} + \frac{\tilde{\omega}_{p,e}^2}{v_{ph}^2 - v_{T,e}^2} \right]. \tag{25}$$

The graph is presented in Figure 6. The domain of negative phase velocities, $v_{ph} < 0$, could be excluded in this case because of the symmetry of $f(v_{ph})$. However, the symmetry will be violated when drift motions are included in the analysis, hence for the convenience of their comparison, all the diagrams are shown for all values of $v_{ph}$. In Figure 6, the straight line $y = 1$ intersects with $y = f(v_{ph})$ at six points corresponding to three pairs of eigenwaves with $v_{ph} = v_{ph,j}$. The Figure also shows another straight line, namely $y = -1$, that will be discussed in the next Subsection in connection with self-gravitating particle beams.

To obtain approximate estimates of $v_{ph,j}$, we adopt the same approach as in Chapter 3, making use of the physically meaningful inequalities. For $v_{ph,1}$ we can assume $\Delta V \ll v_{ph,1} \ll v_{T,e}$ and hence neglect $\Delta V^2$ in the denominator of the first term in Equation (25) and $v_{ph}^2$ in the second. The third term may be totally ignored (see the corresponding inequalities in Chapter 3). Then the modal equation reduces to

$$d^2 \left( \frac{\tilde{\omega}_{p,d}^2}{v_{ph,1}^2} - \frac{\tilde{\omega}_{p,i}^2}{v_{T,i}^2} \right) \simeq 1,$$

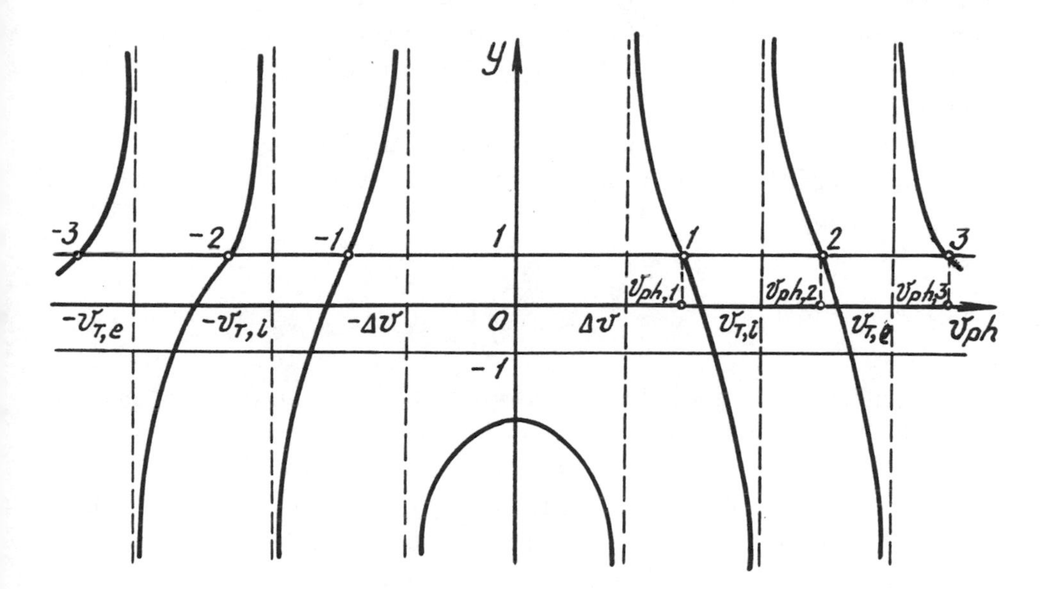


Figure 6: Graphic Analysis of the Dispersion Relation. The Intersection Points of $y = f(v_{ph})$ with the Constant Level $y = 1$ ($y = -1$) Correspond to Eigenwaves in a Thin Beam of a Three-Component Dusty Plasma (Beam of Neutral Gravitating Particles).

whence

$$v_{ph,1}^2 \simeq \frac{n_{0,d} Q_d^2}{n_{0,i} Q_i^2} \cdot \frac{1}{1 + \lambda_{D,i}^2 / d^2} \cdot \frac{T_i}{m_d}, \tag{26}$$

where $\lambda_{D,i} = v_{T,i}/\omega_{p,i}$ is the Debye length of ions. The first factor in Equation (26) is determined by the electric charge distribution between different particle species in the beam, obeying the demand of quasineutrality. The second factor is purely geometric, tending to 1 with $d \to \infty$. Of major importance is the third factor responsible for the $v_{ph} \sim \sqrt{T_i/m_d}$ dependence characteristic of dust-acoustic waves.

The second root of Equation (25) can be estimated in a similar way,

$$v_{ph,2}^2 \simeq \frac{n_{0,i} Q_i^2}{n_{0,e} e^2} \cdot \frac{1}{1 + \lambda_{D,e}^2 / d^2} \cdot \frac{T_e}{m_i}.$$

It represents an ion-acoustic wave.

The phase velocity of the third wave is $v_{ph,3} \simeq \widetilde{\omega}_{p,e}d(1+\lambda_{D,e}^2/d^2)^{1/2} \simeq \widetilde{\omega}_{p,e}d$. (It seems natural to assume that the beam diameter is much greater than the Debye length, otherwise it would be senseless to speak of a plasma beam.) The wave is the high-frequency electron sound characterized by the dispersion law $\omega \simeq kd\omega_{p,e}$ that was briefly mentioned at the end of the preceding Subsection.

Now we consider the instabilities that may arise owing to different regular velocities of the different beam components. If the electrons drift at a velocity $V_0$, then

$$f(v_{ph}) \simeq d^2 \left[ \frac{\widetilde{\omega}_{p,d}^2}{v_{ph}^2 - \Delta V^2} + \frac{\widetilde{\omega}_{p,i}^2}{v_{ph}^2 - v_{T,i}^2} + \frac{\widetilde{\omega}_{p,e}^2}{(v_{ph} - V_0)^2 - v_{T,e}^2} \right]. \qquad (27)$$

The symmetry with respect to $\pm v_{ph}$ has been violated, however the general run of the curve remains the same as in Figure 6 if $V_0 \ll v_{T,e}$. The poles of $f(v_{ph})$ have shifted from $\pm v_{T,e}$ to $v_{T,e}+V_0$ and $-v_{T,e}+V_0$, respectively. The number of intersections with $y = 1$ remains equal to six and all the points $v_{ph,j}$ are real, which implies absence of instability.

The situation is sharply different for $V_0 \gg v_{T,e}$. The two poles at $V_0 \pm v_{T,e}$ shift to the domain $v_{ph} > 0$, lying near $V_0$. The corresponding diagram $y = f(v_{ph})$ is shown in Figure 7. While the number of vertical asymptotes of the function remains the same, curves between them run in a different way. In particular, note the loops $A$ and $B$. The first of these may lie either above or below the $y = 1$ level. In the first case, the number of intersection points of $y = 1$ and $y = f(v_{ph})$ will be four, i.e. two of the six roots of the sixth-order equation should be complex-conjugate. This means that the beam becomes unstable relative excitation of waves with the phase velocities $v_{ph}$ between $v_{T,i}$ and $V_0 - v_{T,e}$. (This is the range of velocities of ion-acoustic waves.) In the second case the number of intersection points, i.e. real solutions of Equation (24) with the $f(v_{ph})$ of Equation (27) remains equal to the equation order and the beam is stable.

To analyze the dust-acoustic waves, we need to take account of the dust grain drift relative the electrons and ions which always occurs in planetary rings off the synchronous orbit. It might be convenient to use the frame of reference in which the dust grains are (on the average) at rest, while electrons and ions drift at the velocity $V_0 = (1/2)[V_0(a_{max}) + V_0(a_{min})]$ (see Subsection 4.1.4). In that frame, it is particularly easy to follow how the $y = f(v_{ph})$ dependence is deformed compared with Figure 6 and Figure 7. Let $V_0$ obey the inequalities $v_{T,i} \ll V_0 \ll v_{T,e}$. Once again, the diagram

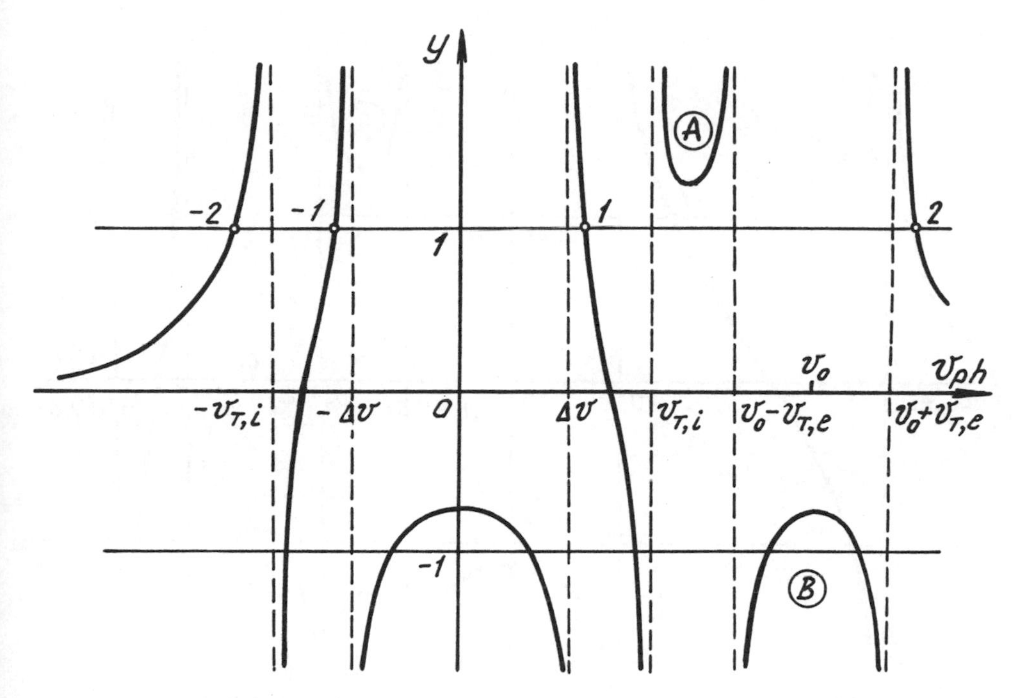


Figure 7: Graphic Analysis of the Dispersion Relation with Electrons Drifting at a Velocity $V_0 > v_{T,e}$. The Intersection Points of $y = f(v_{ph})$ with the Constant Level $y = 1$ ($y = -1$) Correspond to Eigenwaves in a Thin Dusty Plasma Beam (Beam of Neutral Gravitating Particles with a Drifting Light-Weight Component).

contains the loops $A$ and $B$, however loop $A$ associated with the beam instability now lies between $\Delta V$ and $v_{T,i}$, i.e. in the range of existence of dust-acoustic waves (Figure 8). The previous considerations remain valid, namely if the loop descends below the $y = 1$ level, the particle beam is stable. In case it lies above $y = 1$, the beam is unstable against excitation of dust-acoustic waves.

The threshold drift velocity cannot be evaluated through the graphic analysis, nor can growth rates of the ion-acoustic and dust-acoustic waves be compared. Yet it seems plausible to assume that the instability to prevail in the plasma of planetary rings should be that of dust-acoustic waves, because dust grains drift relative electrons and ions throughout the ring, except the synchronous orbit.

The problems that have been touched upon in this Section were also

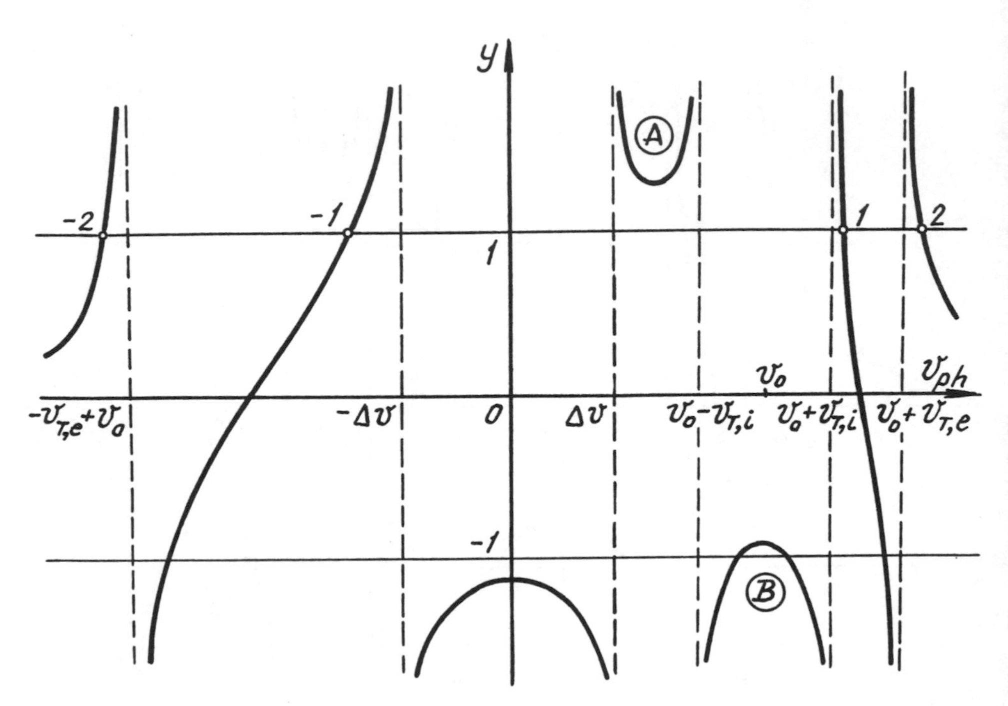


Figure 8: Graphic Analysis of the Dispersion Relation with Electrons and Ions Drifting at a Velocity $V_0$ ($v_{T,i} < V_0 < v_{T,e}$). The Intersection Points of $y = f(v_{ph})$ with the Constant Level $y = 1$ ($y = -1$) Correspond to Eigenwaves in a Thin Dusty Plasma Beam (Beam of Neutral Gravitating Particles with a Drift of Two Lightest-Weight Components).

analyzed in a more rigorous kinetic approach (Rosenberg, 1993), however for a more idealized model (a laterally unbounded beam involving dust particles of a single species with a Maxwellian velocity distribution).

To consider the range of dust-acoustic waves, the distribution functions should be taken as

$$f_{0,e,i} = \left(\frac{m_{e,i}}{2\pi T_{e,i}}\right)^{1/2} \exp\left[-\frac{m_{e,i}(V - V_0)}{2T_{e,i}}\right]^2,$$

which account for the joint drift of ions and electrons relative the dust grains. The dispersion relation takes the form

$$1 = \sum_{\alpha} \frac{\omega_{p,\alpha}^2}{k} \int_{-\infty}^{\infty} \frac{\partial f_{0,\alpha}/\partial V}{kV - \omega} dV,$$

where analysis of the integral leads to certain conditions on the drift velocities. Namely, ion-acoustic waves can be excited if

$$V_0 > c_s\sqrt{\frac{\delta}{1+k^2\lambda_{D,e}^2}}\left\{1+\delta\left(\frac{T_e}{T_i}\right)^{3/2}\cdot\left(\frac{m_i}{m_e}\right)^{1/2}\cdot\exp\left[-\frac{\delta T_e/T_i}{2(1+k^2\lambda_{D,e}^2)}\right]\right\}$$

(Rosenberg, 1993). The dust-acoustic waves are excited if the common drift velocity of electrons and ions exceeds

$$c_s\sqrt{\varepsilon_d Z_d(1-\delta)\frac{m_i}{m_d}}\left(1+\delta T_e/T_i+k^2\lambda_{D,e}^2\right)^{-1/2}.$$

The corresponding growth rate is

$$\left.\frac{\nu}{\omega}\right|_{DAW} \simeq \sqrt{\frac{\pi}{8}}\cdot\frac{V_0}{c_s}\cdot\frac{1}{1+\delta T_e/T_i+k^2\lambda_{D,e}^2}, \qquad (28)$$

where $c_s = (T_e/m_i)^{1/2}$; $\delta = n_{0,i}/n_{0,e}$; $Z_d$ is the dust grain charge number ($Z_d = Q_d/e$), and $\varepsilon_d$ assumes the value of 1 for positively and $-1$ for negatively charged grains.

While the results described were obtained for a dusty plasma with dust grains of a single species (size), they retain their importance for a wider variety of parameters. As was shown in Subsection (4.1.4), the ensemble of different dust grains in a thin planetary ring can be replaced by a beam of identical particles with the average velocity $V_0$, velocity variation $\Delta V$ and some effective number density $n_{0,d}$ appearing in the formula for $\omega_{p,d}^2$. Certainly, the velocity distribution associated with the power-law function $\tilde{n}_d(a) \sim a^{-\beta}$ differs from the Maxwellian but the specific form of the initial distribution function should not affect the general course of instability development.

Possible astrophysical implications of the ion-acoustic and dust-acoustic instabilities were discussed by Rosenberg (1993) for such objects as weakly ionized protostellar clouds and planetary rings. The growth rates estimated after Equation (28) for dust-acoustic waves in Saturn rings $E$ or $G$ were about $\nu \sim 10^{-3}\,\mathrm{s}^{-1}$, which is much greater than the grain—grain collision frequency.

## 4.2 Beam Instability in the Self-Gravitational Dusty Plasma

As was shown in Chapter 3, electric and gravitational disturbances are not independent in the self-gravitational dusty plasma. For the ensemble of particles without average velocities that was considered, the two kinds of disturbances were coupled owing to the thermal motion alone. (Accordingly, the thermal coupling factor

$$K_T = \sum_\alpha \frac{\omega_{p,\alpha}\omega_{G,\alpha}}{\omega^2 - k^2 v_{T,\alpha}^2}$$

turned to zero with $v_T = 0$).

If particles of different species move at regular velocities, then the electric and gravitational perturbations are coupled even in a cold medium. The dispersion relation to describe longitudinal waves in a self-gravitational dusty plasma can be derived in the same way as Equation (28) of Chapter 3, however the linearized equations of motion and continuity should involve extra terms depending on the drift velocities $\mathbf{V}_{0,\alpha}$. The result for a model of unbounded, mutually penetrating cold particle beams would be

$$\varepsilon_p(\omega, \mathbf{k})\varepsilon_G(\omega, \mathbf{k}) + K_V^2(\omega, \mathbf{k}) = 0, \tag{29}$$

with

$$\varepsilon_p(\omega, \mathbf{k}) = 1 - \sum_\alpha \frac{\omega_{p,\alpha}^2}{(\omega - \mathbf{k}\mathbf{V}_{0,\alpha})^2}, \tag{30}$$

$$\varepsilon_G(\omega, \mathbf{k}) = 1 + \sum_\alpha \frac{\omega_{G,\alpha}^2}{(\omega - \mathbf{k}\mathbf{V}_{0,\alpha})^2}, \tag{31}$$

and

$$K_V(\omega, \mathbf{k}) = \sum_\alpha \frac{\omega_{p,\alpha}\omega_{G,\alpha}}{(\omega - \mathbf{k}\mathbf{V}_{0,\alpha})^2}. \tag{32}$$

Perturbations of the two kinds are indeed coupled through the *drift coupling factor* $K_V$. For a more complex model of the medium (e.g., with allowance for both $\mathbf{V}_{0,\alpha}$ and $v_{T,\alpha}$), the coupling factor would be $K = \sum_\alpha \sqrt{f_{p,\alpha} f_{G,\alpha}}$, where $\varepsilon_p = 1 - \sum_\alpha f_{p,\alpha}$ and $\varepsilon_G = 1 + \sum_\alpha f_{G,\alpha}$.

### 4.2.1 Instability of Neutral Self-Gravitating Beams

The beam instability can arise not only in charged particle beams but in neutral beams as well, where particles interact through gravitation alone. The conditions for the growth of oscillation amplitudes are similar to those of the beam instability in conventional plasmas. However, in contrast to the latter the beam instability of gravitating particles develops against the background of the Jeans-unstable ambient medium. The wavelengths characteristic of the two instabilities are the same owing to the resonance condition

$\omega_G \sim k\Delta V$ (Mikhailovsky and Fridman, 1971; Polyachenko and Fridman, 1981), which makes it difficult to distinguish between them. Indeed, let us consider the dispersion relation for disturbances in an ensemble of self-gravitating particle beams,

$$\varepsilon_G = 1 + \sum_\alpha \omega_{G,\alpha}^2/(\omega - kV_{0,\alpha})^2 = 0. \quad (33)$$

Representing Equation (33) as

$$-k^2 = f(v_{ph}), \quad (34)$$

by analogy with the plasma medium, we find

$$f(v_{ph}) = \sum_\alpha \omega_{G,\alpha}^2/(v_{ph} - V_{0,\alpha})^2$$

to coincide with the standard function characteristic of plasmas if $\omega_{p,\alpha}$ were replaced by $\omega_{G,\alpha}$. Hence, we may appeal once again to Figure 1 showing $f(v_{ph})$ diagrammatically for the case of two particle beams. In the present case, however, $y = f(v_{ph})$ can never intersect with $y = -k^2$ lying in the lower halfplane $y < 0$. This means that the dispersion relation Equation (33) does not have real solutions at all, which in fact was clear *a priori* as the sum of positive terms cannot be equal to zero.

Consider the simpler situation of two particle streams with $\omega_{G,1}^2 = \omega_{G,2}^2 = \omega_G^2$. In this case, Equation (33) becomes a biquadratic equation with respect to $\tilde{\omega} = (\omega - kV_0)$, where $V_0 = (1/2)(V_{0,1} + V_{0,2})$ and $\Delta V = (1/2)(V_{0,1} - V_{0,2})$. The solution,

$$\omega_{1-4} = \pm[(-\omega_G^2 + k^2\Delta V^2) \pm \omega_G(\omega_G^2 - 4k^2\Delta V^2)^{1/2}]^{1/2} \quad (35)$$

cannot assume real values for any $\omega_G$ or $k\Delta V$. If the beam instability condition is met,

$$k\Delta V < \omega_G,$$

then all the roots of Equation (33) are imaginary,

$$\begin{aligned} \tilde{\omega}_{1,2} &\simeq \pm ik\Delta V \\ \tilde{\omega}_{3,4} &\simeq \pm i\omega_G. \end{aligned} \quad (36)$$

The beam instability of two gravitating particle streams is represented by the root $\omega_1 = kV_0 + ik\Delta V$, while $\omega_3 = kV_0 + i\omega_G$ is characteristic of the simultaneously developing Jeans instability.

In the case of an opposite inequality,

$$k\Delta V > \omega_G,$$

two solutions from Equation (35) represent growing disturbances, i.e.

$$\widetilde{\omega}_{1,2} \simeq \pm k\Delta V + i\omega_G,$$

while the other two are attenuated,

$$\widetilde{\omega}_{3,4} \simeq \pm k\Delta V - i\omega_G.$$

In contrast to the beam instability of charged particles, the growth rate is determined by the Jeans frequency, i.e. $\widetilde{\omega}_{1,2}$ represents a Jeans instability drifting with each of the particle streams at the velocity $V_0$. As was shown in Section 4.1, the model of laterally unbounded dusty plasma beams is but poorly fit to describe waves in planetary rings. For this reason, we rather discussed the waves propagating in thin filamentary beams of charged particles. The natural question is whether this theoretical model can be applied to waves in streams of neutral particles interacting through gravitation. In fact, the choice of a specific model for the particle beam as such is not dependent on the kind of particle interaction. Of importance is the model applicability to waves in planetary rings. The velocity of a neutral particle at a given orbit happens to be independent of its size (i.e. mass). Therefore, no drifts characterized by the velocities $V_{0,\alpha}$ can appear in a filamentary beam. True, the velocities $V_0$ are different for orbits of different radii $R$ but it might seem that the particles from different orbits do not form mutually penetrating streams. In fact, it can be shown (as we will do in the next Chapter) that the waves developing in particle beams at orbits $R_1$ and $R_2$ are subject to a strong interaction if $\Delta R = R_2 - R_1$ is shorter than the wavelength $\lambda$, and hence may be regarded as mutually penetrating. The beams propagating along such orbits that $k\Delta R \ll 1$ are "filamentary" and "mutually penetrating" at the same time. If the orbit radius $R$ obeys, as before, $kR \gg 1$, then the geometrical curvature of the particle beam may be disregarded (at least for wavelike disturbances of sufficiently short wavelength). All the conditions formulated would be consistent for $\Delta R/R \ll 1$.

Thus, we can consider the model of "filamentary" beams of neutral particles, however both the mean velocity $V_0$ and the scatter $\Delta V$ should be found from the purely Keplerian equations of motion,

$$V_0 = \frac{1}{2}\left[V_K(R_1) + V_K(R_2)\right] \simeq \Omega_K(R_0)R_0$$

and

$$\Delta V = \frac{1}{2}\left[V_K(R_1) - V_K(R_2)\right] \simeq \frac{1}{4}\Omega_K(R_0)\Delta R,$$

where

$$\Delta V \simeq -\frac{1}{2} \cdot \left.\frac{dV_K}{dR}\right|_{R=R_0} \cdot \Delta R \text{ and } R_1 < R_0 < R_2.$$

At this point, we can directly use the results of the graphic analysis of the preceding Section. It is sufficient to find intersections of the curves $y = f(v_{ph})$ with $y = -1$ in the same diagram as before.

First, consider Figure 6 relating to a stream of three kinds of particles without a regular drift. The common velocity $V_0$ of all the particles is of no importance as we can always use a frame of reference in which $V_0 = 0$. As for $\Delta V$ and $v_{T,\alpha}$, the thermal velocity can be ignored for large grains with $v_{T,\alpha} \ll \Delta V$. Contrary to this, the scatter in Keplerian velocities is of no importance for the lighter (smaller) particles with $v_{T,\alpha} \gg \Delta V$. Equation (25) represented in the Figure should be modified to describe neutral particles, namely $\omega^2_{p,\alpha}$ replaced by $\omega^2_{G,\alpha}$ and the beam cross-section area $\pi d^2/4$ by $h\Delta R$ (where $h$ is the ring thickness). The graph of $f(v_{ph})$ remains unchanged if the subscripts $e$ and $i$ are related to the neutral macroparticles rather than electrons and ions.

The curves of Figure 6 cross the $y = -1$ level at four points, which implies existence of two complex-conjugate roots of the modal equation, i.e. instability of the particle beam. The loop near $v_{ph} = 0$ moves up if the particle mass density decreases, and as soon as its upper point crosses the level $y = 1$, the number of intersection points becomes equal to six. This is the situation where all the roots of the modal equation are real and the system is stable.

An approximate analytical description of the roots of the modal equation is also possible if we note that for $v_{ph} \ll \Delta V$ it suffices to retain just one term, relating to the heaviest particles. All other terms of Equation (25) become important only in the vicinity of proper poles of $f(v_{ph})$, i.e. near $v_{ph} = \pm v_{T,\alpha}$. Thus, the simplified form of the dispersion relation is

$$\frac{\omega^2_{G,d} h\Delta R}{v^2_{ph} - \Delta V^2} \simeq -1,$$

with the solutions $v_{ph} = \pm\sqrt{\Delta V^2 - h\Delta R\omega^2_{G,d}}$,

$$\text{or } \omega = \pm k\sqrt{\Delta V^2 - h\Delta R\omega^2_{G,d}}. \tag{37}$$

These values are real with $h\Delta R\omega_{G,d}^2 < \Delta V^2$ and imaginary in the opposite case. Equation (37) takes a more familiar form for a "hot" unbounded medium where $k^2 h\Delta R\omega_{G,d}^2$ is replaced by $\omega_{G,d}^2$ and $\Delta V^2$ by $v_T^2$, i.e.

$$\omega^2 = k^2 v_T^2 - \omega_{G,d}^2. \tag{38}$$

This equation illustrates the statement that a self-gravitating ensemble of particles with thermal velocities $v_T$ is stable with respect to short-wave perturbations $k > \omega_G/v_T$ and unstable against the excitation of long waves $k < \omega_G/v_T$, i.e. $2\pi v_T/\omega_G$ is a critical wavelength.

In the case of planetary rings and Equation (37) ($\Delta V \simeq (1/4)\Omega_K(R_0)\Delta R$) the stability condition, $h\Delta R\omega_G^2 < \Delta V$, can be written as

$$\omega_G^2 < \Omega_K^2(R_0)\Delta R/16h. \tag{39}$$

Thus, stability of a ring against formation of periodic condensations (longitudinal waves) is determined by the ratio of the Jeans and Kepler frequencies (with account of the geometric correction factor $\Delta R/h$). With all the parameters unchanged, a ring would grow unstable if the orbit radius $R_0$ increased to such a value as to violate the condition of Equation (39).

The waves in beams with regular grain velocities (see Figures 7 and 8) can be analyzed almost as simply, however the orbit radii are different for different particle species $\alpha$, $R_\alpha \neq R_0$ (while $k|R_0 - R_\alpha| \ll 1$). To decide if the system is stable, it suffices to follow the behavior of loops in the lower part of the diagrams. Each loop lying below the level $y = -1$ suggests instability of the particle beam against excitation of waves with the phase velocity $v_{ph}$ given by the position of the loop apex. An approximate analytical description can be developed within the same approach as before (i.e., the dispersion relation retains just one term corresponding to the particular loop). E.g., the three-component particle stream of Figure 7 is stable. It can support four waves propagating in the same direction $\mathbf{V}_0$ as the drifting particles, and two waves traveling in the opposite direction. All the waves belong to the acoustic mode; however the part of thermal velocities is played for the slow waves by the scatter $\Delta V$ of Keplerian velocities.

If the mass density of the lighter component increased, loop $B$ would move down rendering the beam unstable against wavelike disturbances with $v_{ph} \simeq V_0$. The same effect occurs near $v_{ph} < \Delta V$ if the mass density of the heaviest component increases, in which case very slow waves are excited.

### 4.2.2 Two-Beam Instability of the Self-Gravitational Dusty Plasma. Small Amplitudes

Until now we have considered the instabilities developing in beams of neutral particles characterized by a single type of interaction. The simultaneous action of electric and gravitational forces in a plasma medium can alter the conditions for the development of an instability or change its growth rate. To analyze the instability of self-gravitational plasma flows, let us return to the dispersion relation Equation (29) with $\varepsilon_p$, $\varepsilon_G$ and $K_V$ given by Equations (30), (31) and (32), respectively. As a result, the dispersion relation becomes

$$1 - \sum_{\alpha} \frac{\omega_{p,\alpha}^2 - \omega_{G,\alpha}^2}{(\omega - kV_{0,\alpha})^2} - \frac{1}{2} \sum_{\substack{\alpha,\beta \\ \alpha\neq\beta}} \frac{(\omega_{p,\alpha}\omega_{G,\beta} - \omega_{p,\beta}\omega_{G,\alpha})^2}{(\omega - kV_{0,\alpha})^2(\omega - kV_{0,\beta})^2} = 0. \tag{40}$$

Once again, we will consider the simplest case of two particle streams, $\alpha;\beta = 1,2$. Then in the frame of reference moving at the mean velocity $V_0 = (1/2)(V_{0,1} + V_{0,2})$ Equation (40) turns into a fourth-order equation in $\tilde{\omega} = \omega - kv_0$, *viz.*

$$1 = \frac{\omega_{p,1}^2 - \omega_{G,1}^2}{(\tilde{\omega} - k\Delta V)^2} + \frac{\omega_{p,2}^2 - \omega_{G,2}^2}{(\tilde{\omega} + k\Delta V)^2} + \frac{\omega_{p,G}^4}{(\tilde{\omega}^2 - k^2\Delta V^2)^2}, \tag{41}$$

where $\omega_{p,G}^2$ denotes $|\omega_{p,1}\omega_{G,2} - \omega_{p,2}\omega_{G,1}|$. Like before, the dispersion relation can be analyzed graphically. To do so, consider the right-hand side of Equation (41) as a function of frequency, $\eta = f(\tilde{\omega})$. The solutions to Equation (41) are given by intersections of this graph with the straight line $\eta = 1$. To analyze the relative importance of the two kinds of particle interaction, let us introduce dimensionless parameters $y_\alpha = \omega_{p,\alpha}/\omega_{G,\alpha} = Q_\alpha/(m_\alpha G^{1/2})$ and re-write $f(\tilde{\omega})$ as

$$f(\tilde{\omega}) = \frac{\omega_{G,1}^2(y_1^2 - 1)}{(\tilde{\omega} - k\Delta V)^2} + \frac{\omega_{G,2}^2(y_2^2 - 1)}{(\tilde{\omega} + k\Delta V)^2} + \frac{\omega_{G,1}^2\omega_{G,2}^2(y_1 - y_2)^2}{(\tilde{\omega}^2 - k^2\Delta V^2)^2}. \tag{42}$$

If $y_1$ or $y_2$ happens to be equal to 1, then the corresponding term in Equation (42) vanishes but parameters of that beam are still represented in the dispersion relation owing to the third term. With any $y_\alpha$, the graph of $f(\tilde{\omega})$ is characterized by two vertical asymptotes at $\tilde{\omega} = \pm k\Delta V$, and four different types of the curves are possible (see Figure 9):

a) $y_\alpha > 1$. Electrical interaction prevails in both beams. The appearance of four intersection points with $\eta = 1$ suggests existence of four eigenmodes,

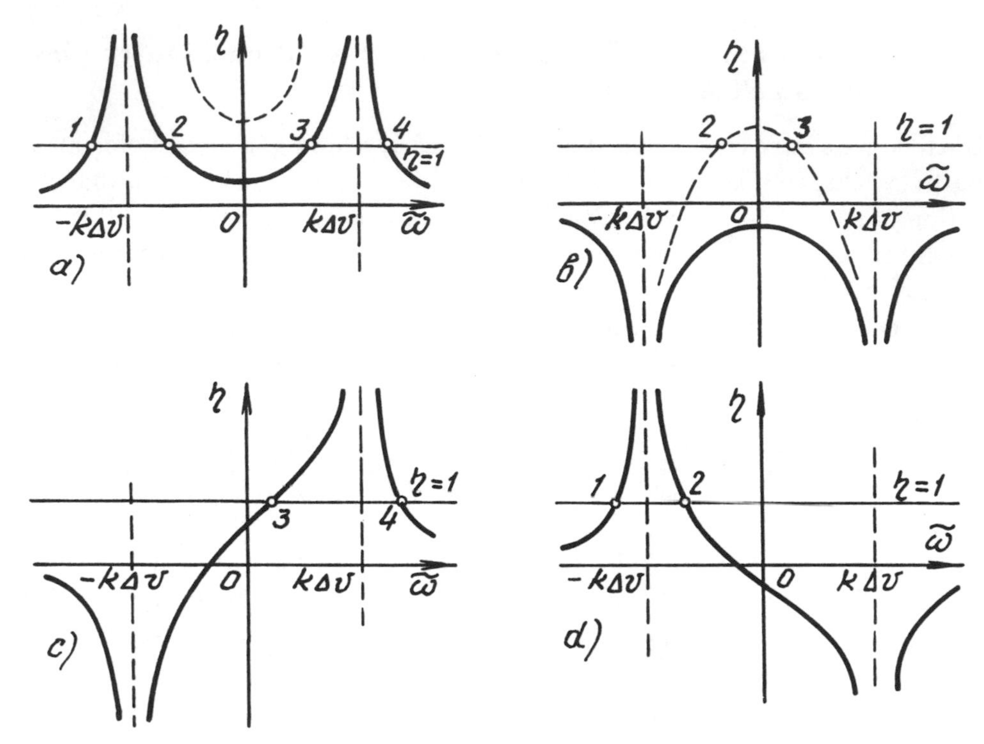


Figure 9: Four Possible Variants of the $\eta = f(v_{ph})$ Dependence: a) $y_1 > 1$ and $y_2 > 1$, the Electric Interaction Prevails in Both Beams; b) $y_1 < 1$ and $y_2 < 1$, the Gravitational Interaction Prevails in Both Beams; c) $y_1 > 1$ and $y_2 < 1$, the Electric Interaction is Dominant in the First Beam, while the Gravitational in the Second; d) $y_1 < 1$ and $y_2 > 1$, the Gravitational Interaction is Dominant in the First Beam and the Electric in the Second.

of which two propagate at phase velocities close to the speed $V_{0,1}$ of one beam and the other two at velocities close to $V_{0,2}$. As the loop in the area $\tilde{\omega} < |k\Delta V|$ moves above the level $\eta = 1$, intersection points 2 and 3 merge and then disappear, giving rise to two complex-conjugate roots of Equation (42). The system becomes unstable with respect to electrical interaction of the beam particles.

b) $y_\alpha < 1$. Gravitation interaction prevails in both beams. The system is unstable against excitation of four wavelike disturbances propagating as two pairs at phase velocities close to $V_{0,1}$ and $V_{0,2}$. In case the beam parameters are such that the loop lying in the domain $\tilde{\omega} < |k\Delta V|$ moves above $\eta = 1$, then two waves with the phase velocities close to the mean velocity of the two beams are stable, however the system as a whole remains unstable.

c) $y_2 < 1$ and $y_1 > 1$. Gravitation prevails in the second beam with $V_{0,2} = V_0 - \Delta V$. Owing to that beam, the system is unstable against gravitational interactions. The waves with $v_{ph} \simeq V_{0,1} = V_0 + \Delta V$ remain stable.

d) $y_1 < 1$ and $y_2 > 1$. The system is unstable as in c), however the growing wave is controlled by gravitation forces in the first beam. Besides, constant-amplitude waves are possible in the system, traveling at a velocity close to $V_{0,2} = V_0 - \Delta V$.

Strictly speaking, the run of $f(\tilde{\omega})$ is not determined by the parameters $y_1$ and $y_2$ alone (altogether, there are five parameters of the frequency dimension, i.e. $\omega_{p,1}$; $\omega_{p,2}$; $\omega_{G,1}$; $\omega_{G,2}$; and $k\Delta V$). Therefore, the inequalities such as $y_{1,2} > 1$ or $y_{1,2} < 1$ should be complemented by other conditions, namely the sign of $f(\tilde{\omega})$ should be determined by that of the first term in the vicinity of $\tilde{\omega} = k\Delta V$ and by the second term of Equation (42) in the vicinity of $\tilde{\omega} = -k\Delta V$.

Investigating the system stability analytically is quite difficult as it involves analysis of a fourth-order algebraic equation. By way of example, we will just consider a restricted model. Let a beam of heavy particles ($\omega_{G,2} \gg \omega_{p,2}$) move through an ensemble of fine dust grains $\omega_{G,1} \ll \omega_{p,1}$. To further simplify the analysis, we assume $\omega_{p,1}^2 - \omega_{G,1}^2 = \omega_{G,2}^2 - \omega_{p,2}^2 = \omega_0^2$. The general run of $f(\tilde{\omega})$ is determined by the magnitude of $k$. For short wave disturbances, $k\Delta V > \omega_{p,G}^2/2\omega_0$, the growth rate is $\nu \approx 2k\Delta V\omega_0/(4k^2\Delta V^2 - \omega_0^2)^{1/2}$. For the disturbances satisfying the opposite inequality, i.e. $k\Delta V < \omega_{p,G}^2/2\omega_0$, $f(\tilde{\omega})$ is symmetrical at $\omega > 0$ and $\omega < 0$ and tends to infinity at $\tilde{\omega} \to \pm k\Delta V$. It reaches the minimum of $\omega_{p,G}^4/k^4\Delta V^4 = f(0)$ at $\tilde{\omega} = 0$. In case $f(0)$ lies below the unit level, which is possible with $\omega_{p,G} < k\Delta V < \omega_{p,G}^2/2\omega_0$, the self-gravitating plasma streams are stable with respect to such disturbances.

In the case of long wave disturbances, $k\Delta V < \omega_{p,G}$, such that $f(0) = \omega_{p,G}^4/k^4\Delta V^4 > 1$, there are two unstable branches with respective growth rates $\nu_{1,2} \approx \omega_{p,G}^2/(\omega_0^2 \pm 4k^2\Delta V^2)^{1/2}$. Other particular cases of interest have been considered by Yaroshenko (1985).

### 4.2.3 Two-Beam and Gravitational Instabilities. Numerical Modeling of the Nonlinear Stage

The dispersion relation Equation (41) allowing for both kinds of the particle interaction is capable of predicting one of the four possible versions (see Figure 9) of instability development in a two-component plasma beam system. Meanwhile, the linear theory describes just the initial stage of the growing disturbance. The important question is, how the process would develop further. If we discussed a "pure" system, e.g. several beams of charged particles not subject to self-gravitation, then the answer would be well known. The disturbance amplitude would ultimately saturate as a result of thermalization of the particle beams whose initial kinetic energy would be converted into the energy of random (thermal) motion of the particles and electrostatic energy of bunches in the beam. The situation is different with the "mixed" system where an instability arising from one kind of the particle interaction can give rise, at a sufficiently high amplitude, to disturbances associated with the other interaction. Such processes are all nonlinear, hence they are not described by the dispersion relation Equation (41) but can be analyzed through numerical modeling of the particle dynamics. Apparently, the first paper of the kind treated instability development in a two-component system without regular drifts (Gisler and Wollman, 1988). The only instability that appeared in the system was of the gravitational (Jeans) type, however accompanied by electric polarization of the medium. When a regular relative drift of the two particle species was also included in the analysis (Gisler *et al.*, 1992), the system became subject to two kinds of instability. The problem formulation was essentially similar to that of the preceding Subsection, however thermal velocities of the heavier species were also taken into account in the initial state. Considered were dust grains of two sorts, moving at a relative mean velocity $V_0$. The masses were assumed greatly different ($m_1 \gg m_2$), whereas the electric charges, $Q_1$ and $Q_2$, were equal. The parameters $y_1$ and $y_2$ introduced to characterize the relative importance of electric and gravitation forces obviously were also very different. For the numerical simulation, the magnitude of $y_2$ was fixed at $y_2 = 3.5 \times 10^3$ (i.e. electric interaction prevailed in the medium of lighter grains), while $y_1$ could assume different values between 0.05 and 100 owing to variation of $m_1$. The growth rates of the unstable waves were evaluated from the linear theory for "pure" media. The Jeans instability (not involving the electric interaction) was characterized by the growth rate $\nu_J \simeq \omega_{G,1}$, whereas the wavelength $\lambda$ could assume arbitrary values longer than the critical, $\lambda_J$ (see Equation (38)). The maximum growth rate characteristic of the two-beam (Buneman) instability in the absence of self-gravitation,

$$\nu_B = \frac{\sqrt{3}}{2}\left(\frac{m_2}{2m_1}\right)^{1/2} \omega_{p,2},$$

occurred for $\lambda_B \simeq 2\pi V_0/\omega_{p,2}$ (see Equations (10) and (11)). With small values of $y_1$, $y_1 \ll 1$ the Jeans instability develops at a much greater rate than the Buneman, $\nu_J \gg \nu_B$, and the gravitational condensation is not accompanied by electric polarization of the medium (recall the results of Section 3.1 that predicted electric-field-free disturbances in a medium of self-gravitational charged particles). Contrary to this, at $y_1 \gg 1$ the developing instability follows the plasma-beam scenario with $\nu_B \gg \nu_J$. Close to the saturation stage, particles of both species are thermalized and gravitational condensation practically does not occur within the wavelength range taken for the numerical simulation (disturbances of the kind have also been discussed in Subsection 3.1.3).

Of greatest interest is the intermediate case $y_1 \leq 1$ and $\nu_B \geq \nu_J$. First, the Buneman growth rate is somewhat higher than the Jeans rate (up to times $t \sim 200\,\omega_{p,2}^{-1}$) and the electrostatic energy per unit volume is greater than the gravitation energy. The Bumeman-unstable disturbances can initiate the Jeans instability if $\lambda_B > \lambda_J$.

Numerical modeling of the particle dynamics for the two species allowed investigating the energy redistribution in the course of instability development up to the saturation stage. Figure 10 shows the final energy stores of the heavier and lighter particles, electrostatic potential and total gravitation energy at the saturation stage of the instability (all the values are normalized to the initial kinetic energy of the particles). The independent parameter is $y_1$, with the rest of the values involved being fixed. The diagrams confirm the qualitative scenario of instability development outlined above. At small values of $y_1$ (i.e. great masses $m_1$) the major role belongs to gravitation. The kinetic energy of heavy particles is comparable in magnitude with the potential gravitation energy of opposite sign. At high values of $y_1$ the gravitational condensation does not occur, the energy distribution being controlled by the Buneman instability. All the energies are comparable at $y_1 \simeq 0.7$.

## 4.3 Parametric Instability in a Dusty Plasma with a Periodically Varying Charge of Dust Grains

While analyzing the instabilities of dusty or self-gravitational plasmas we have tacitly assumed all the unperturbed parameters to remain unchanged. In the present Section, we shall consider the parametric instability developing in a circular plasma flow of small cross-section as a result of periodic variations in the charge density of grains. The reason for such variations is

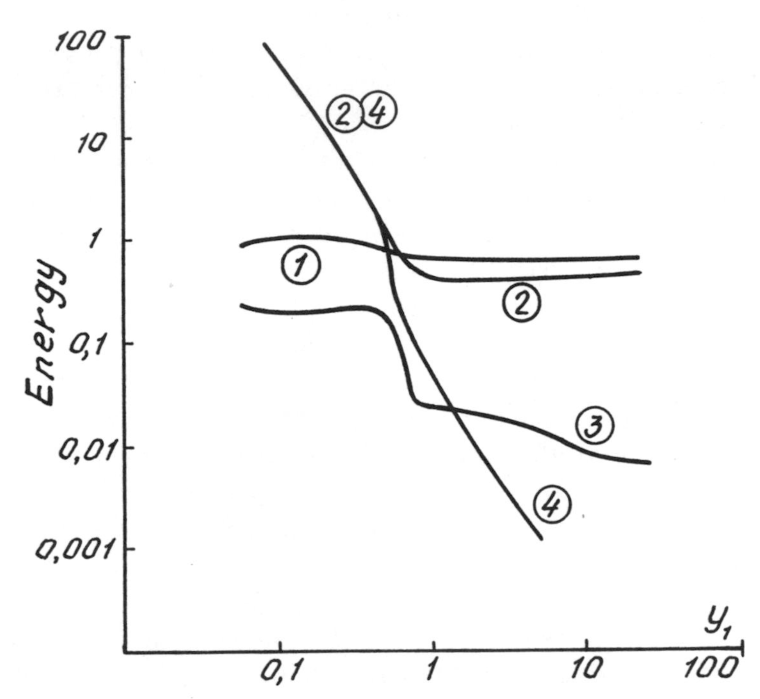


Figure 10: The Final Energy Distribution as Resulting from Numerical Simulation of the Particle Dynamics: Curve 1 Represents the Kinetic Energy of Light-Weight Particles; Curve 2 is the Kinetic Energy of Heavier Particles; Curve 3 is the Electrostatic Energy, and Curve 4 the Absolute Magnitude of Gravitation Energy (after Gisler *et al.*, 1992).

the repeated transit of the particles through light and shadow regions (see Figure 11). This causes changes in the intensity of the photoelectric effect, and hence in the charge of dust grains. The question whether the photoelectric current plays any important part in the general balance of electric currents responsible for grain charging ought to be specially analyzed (see Chapter 1). While *in situ* measurements of the photoelectric current are not possible so far, the effect can be evaluated indirectly through other related phenomena that are detectable in observations.

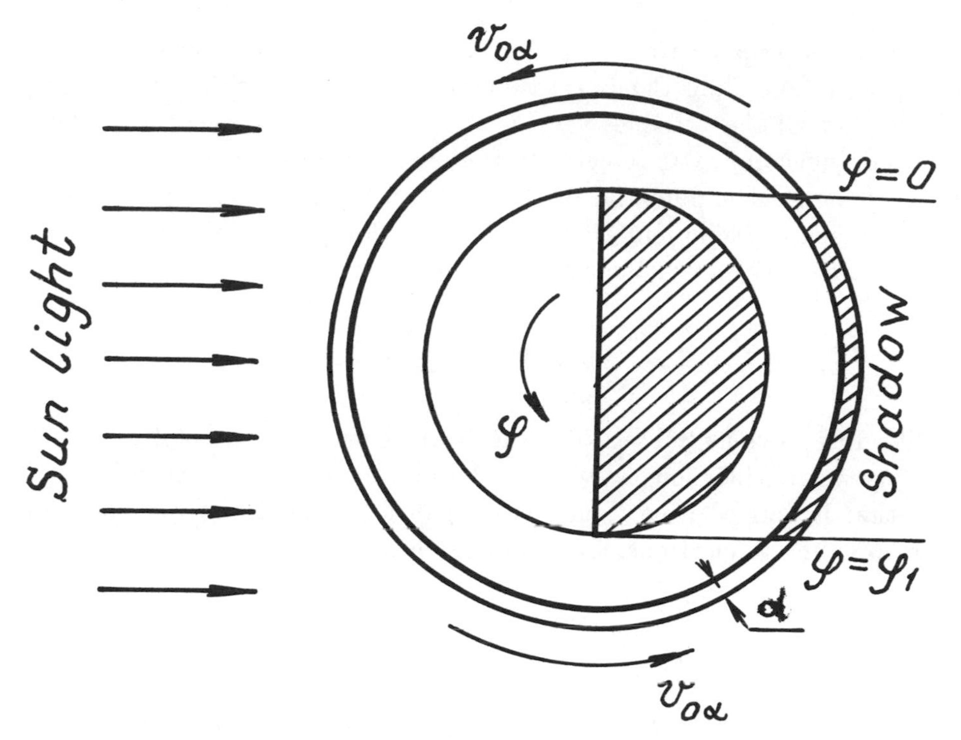


Figure 11: Periodic Light-to-Shadow Transitions Due to Partial Shadowing of a Planetary Ring.

### 4.3.1 Long Wave Disturbances in a Thin Circular Beam (Model of a Narrow Planetary Ring)

The parametric excitation of wavelike disturbances will be discussed for the example of the above described thin circular beam. Along with electrically charged dust grains, the beam involves plasma ions and electrons. It is assumed that the beam is quasineutral and the net electric current is constant along the ring. The disturbances that may arise are described by the linearized equation set

$$\begin{aligned}\frac{\partial \mathbf{V}_\alpha}{\partial t} &+ (\mathbf{V}_{0,\alpha}\cdot\nabla)\mathbf{V}_\alpha = -\frac{Q_\alpha}{m_\alpha}(\nabla\psi_E + c^{-1}\mathbf{V}_\alpha\times\mathbf{B}_0),\\ \frac{\partial n_\alpha}{\partial t} &+ \nabla\cdot(n_{0,\alpha}\mathbf{V}_\alpha + n_\alpha\mathbf{V}_{0,\alpha}) = 0\\ \Delta\psi_E &= -4\pi\sum_\alpha Q_\alpha n_\alpha.\end{aligned} \tag{43}$$

The zero subscripts refer to unperturbed magnitudes, $\alpha$ stands for the particle species and $\mathbf{B}_0$ is the external magnetic field perpendicular to the ring plane. Let all the variables vary as $\exp[-i\omega t + im\varphi]$, where $\varphi$ is the azimuthal angle and $m$ an integer. In the case of long wave disturbances (i.e. $m \sim 1$) the electric potential $\psi_E$ takes the same form as the potential of a uniform ring (Landau and Lifshitz, 1982) with a parametric dependence of the charge density upon the azimuth,

$$\psi_E \approx \frac{\pi d^2}{2} \log \frac{16R}{d} \sum_\alpha n_\alpha Q_\alpha(\varphi).$$

Unlike the previous Subsections, the beam curvature cannot be neglected here as we consider long wave disturbances $\lambda \sim R$. The dispersion relation that is derived from Equation (43) in a standard way involves terms corresponding to electrons, ions and dust grains, i.e.

$$1 = \frac{m^2 d^2}{8R^2} \log \frac{16R}{d} \sum_\alpha \frac{\omega_{p,\alpha}^2}{(\omega - mV_{0,\alpha}/R)^2 - \omega_{B,\alpha}^2}. \tag{44}$$

Note the gyrofrequencies $\omega_{B,e}$, $\omega_{B,i}$ and $\omega_{B,d}$ to be greatly different (e.g., in Saturn's rings the respective numerical values are $\omega_{B,e} \sim 10^5\,\mathrm{s}^{-1}$, $\omega_{B,i} \sim 10^2\,\mathrm{s}^{-1}$ and $\omega_{B,d} \sim 10^{-5}\,\mathrm{s}^{-1}$). Once again, we will consider the disturbances whose frequencies obey $\omega_{B,d}^2 \ll \omega^2 \ll \omega_{B,i}^2, \omega_{B,e}^2$, which implies that the major contribution to the right-hand part of Equation (44) is given by macroscopic particles,

$$1 \approx \frac{m^2 d^2}{8R^2} \log \frac{16R}{d} \sum_{\alpha=1}^{N_d} \frac{\omega_{p,\alpha}^2}{(\omega - mV_{0,\alpha}/R)^2}. \tag{45}$$

In what follows, we shall consider, for the sake of simplicity, a single sort of dust grains.

### 4.3.2 Dispersion Relation with the Grain Charge Abruptly Changing at the Light-Shadow Boundary

The electric charge of the dust grains will be different in the illuminated part of the circular beam and in the shadow. Let the dust charge density be $n_{0,d}Q_1$ in the illuminated zone (the range of angles $0 < \varphi < \varphi_1$) and $n_{0,d}Q_2$ in the shadow ($\varphi_1 < \varphi < 2\pi$), as is shown in Figure 12a. The dispersion relation Equation (45) is valid separately in each of the two regions and

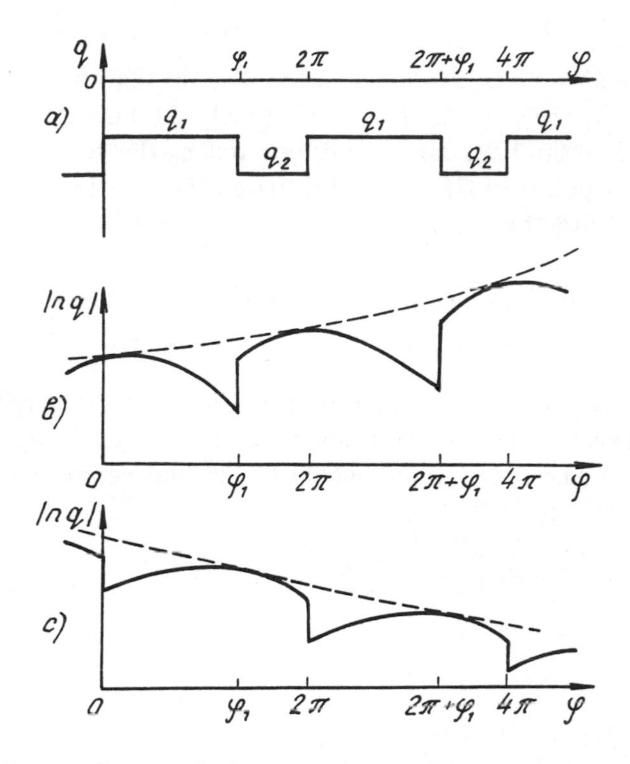


Figure 12: Grain Charge Variations along a Planetary Ring (Curve a), and Increasing (b) and Decreasing (c) Charge Density Waves (the Frame of Reference Moves with the Beam).

provides solutions for the two angular domains. Accordingly, the current and velocity disturbances in the illuminated (subscript 1) and shadow (subscript 2) zone are

$$\begin{aligned} J_{1,2} &= \exp\left(-i\omega t + \frac{i\omega R\varphi}{V_0}\right)\left[A_{1,2}\exp\left(-i\tau_{1,2}\frac{\omega R\varphi}{V_0}\right) + \right. \\ &\left. + B_{1,2}\exp\left(+i\tau_{1,2}\frac{\omega R\varphi}{V_0}\right)\right], \\ V_{1,2} &= \frac{\tau_{1,2}}{n_{0,d}Q_{1,2}}\exp\left(-i\omega t + \frac{i\omega R\varphi}{V_0}\right)\left[A_{1,2}\exp\left(-i\tau_{1,2}\frac{\omega R\varphi}{V_0}\right) - \right. \\ &\left. - B_{1,2}\exp\left(+i\tau_{1,2}\frac{\omega R\varphi}{V_0}\right)\right]. \end{aligned} \tag{46}$$

with $\tau_{1,2} = (d/2)\omega_{p,1,2}(\log 16R/d)^{1/2}/V_0$. These solutions are to be matched at the boundary of the regions 1 and 2 to yield the modal equation corresponding to the specific $Q(\varphi)$ dependence. This is the approach typical of the theory of waves in periodic structures (Brillouin and Parodi, 1956). It can be shown that the boundary conditions reflect the continuity of $J$ and $V$ at the singular points of $Q(\varphi)$. By definition, the solutions should also be periodical in $\varphi$ (with the period of $2\pi$), and hence the boundary conditions are

$$J_1(\varphi_1) = J_2(\varphi_1) \qquad J_1(0) = J_2(2\pi)$$
$$V_1(\varphi_1) = V_2(\varphi_1) \qquad V_1(0) = V_2(2\pi)$$

By imposing these on the solutions of Equation (46) we arrive at a set of four homogeneous algebraic equations for the unknown amplitudes $A_1$, $A_2$, $B_1$ and $B_2$. The determinant yields the modal equation (Bliokh and Yaroshenko, 1984; 1994)

$$\begin{aligned}\cos\Gamma &= \cos\left(\frac{\tau_1\omega R\varphi_1}{V_0}\right)\cos\left(\frac{\tau_2\omega R\varphi_2}{V_0}\right) - \frac{1}{2}(S+S^{-1})\sin\left(\frac{\tau_1\omega R\varphi_1}{V_0}\right)\times \\ &\times \sin\left(\frac{\tau_2\omega R\varphi_2}{V_0}\right) \end{aligned} \qquad (47)$$

with $\Gamma = 2\pi\omega R/V_0$ and $S = \tau_1 Q_2/\tau_2 Q_1$.

### 4.3.3 Parametric Resonance and Azimuthal Modulation of the Planetary Ring Brightness

If the right-hand side of Equation (47) is greater than one (in absolute magnitude), then $\Gamma$, and hence $\omega$, become complex values ($\omega = \omega_r + i\nu$), which implies excitation of an exponentially growing wave in the ring. The growth rate is the higher, the greater the difference between the right-hand part of Equation (47) and unity. The maximum rate can be found to correspond to the conditions $\tau_1\omega_r R\varphi_1/V_0 = \tau_2\omega_r R\varphi_2/V_0 = (2n_1+1)\pi/2$ and $\omega_r R/V_0 = (2n_2+1)/2$, with $n_1 = 0, 1, 2,$ *etc.* and $n_2 = 1, 2,$ *etc.*, $n_2 > n_1$. The magnitude of $\nu$ is determined by the beam parameters (specifically, $R$ and $V_0$), the ordinal number $n_1$ of the unstable zone and the real frequency $\omega_r$, *viz.*

$$\nu \approx \pm\frac{\eta V_0(2n_2+1)}{2\pi R[(2n_2+1)^2-(2n_1+1)^2]^{1/2}},$$

where $\eta = |\tau_1 Q_2 - \tau_2 Q_1|/(\tau_1\tau_2 Q_1 Q_2)^{1/2}$. The highest value of $\nu$ is reached in the first unstable zone ($n_1 = 0$) at $n_2 = 1$. The variations of $J, V$ and the charge density $Q(\varphi)n_d(\varphi)$ along the ring are all characterized by the same phase factor $\exp(-i\omega t + i\omega R\varphi/V_0)$, however with different laws for the amplitudes. The charge density disturbances are at the same time mass density waves of the dust grains, which allows suggesting they could result in some observable effects. The spatial variations of $n_dQ$ for the growing and the attenuated mode are shown diagrammatically in Figure 11,

$$|n_dQ|_g \sim \begin{cases} e^{\nu t}\left[\cos^2\left(\frac{\omega_r R}{V_0}\tau_1\varphi\right) + \tau_1^2 \sin^2\left(\frac{\omega_r R}{V_0}\tau_1\varphi\right)\right]^{1/2} & \text{a)} \\ e^{\nu t}\left[\cos^2\left(\frac{\omega_r R}{V_0}\tau_2(\varphi-2\pi)\right) + \tau_2^2 \sin^2\left(\frac{\omega_r R}{V_0}\tau_2(\varphi-2\pi)\right)\right]^{1/2} & \text{b)} \end{cases}$$

$$|n_dQ|_a \sim \begin{cases} e^{-\nu t}\left[\sin^2\left(\frac{\omega_r R}{V_0}\tau_1\varphi\right) + \tau_1^2 \cos^2\left(\frac{\omega_r R}{V_0}\tau_1\varphi\right)\right]^{1/2} & \text{a)} \\ e^{-\nu t}\left[\sin^2\left(\frac{\omega_r R}{V_0}\tau_2(\varphi-2\pi)\right) + \tau_2^2 \cos^2\left(\frac{\omega_r R}{V_0}\tau_2(\varphi-2\pi)\right)\right]^{1/2}. & \text{b)} \end{cases} \quad (48)$$

where a) is for $0 \le \varphi \le \varphi_1$ and b) for $\varphi_1 \le \varphi \le 2\pi$. After some time, only the growing disturbances $n_d(\varphi)Q(\varphi)$ remain, which are characterized by a minimum at the light-shadow transition and an increase in the shadow.

The angular dependences of $n_dQ$ have been calculated for a quite definite (negative) charge of the dust grains. In the case of a different sign of the grain charge, the phase of $Q(\varphi)$ variation would be also different. That would result in different dependences of $|n_dQ|_g(\varphi)$ and $|n_dQ|_a(\varphi)$. The parametric excitation mechanism would remain active, however result in a lower growth rate. Indeed, the conditions Equation (48) can be met only with $Q_1 < Q_2$ (or $\tau_1 < \tau_2$), which inequality corresponds to negatively charged dust grains.

According to some data that can be found in the literature, the role of photoeffect is insignificant at great heliocentric separations, and hence the grain charge is controlled mainly by the ion and electron currents from the ambient plasma (Grün *et al.*, 1984). Yet this does not resolve the question whether or not small variations in charge, owing to the photoeffect, can result in parametrically excited wavelike perturbations in the ring. The answer will be ultimately found through observations. The above discussed theoretical model is in principle capable of providing such an opportunity. Suppose, azimuthal variations of the ring brightness have been accurately measured. Since brightness is determined by the material density, the measured data can be confronted with calculations for the density modulation (Figure 12). Detection of similar regularities will be evidence for the parametric instability in the planetary ring. Unfortunately, detailed observational data on azimuthal variations of the material density in the rings of Saturn or Jupiter are not currently available.

Considering the instabilities of dusty and self-gravitational plasmas, we have concentrated on the instability mechanisms associated with a regular drift of individual plasma components, having in mind such astrophysical objects as planetary rings. Meanwhile, regular drifts are also shown by comet environments where the beam-plasma instability can be excited by flows of solar wind particles passing through the comet's dusty envelope (Havnes, 1988).

Bharuthram *et al.* (1992) discussed the two-beam instability arising in beams of dust grains and ions.

Other mechanisms have also been discussed in the literature, e.g. the Kelvin-Helmholtz instability produced by the velocity gradient in adjacent liquid layers. With regard to the comet environment, it was considered by Ershkovich and Mendis (1986). D'Angelo and Song (1990) analyzed the same instability mechanism for a low-pressure magnetized plasma with shear in the ion flow, using a multi-fluid hydrodynamic approach. The effect produced by the dust component proves to depend on the sign of the grain charge, which is quite understandable if we recall that the phase velocity of dust-acoustic waves is determined by the balance of electron and ion charge density in the medium (cf. Subsection 3.2.1).

## References

Bharuthram, R., Salem, H. and Shukla, P.K. 1992, *Phys. Scripta* **45**, 512.

Bliokh, P.V. and Yaroshenko, V.V. 1984, *Sov. Radio Phys. and Quantum Electron.* **27**, 1471 (*in Russian*).

Bliokh, P.V. and Yaroshenko, V.V. 1994. In H. Kikuchi (Ed.). *Dusty and Dirty Plasmas, Noise, and Chaos in Space and in the Laboratory* (Plenum Publ. Corp.: New York), p. 23.

Brillouin, L. et Parodi, M. 1956. Propagation des ondes dans les milieux périodiques (Dunod et C[ie]: Paris).

D'Angelo, N. and Song, B. 1990, *Planet. Space Sci.* **38**, 1577.

Dawson, J.M. 1960, *Phys. Rev.* **118**, 381.

Ershkovich, A.I. and Mendis, D.A. 1986, *Astrophys. J.* **302**, 849.

Gisler, G.R. and Wollman, E.R. 1988, *Phys. Fluids* **31**, 1101.

Gisler, G.R., Q. Rushdy Ahmand and Wollman, E. 1992, *IEEE Trans. on Plasma Sci.* **20**, 922.

Grün, E., Morfill, G.E. and Mendis, D.A. 1984. In R. Greenberg and A. Brahic (Eds.) *Planetary Rings*, (Univ. of Ariz. Press: Tucson, AZ), p.275.

Havnes, O. 1988, *Astron. Astrophys.* **193**, 309.

Havnes, O., Aslaksen, T.K. and Melandsø, F. 1990, *J. Geophys. Res.* **95A**, 6581.

Krall, N.A. and Trivelpiece, A.W. 1973. Principles of Plasma Physics (McGraw-Hill Inc).

Landau, L.D. and Lifshitz, E.M. 1982. Electrodynamics of Solids (Nauka: Moscow), p. 621 (*in Russian*).

Mendis, E.A. and Rosenberg, M. 1994, *Ann. Rev. Astron. Astrophys.* **32**, (*in press*).

Mikhailovsky, A.B. 1975. Theory of Plasma Instabilities, vol. 1 (Atomizdat: Moscow) (*in Russian*).

Mikhailovsky, A.B. and Fridman, A.M. 1971, *Sov. Phys.—JETP* **61**, 13.

Polyachenko, V.L. and Fridman, A.M. 1981, *Sov. Phys.—JETP* **81**, 457.

Rosenberg, M. 1993, *Planet. Space Sci.* **41**, 229.

Yaroshenko, V.V.1985. Electrostatic Waves in Planetary Rings: PhD Thesis (Institute of Radiophysics and Electronics: Kharkov, Ukraine) (*in Russian*).

# 5 Dusty and Self-Gravitational Plasmas of Planetary Rings

The principal force responsible for the structure and the very existence of planetary rings is gravitation which includes the gravity field of the planet as the major component, perturbations on the part of satellites, and self-gravitation in the ring. From the electromagnetic point of view the ring is a set of charged particles moving through a magnetic field. Clearly, the system should show some electromagnetic effects. These vague general considerations were transformed into certainty when space probes revealed such structural details of Saturn and Jupiter rings that could not be explained by the action of gravitation alone.

In this Chapter, we will consider only those effects in planetary rings (pricipally in Saturn's rings) that are associated with the dusty plasma. First, we shall analyze the dust concentration profile in the ring. The profile is formed with the participation of electrostatic pressure. Vertical density oscillations arise here, which can increase owing to resonance effects. The radial profile of particle number density is also influenced by electromagnetic forces probably responsible for the extremely complex radial variation of density that is observed. Further on, we will discuss electrostatic waves in narrow and broad rings, marked by rather unusual behavior because of the special geometry and differential rotation in the system. To conclude the Chapter, we analyze several interpretations of the interesting effect known as "spokes" in Saturn's ring.

## 5.1 Planetary Rings

The ring around Saturn was first seen by man through Galileo's telescope in 1610 and the word itself was not enunciated until almost half a century later (Huygens, 1659). Huygens' hypothesis concerning the structure did not result from observations with improved telescopes but rather appeared from a purely theoretical model of invisible vortices filling the space. While that model (suggested by Descartes in 1644) proved untenable, it had inspired Huygens to put forward his idea.

As an outstanding celestial object, Saturn's ring has at all times attracted the brilliant minds of mankind. Laplace and Maxwell have shown that a continuous solid ring (as imagined by Huygens) would be unstable. Later spectral observations showed the ring in fact to consist of a multitude of individual bodies moving along Keplerian orbits. Poincaré's papers laid the foundation of the dynamical theory of planetary rings that seemed to be in a perfect agreement with observations.

The smooth development of events was interrupted in 1977 by an impetuous flow of new data. That year, nine rings of Uranus were discovered and Saturn lost its status of the only ringed planet. Then followed the discovery of Jupiter's rings (with the aid of space probes) and, quite unexpectedly, of the fine structure of Saturn's rings and "spokes" in its broad ring $B$. Later on, the family of "ringed" planets was joined by the fourth member, Neptune.

Having lost its exclusive position, Saturn still continues to attract the attention of plasma physicists and specialists in celestial mechanics. Their joint efforts have resulted in considerable progress in understanding the behavior of rings; however many details of the ring structure and dynamics have not been given any interpretation.

The properties of Saturn's rings are largely determined by the physics of the dusty plasma. That is why we have selected them to illustrate many of the theoretical concepts of the preceding Chapters. At present, various aspects of the planetary ring physics are widely discussed in the vast literature. We will only mention the book edited by Greenberg and Brahič (1984) which is an excellent state of the art review for the period of mid-1980s.

Some parameters of the Saturn and Jupiter ring were given in Chapter 2 (see Figure 3 of Chapter 2). The different rings are separated owing to gravitational resonances with the satellites and, in some cases, to gravitoelectrodynamic effects (see Subsection 2.3.5). Saturn rings $A$, $B$ and $C$ are of greater density than $D$, $E$ and $G$. Ring $F$ is rather dense and shows peculiar curvatures. Parameters of the dusty plasma in planetary rings are listed in Table 1 taken from the reviews of Hartquist *et al.* (1992) and Goertz (1989) ($n_0$ is the electron number density; $T$ – temperature; $z_m$ – half-thickness of the ring; $a$ – scale size of dust grains, and $n_d$ the dust grain number density).

## 5.2 Electromagnetic Effects and Grain Number Profile in the Ring

### 5.2.1 The Dust Grain Distribution Across the Ring Produced by Electrostatic Forces

We shall consider dust grain distributions in wide rings of low thickness, such that the problem can be formulated in one dimension. A dust layer in

Table 1: Some parameters of the dusty plasma of planetary rings

| Ring | $n_0\,[\mathrm{cm}^{-3}]$ | $T\,[\mathrm{eV}]$ | $z_m\,[\mathrm{cm}]$ | $a\,[\mathrm{cm}]$ | $n_d\,[\mathrm{cm}^{-3}]$ |
|---|---|---|---|---|---|
| $A,\ B$ | 0.1 | 2 | $5 \times 10^4$ | $(1 \div 5) \times 10^3$ | $2.5 \times 10^{-8}$ |
| | | | | $10^{-5} \div 10^{-3}$ | $6.3 \times 10^{-7}$ |
| Spokes | $0.1 \div 10^2$ | 2 | $3 \times 10^6$ | $10^{-4}$ | 1 |
| $F$ | 10 | 100 | $10^5$ | $10^{-4} \div 1$ | $10^{-12}$ |
| $G$ | 10 | 100 | $6 \times 10^8$ | $10^{-4} \div 3 \times 10^{-2}$ | |
| $E$ | 20 | 10 | $2 \times 10^8$ | $10^{-4}$ | |
| Jupiter ring | 100 | 100 | $10^5$ | $2 \times 10^{-4}$ | |
| Jupiter halo | 100 | 100 | $10^9$ | $10^{-5}$ | |
| Uranus $\alpha$ | $50 \div 200$ | 30 | $10^4$ | 5 | $5 \times 10^{-7}$ |
| Uranus $\varepsilon$ | $50 \div 200$ | 30 | $10^4$ | 20 | $10^{-7}$ |

a plasma environment can be characterized by six functional dependences, namely the dust grain charge, $Q(z)$, and potential, $\psi(z)$, the ambient plasma potential, $\Phi(z)$, and the electron, ion and dust grain number densities $n_e(z)$; $n_i(z)$, and $n_d(z)$, where the $z$-axis is perpendicular to the ring plane.

A similar problem was discussed in Section 1.5 treating the charge of grains in a dust cloud. The density profile $n_d(z)$ was assumed known, while the five resting functions obeyed the set of five equations which we write here again:

1. The charge $Q$ of a dust grain of radius $a$ is related to the potential as

$$Q = aU \qquad (U = \psi - \Phi). \tag{1}$$

2. The currents flowing on the dust grain from the ambient plasma and the photoelectric current $J_{ph}$ are balanced as

$$n_e \left(\frac{8T}{\pi m_e}\right)^{1/2} \exp(eU/T) = n_i \left(\frac{8T}{\pi m_i}\right)^{1/2} [1 - eU/T] + J_{ph}. \tag{2}$$

3. If the scale length of variations in the dust grain density is much greater than the Debye length, the Poisson equation for the potential is replaced by the quasineutrality condition,

$$(n_e - n_i)e - a n_d U = 0. \tag{3}$$

4. The electron number density obeys the Boltzmann distribution law,

$$n_e = n_0 \exp(e\Phi/T). \tag{4}$$

5. The ion number density obeys the Boltzmann distribution law,

$$n_i = n_0 \exp(-e\Phi/T). \tag{5}$$

For our present purposes, the equation set Equation (1-5) is insufficient as the dust grain density $n_d(z)$ should be established in a self-consistent way. The necessary sixth equation will represent the balance of electrostatic and gravitational forces (Havnes and Morfill, 1984). The electric forces,

$$f_E = -Q\frac{d\Phi}{dz},$$

acting similarly to gas-kinetic pressure, tend to oust the grain from the area of higher density. Contrary to this, the gravitation force

$$f_g = GM_p m_d z/R^3$$

attempts to bring the grain back to the equator plane $z = 0$ (which is the middle plane of the ring). The equilibrium $n_d(z)$ profile is given by

$$\Omega_K^2 z m_d = -Q(z)\frac{d\Phi}{dz}, \tag{6}$$

where $\Omega_K = (GM_p/R^3)^{1/2}$ is the Kepler frequency for the orbit of radius $R$. The conditions governing the $n_d(z)$ profile formation may be different in the illuminated parts of the ring and those shadowed by the planet. Equations (1-6) are a nonlinear set that generally permits numerical analysis only. In some limiting cases analytical solutions have been found by means of linearization. The parameter of major importance is the dust layer density characterized by the dimensionless value $p = Tan_d/(e^2 n_0)$. It is often defined for the middle plane of the ring where $n_d = n_d(0)$, while $a$ and $T$ are expressed in specific units, namely

$$p = a\,[\mu]\,T\,[\text{eV}] n_d(0)/n_0.$$

The ring is considered dense if $p \gg 10^{-4}$ (e.g., the value for Saturn's $B$-ring is $p \geq 10^{-2}$ in the region of "spokes"). In the case of a dense ring the electric field concentrates mainly at the boundary, therefore the charge of dust grains is small in the interior, and $\psi - \Phi \simeq 0$. The corresponding magnitudes of $\Phi = \Phi_m = \psi = \psi_m$ can be found from Equations (2), (4) and (5). Neglecting the photocurrent, we have

$$\Phi_m = \psi_m = -\frac{1}{2}\left(\frac{T}{e}\right)\log\left(\frac{m_i}{m_e}\right). \tag{7}$$

Taking the photocurrent into account in Equation (2), one arrives at a more complicated formula (Havnes, 1986) where $\Phi_m$ and $\psi_m$ are expressed in terms of $F = (1/2)[X + (X^2 + 4m_e/m_i)^{1/2}]$ and

$$X = 1.5 \times 10^{-8} J_{ph} / \left(n_0\sqrt{T\,[\mathrm{eV}]}\right), \tag{8}$$

*viz.*

$$\Phi_m = \psi_m = (T/e)\log F(X). \tag{9}$$

By linearizing the potentials about $\Phi_m$ and $\psi_m$ ($\Phi = \Phi_m + \Delta\Phi$ and $\psi = \psi_m + \Delta\psi$) we can derive a parabolic profile for the dust grain density,

$$n_d(z) \simeq n_d(0)(1 - z^2/z_m^2). \tag{10}$$

The half-thickness $z_m$ of the ring can be expressed in terms of known parameters, i.e.

$$z_m = z_m^{(0)} \simeq 4 \times 10^{-2}\left(\frac{m_i}{m_e}\right)^{1/2} n_0/(\Omega_K a\,[\mu] n_d(0)) \qquad \text{with } J_{ph} = 0 \tag{11}$$

or

$$z_m \simeq z_m^{(0)}\left(\frac{1-F^2}{F}\right)\left(\frac{m_e}{m_i}\right)^{1/2} \qquad \text{with } J_{ph} \neq 0. \tag{12}$$

The parabolic representation of Equation (10) is not valid for the outer parts of the ring where the dust concentration is very low.

In the other limiting case of small $p$ ($p \ll 10^{-4}$) electrostatic screening inside the ring is very weak, such that the potentials $\psi$ and $\Phi$ remain close to their respective values in free space, where $\Phi \simeq \Phi_0 = 0$, while $\psi_0$ depends on the chemical composition of the plasma. In an electron-proton plasma it is $\psi_0 \simeq -2.5\,T/e$, while for oxygen ions $\psi_0 \simeq -3.6\,T/e$ (see Chapter 1). By linearizing the potentials near $\Phi_0$ and $\psi_0$ one would obtain a parabolic profile once again, however with a different half-thickness parameter (Havnes and Morfill, 1984).

Allowance for the photoelectron emission in the balance of currents equation permits analyzing the different density profiles arising in the dark and illuminated parts of the ring (Havnes, 1986). The major difficulty is the lack of reliable data on the magnitude of $J_{ph}$. If the intensity of solar $UV$ radiation may be assumed known, then the major sources of uncertainty are the effective absorption cross-section $q_{abs}$ for the radiation and the photoelectron emission factor $Y$. Evaluating $Y$ from laboratory measurements for different materials leads to relatively low values of the parameter $X$ involved in Equation (9). Taking $n_0 = 10^2\,\mathrm{cm}^{-3}$ and $T = 2\,\mathrm{eV}$ in Equation (8) one can obtain, with account of the possible scatter in $Y$, $X \simeq 0.002 \div 0.04$. The distinction between the dark and the light part of the ring is insignificant. In fact, the estimates of the photoelectron current from small dielectric grains in space derived through laboratory measurements of $Y$ may prove greatly underrated (Gail and Sedlmayr, 1980). With this in mind, Havnes (1986) indicated that the parameter $X$ might reach as high a value as 0.4 with the same magnitudes of $n_0$ and $T$, which would result in a noticeable difference between physical properties of the ring on two sides of the light-shadow boundary (Figure 1).

### 5.2.2 Profile of Dust Concentration with Account of Different Grain Sizes and Thermal Motion

The profile derived in the preceding Subsection has appeared as a result of significant simplifications. While mentioning the gas-kinetic pressure $f_p$, we failed to include it in the balance of forces Equation (6). Besides, all the dust grains were assumed to be of the same size $a$ which certainly is not true. Making the appropriate amendments, one should be able to derive more realistic profiles $n_d(z)$ of the dust concentration and establish limits of validity for these or other simplifications (Aslaksen and Havnes, 1992). The functions to be found and, accordingly, the set of equations remain the same, however two of the six equations need to be modified. The quasineutrality condition Equation (3) becomes, with allowance for the different grain sizes,

$$(n_e - n_0)e - U \int_{a_{min}}^{a_{max}} a\tilde{n}_d(z,a)\,da = 0 \tag{13}$$

(once again, $\tilde{n}_d(z,a)$ is the spatial density of $n_d(z)$). The balance of forces Equation (6) acquires the kinetic pressure term in its right-hand side, *viz.*

$$\Omega_K^2 z m_d(a) = -aU\frac{d\Phi}{dz} - \frac{T_d}{n_d}\cdot\frac{dn_d}{dz}. \tag{14}$$

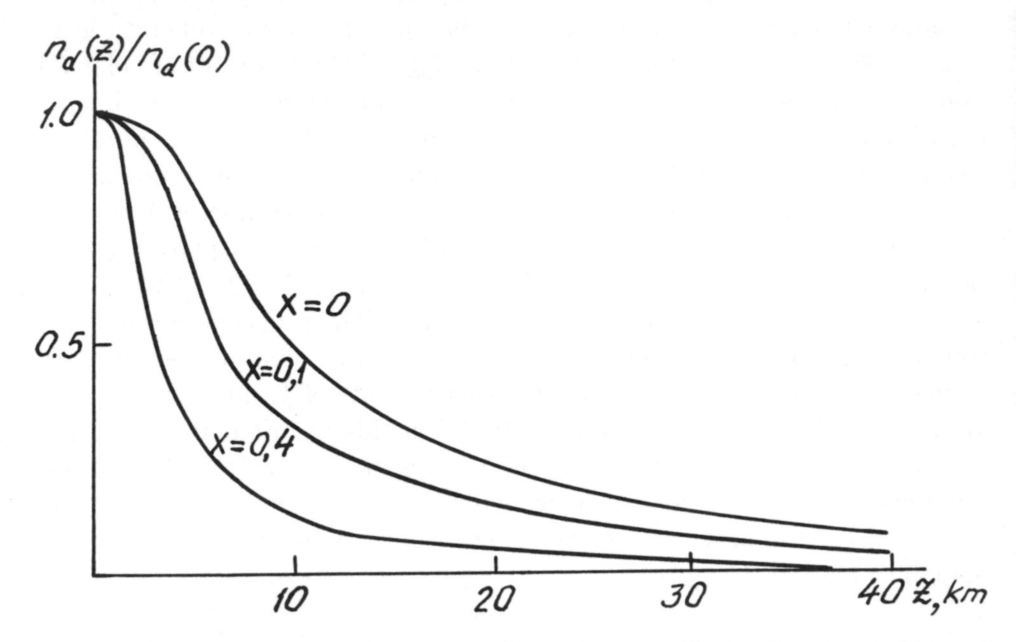


Figure 1: Normalized Vertical Profiles of Dust Grain Density: $X = 0$ Corresponds to Nightside Conditions; $X = 0.1$ and $X = 0.4$ Correspond to Dayside Conditions with Different Photoelectric Current Magnitudes; $n_0 = 100\,\text{cm}^{-3}$, $T_e = 2\,\text{eV}$ and $a = 1\,\mu$ for all the Cases (after Havnes, 1986).

Integrating Equation (14) gives

$$\tilde{n}_d(z,a) = \tilde{n}_d(0,a)\exp\left[-\left(\Omega_K^2 m_d z^2/2 + a\int_{\Phi_0}^{\Phi(z)}(\psi - \Phi')\,d\Phi'\right)/T_d\right], \quad (15)$$

where $\Phi_0$ is the electric potential at $z = 0$ in the dusty plasma layer. At this point we have to specify the size distribution of dust grains. While in the preceding Sections we have often used the power law distribution $\tilde{n}_d(a) = n_{0,d}a_0^{\beta-1}a^{-\beta}$, now it seems reasonable to specify the total number of grains of a given size,

$$N_d(a) = \int_0^\infty \tilde{n}_d(z,a)\,dz.$$

The integrated value $N_d(a)$ retains a power-law form,

$$N_d(a) = Aa^{-\beta},$$

with the normalizing factor $A$ being determined by the fixed total number of grains. The exponent $\beta$ lies, in the case of planetary rings, between 0.9 and 4.5 (Burns *et al.*, 1984).

First, consider cold dust ($T_d = 0$) in which the grains of different sizes occupy their equilibrium positions at different specific levels $z$. The reason can be easily understood if we recall that $f_E \sim a$, while $f_g \sim a^3 z$, and hence the equilibrium condition $f_E = f_g$ in the field $\Phi = \Phi(z)$ "selects" a specific size $a$ for every level $z$. The larger (heavier) are the grains, the closer they lie to the plane $z = 0$. A detailed analysis was performed by Aslaksen and Havnes (1992) who derived both the $a = a(z)$ and $n_d = n_d(z)$ dependences and the self-consistent potential distributions $\psi(z)$ and $\Phi(z)$.

The situation is markedly different for the "hot" dust, $T_d \neq 0$. The thermal motion of dust grains is able to prevent the system from gravitational collapse, resulting in a finite thickness of the ring even with electrically neutral grains. If $Q = 0$, then the potentials $\Phi$ and $\psi$ are also zeros, and Equation (15) yields

$$n_d(z) = n_d(0) \exp(-z^2/H_p^2), \tag{16}$$

where $H_p = (2T_d/\Omega_K^2 m_d)^{1/2}$ is the scale thickness for the neutral ring (Burns *et al.*, 1984). It is of interest to compare the value with the ring half-thickness $z_m$ resulting from a purely electrostatic equilibrium (Equation (11)). In terms of the density parameter $p$, the ratio is (Aslaksen and Havnes, 1992)

$$\frac{z_m}{H_p} \sim 17 \left(T^2\,[\text{eV}]\, a(\mu) p / T_d\,[\text{eV}]\right)^{1/2},$$

and the condition $z_m/H_p \leq 1$ yields an estimate of the dust temperature $T_d$ at which the kinetic effects prevail,

$$T_d\,[\text{eV}] \geq 3 \times 10^2 T^2\,[\text{eV}]\, a(\mu) p.$$

In tenuous dust layers ($p \leq 3.5 \times 10^{-3}/T\,[\text{eV}] a(\mu)$) this happens at $T_d \sim T$, while in layers of higher density kinetic effects are noticeable at high dust temperatures.

Taking into account all the complicating effects simultaneously seems possible only numerically. E.g., the integral in Equation (13) can be substituted by a discrete sum,

$$\int_{a_{min}}^{a_{max}} a\tilde{n}_d(z,a)\,da \simeq \sum_{j=1}^{N} a_j n_d(z,a_j),$$

where $a_j$ are given by

$$a_j = \frac{a_{max}}{2}\left[\left(\frac{a_{min}}{a_{max}}\right)^{(j-1)/N} + \left(\frac{a_{min}}{a_{max}}\right)^{j/N}\right], \qquad j = 1, 2, \ldots, N.$$

The results for a proton-electron plasma of constant temperature $T$ are shown in Figures 2 and 3 for $N = 10$, $a_{min} = 0.1\,\mu$ and $a_{max} = 10\,\mu$. The dust temperature $T_d$ is a variable parameter. If it were zero, the dust grains would be sorted out electrostatically according to their size, with the larger ones concentrating near $z = 0$ and the smaller being dispersed at the ring periphery. With $T_d \neq 0$, sharp boundaries between the strata are smeared out, yet the small size grains fail to reach the inner ring levels. In contrast to them, the larger grains remain at low heights (Figure 2, $T_d = 0.1T$). Kinetic effects become increasingly more important at higher dust temperatures and the areas of concentration of dust grains of different sizes overlap (Figure 3 with $T_d = T$). Finally, the stratification almost vanishes at $T_d = 10T$ and all the moving grains can reach and cross the central plane. The density profile approaches that of Equation (16).

Along with the effects due to a nonzero temperature $T_d$ equal for dust grains of all sizes, kinetic effects of a different kind can be analyzed, namely a constant thermal scatter $v_T$ of grain velocities (Havnes *et al.*, 1992). Apparently, such a condition suggests the presence of mechanisms to reduce the statistical variance of smaller grain velocities. Of considerable importance among such mechanisms is the collisionless damping associated with inertia of grain charge variations $Q(t)$ of the grains vibrating in a dust layer.

### 5.2.3 Grain Vibrations Across the Ring and Resonances

A natural continuation to the analysis of the static density distributions across the ring is the discussion of grain vibrations about the equilibrium (Melandsø and Havnes, 1991). The equation set of Subsection 5.2.1 remains

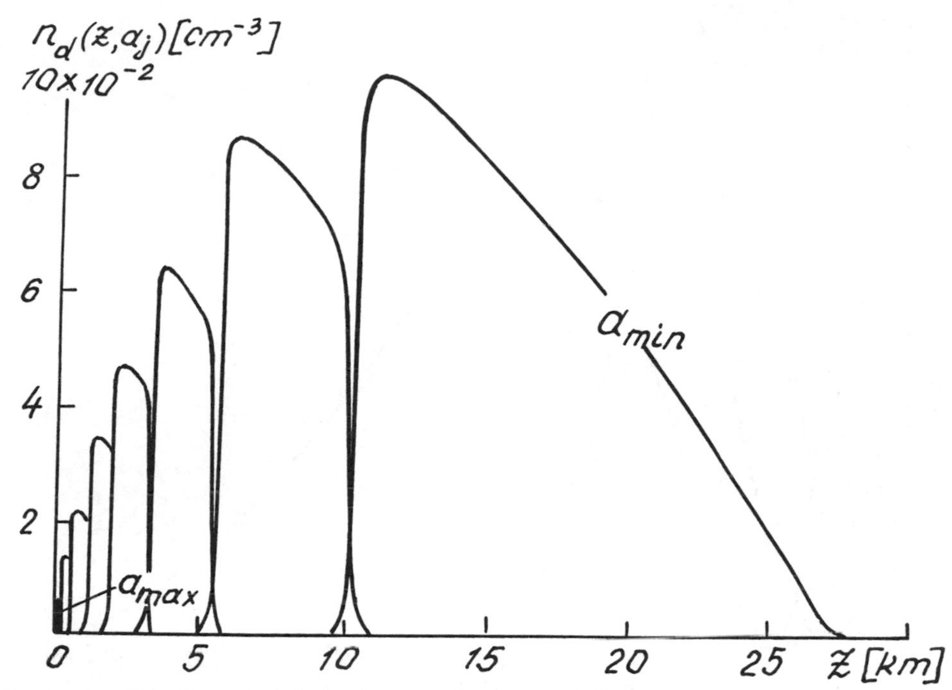


Figure 2: The Distribution of Dust Grains of Different Sizes between $a_{max} = 8.2\,\mu$ and $a_{min} = 0.13\,\mu$ in Saturn's Ring. The Largest Grains Are Closest to the Ring Plane at $z = 0$, while the Smallest Are the Farthest Away. The Temperature of the Electron-Proton Plasma is $T = 1\,\text{eV}$ and the Number Density $n_0 = 10^2\,\text{cm}^{-3}$. The Kinetic Temperature of Dust Grains is $T_d = 0.1\,\text{eV}$ and the Optical Depth $\tau = 10^{-3}$. The Ring is at $L = 5$ from the Planet (after Aslaksen and Havnes, 1992).

generally valid, however the balance of forces Equation (6) now should involve dynamical terms, *viz.*

$$\frac{\partial v_d}{\partial t} + v_d \frac{\partial v_d}{\partial z} = -\frac{Q}{m_d} \cdot \frac{\partial \Phi}{\partial z} - \Omega_K^2 z. \tag{17}$$

Introduction of the new unknown function (the grain velocity $v_d$) brings forth one more equation, namely, that of continuity

$$\frac{\partial n_d}{\partial t} + \frac{\partial}{\partial z}(n_d v_d) = 0. \tag{18}$$

Small variations $\Delta n_d(z,t)$ of the dust grain density about its equilibrium value can be described analytically by means of linearizing the equations

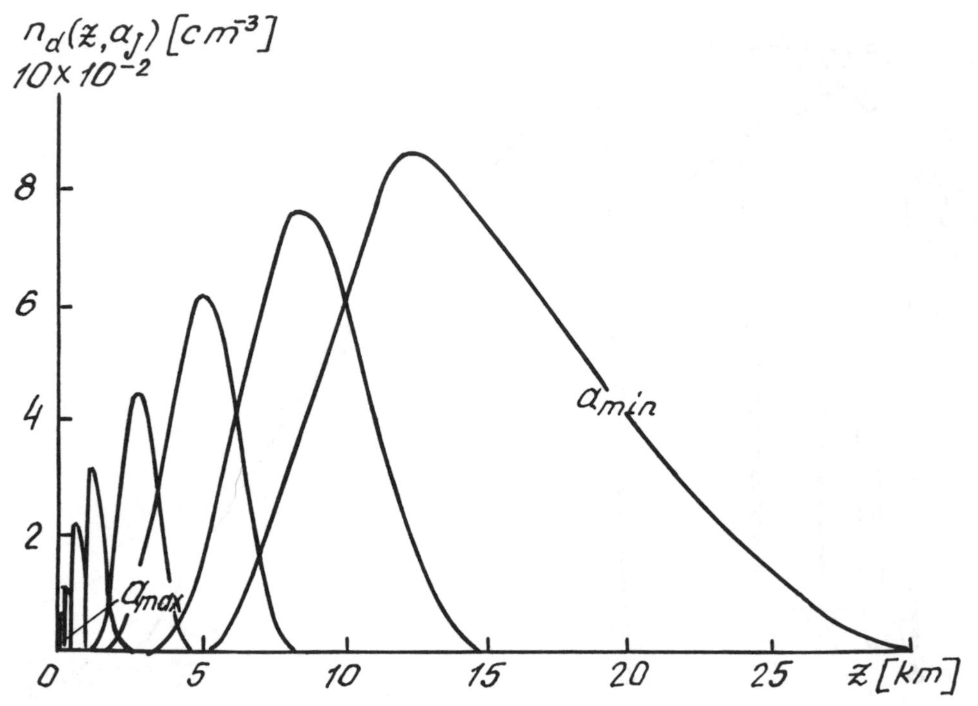


Figure 3: Same as in Figure 2 but with the Dust Kinetic Temperature $T_d = 1\,\text{eV}$.

and setting

$$\begin{aligned} n_d(z,t) &= n_{0,d}(z) + \Delta n_d(z,t) \\ U(z,t) &= U_0(z) + \frac{\partial U_0}{\partial n_{0,d}} \Delta n_d(z,t) \\ \Phi(z,t) &= \Phi_0(z) + \frac{\partial \Phi_0}{\partial n_{0,d}} \Delta n_d(z,t) \end{aligned} \tag{19}$$

and $v_d(z,t) = \Delta v_d(z,t)$. (The equilibrium values of the variables were found in the preceding Sections where they did not carry the subscript "0").

By combining Equations (17,18) and (19) we can obtain a linear equation for $\Delta n_d(z,t)$,

$$\frac{\partial^2 \Delta n_d}{\partial t^2} - \frac{a}{m_d} \cdot \frac{\partial}{\partial z} \left[ n_{0,d} \frac{\partial}{\partial z} \left( U_0 \frac{\partial \Phi_0}{\partial n_{0,d}} \Delta n_d \right) \right] = 0. \tag{20}$$

The coefficients being functions of $z$ alone, this is a separable equation.

Taking $\Delta n_d(z,t) = Z(z)\cos\omega t$ we arrive at

$$\frac{d}{dz}\left[n_{0,d}(z)\frac{d}{dz}\left(\frac{zZ}{\partial n_{0,d}/dz}\right)\right] - \frac{\omega^2}{\Omega_K^2}Z = 0, \tag{21}$$

which can be brought to the form of a Legendre equation,

$$\frac{d}{dx}\left[(1-x^2)\frac{dZ}{dx}\right] + s(s+1)Z = 0$$

with $s(s+1) = 2\omega_s^2/\Omega_K^2$, through the substitutions $x = z/z_m$ and $n_{0,d}(z) = n_{0,d}(0)[1-(z/z_m)^2]$ (*cf.* Equation (10)). We seek bounded solutions, symmetrical with respect to $\pm x$, hence the constant value $s$ should be an even integer number. $s = 0$ should be excluded as it corresponds to static "displacements" $\Delta n_d(z)$. Thus, the lowest oscillation frequency $\omega_s$ is given by $s = 2$,

$$\omega_2 = \sqrt{3}\Omega_K, \tag{22}$$

and the corresponding spatial dependence of the density perturbation is given by the Legendre polynomial $P_2(x) = (1/2)(3x^2-1)$. Finally, the fundamental mode is described by

$$\Delta n_d(z,t) = \Delta n_m \cos(\sqrt{3}\Omega_K t)\left[3\left(\frac{z}{z_m}\right)^2 - 1\right], \tag{23}$$

which equation represents a standing wave with a node at $z = z_m/\sqrt{3}$. The amplitude is determined by the driving force that may be related to variable external conditions (e.g., parameters of the ambient plasma). Periodical variations of the external force can result in a resonant enhancement of the vibration amplitude. A periodicity can arise from azimuthal nonuniformity of the plasma corotating with the planet at the frequency $\Omega_p$. Meanwhile, the orbital frequency of dust grains is close to $\Omega_K$. Should the difference of these frequencies and the oscillation frequency $\sqrt{3}\,\Omega_K$ make a rational or an integer ratio, the result would be a resonance at some orbit $R = LR_p$:

$$\mp\, n\,(\Omega_K(L) - \Omega_p) = m\sqrt{3}\,\Omega_K(L). \tag{24}$$

Here $n$ and $m$ are integer numbers; the upper and the lower signs relate to resonances within and without the synchronous orbit $L = L_{co} = R_{co}/R_p$. Using the explicit $\Omega_K = \Omega_K(L)$ dependence one can easily find the resonance values of $L$,

$$L_{n,m} = L_{co}\left(1 \pm \frac{m}{n\sqrt{3}}\right)^{3/2}. \tag{25}$$

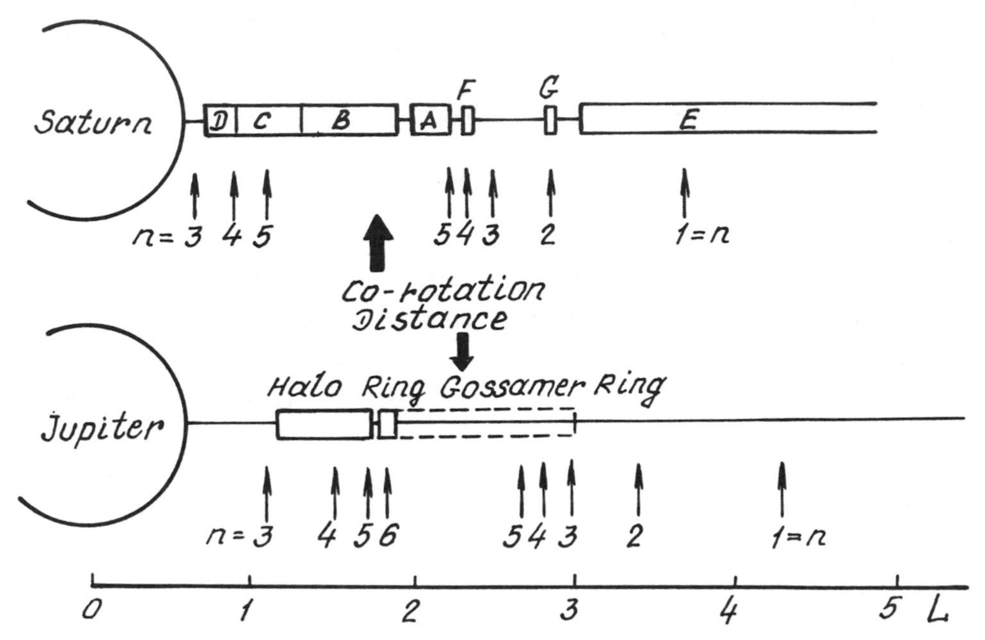


Figure 4: Radial Distances of the Inner and Outer Major Resonances ($m = 1$; $n = 1, 2, \ldots$) in Jupiter and Saturn and Major Ring Features (after Melandsø and Havnes, 1991).

These are shown in Figure 4 for Jupiter and Saturn. The principal resonances correspond to $m = 1$ and $n = 1, 2$. Note the three "internal" resonances ($n = 3, 4, 5$) to nearly span the region of Jupiter's halo ($1.3 \leq L \leq 1.7$), while the next resonance ($n = 6$) falls in the range $L \simeq 1.7 \div 1.8$. It seems therefore quite plausible that resonance effects have taken part in the ring formation.

Interesting coincidences can be noted for the rings of Saturn. The inner resonances $n = 3, 4$ and 5 correspond to the tenuous rings $D$ and $C$, while the outer resonances fall into the regions of ring $E$ ($n = 1$) and ring $G$ ($n = 2$). The $n = 3$ resonance at $L = 2.82$ is not associated with a dust structure, while the resonance $n = 4$ almost coincides with ring $F$ (Melandsø and Havnes, 1991; Hartquist *et al.*, 1992; Havnes, 1993).

As is well known, the intensity of a resonant response is determined not by the driving force alone but also by losses in the system. The losses due to dust grain collisions can be estimated through including friction in the equation of motion Equation (17). Meanwhile, a different, collisionless damping mechanism may prove even more efficient (Havnes *et al.*, 1992; Havnes, 1993). We have already discussed this in Chapter 3 in connection with the attenuation of low frequency waves in a dusty plasma. One and the same physical effect may be responsible for the attenuation of quasiperiodic density variations in a dust layer near a large body (*cf.* Subsection 1.6.3), gyrophase drift of particles (Subsection 2.3.3) and attenuation of low frequency waves (Subsection 3.2.4).

The physics is as follows. A negatively charged particle moving into a dust layer from its exterior is subjected to the braking force on the part of the screening electric field **E** (see Figure 10 of Chapter 1). If the layer is of moderate density and thickness, the particle will be able to fly through it without changing its electric charge. Then the energy lost by the particle when entering the layer can be estimated as $\Delta W_{in} \simeq Q_0\Phi$, where $Q_0$ is the particle charge outside the layer and $\Phi$ the electric potential in the layer. The same amount of energy will be given back to the particle when it leaves the dust layer. In other words, there is no energy exchange between the particle and the layer.

The matter does not change if the particle charge becomes dependent on its location $z$ in the layer. It is only essential that the particle acquired the charge $Q(z)$ matching the ambient plasma without delay. Indeed, the work done by the electrostatic field **E** at the entrant part of the particle trajectory from $z$ to $z = 0$ is

$$W_{in} = \int_z^0 Q(z)E(z)\,dz,$$

while the work done on the departing particle between 0 and $-z$ is

$$\begin{aligned} W_{out} &= \int_0^{-z} Q(z)E(z)\,dz = -\int_0^z Q(-z)E(-z)\,dz = \\ &= -\int_z^0 Q(z)E(z)\,dz = -W_{in}. \end{aligned}$$

Energy exchange between the particle and the medium is absent again, and hence the density vibrations are not attenuated.

Allowing for delays in "matching" the charge $Q(z)$ to the potential $\Phi(z)$ in the layer, we would see $Q(z)$ to depend on the particle history, in particular on the way it reached the level $z$. If it moved toward inner parts of the layer, then its charge would be somewhat greater than the matched value

$Q$ (as the charge outside the layer was greater in magnitude). Accordingly, the new value of the field work along the entrant trajectory,

$$W_{in} \simeq \int_z^0 [Q(z) + \delta_1 Q] E(z)\, dz,$$

is somewhat higher than the old one. Let the particle stay in the inner part of the layer sufficiently long to have ultimately acquired the equilibrium ("matched") charge $Q(z)$. Then its charge at the symmetrically located level $-z$ on the outgoing trajectory will be somewhat smaller in absolute magnitude than $Q(z)$, such that

$$W_{out} \simeq \int_0^{-z} [Q(z) - \delta_2 Q] E(z)\, dz.$$

The energy lost by the particle retarded at the entrance will not be compensated for by acceleration at the exit. There is a net loss of kinetic energy of the vibrating particle, i.e. the oscillation is damped.

This mechanism of particle deceleration is electrostatic in nature and apparently may be regarded as collisionless, despite the fact that variations of the grain charge $Q$ result from collisions with the plasma electrons and ions. Indeed, the grain is not involved in direct exchange of momentum with the plasma microparticles over times comparable with the oscillation period.

The qualitative considerations presented can be replaced by a more rigorous analysis through a slight modification of the basic equation set. The balance of currents Equation (2) is substituted by

$$\frac{dQ}{dt} = J_e + J_i$$

(the photocurrent has been omitted for simplicity). With account of Equations (1),(4) and (5), the explicit form of this equation becomes (Havnes *et al.*, 1992)

$$\begin{aligned} a\frac{dU}{dt} &= \pi a^2 e n_0 \left(\frac{8T}{\pi m_i}\right)^{1/2} \left[\exp\left(-\frac{e\Phi}{T}\right)\left(1 - \frac{eU}{T}\right) - \right. \\ &\left. - \left(\frac{m_i}{m_e}\right)^{1/2} \exp\left(\frac{e\Phi + eU}{T}\right)\right]. \end{aligned} \tag{26}$$

Linearized equations can be treated analytically to describe rings of moderate density, while numerical techniques permit lifting the limitations on

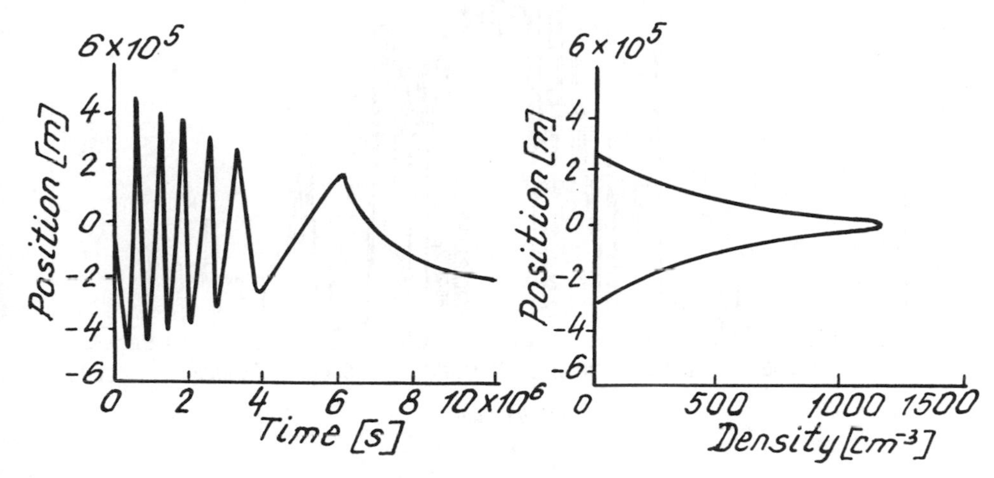


Figure 5: Position of a Probe Grain Injected at $t = 0$ from the Profile Edge at a Downward Velocity $V(t = 0) = -5 \times 10^2\,\mathrm{cm/s}$ (a) and the Dust Density Profile (b). The Ambient Plasma Density is $n_e = 1\,\mathrm{cm}^{-3}$, Electron Temperature $T = 10\,\mathrm{eV}$, and Dust Grain Radius $a = 1\,\mu$ (Havnes *et al.*, 1992).

particle concentration and other parameters. Numerical results concerning grain vibrations and ring thickness are shown in Figure 5. The effective ring thickness would have been much greater if the oscillations had not been attenuated.

### 5.2.4 Radial Profile of Grain Concentration in the Ring

A theoretical description of the radial distribution of grains in a planetary ring is much more complex than interpretation of the vertical profile. This has become evident when the data collected by the *Voyager* spacecraft demonstrated the queer radial structure peculiar to broad rings (Saturn's ring $B$ especially, Figure 6). The sharp variations of optical depth along the radial certainly cannot be ascribed to satellite-produced gravity variations

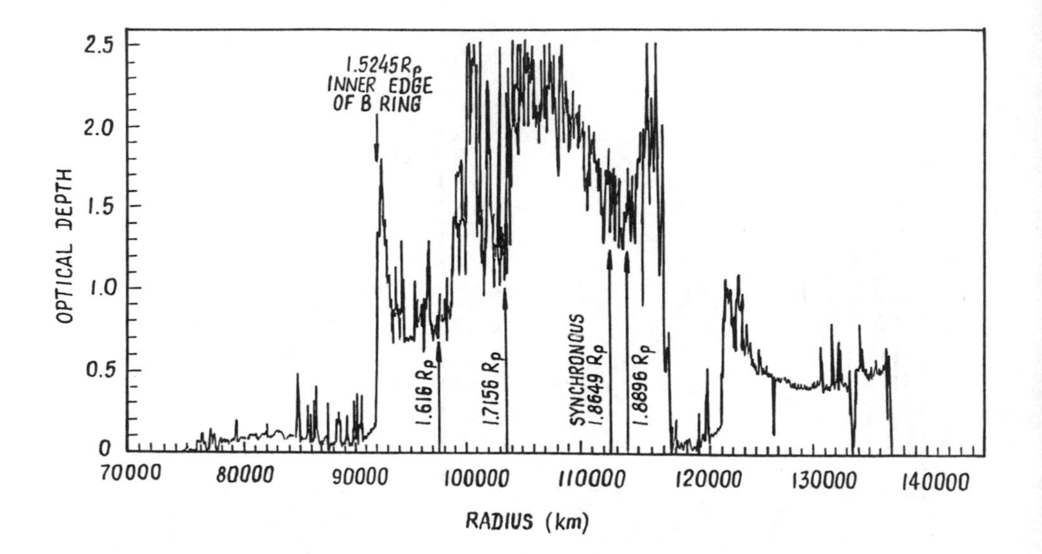


Figure 6: Optical Thicknesses of Saturn Rings $A$, $B$ and $C$ ($A$-Ring is between 122,000 and 137,000 km; $B$-Ring is between 92,000 and 118,000 km, and $C$-Ring from 75,000 to 92,000 km) (Northrop and Hill, 1983).

alone. No wonder a number of theories have appeared appealing somehow to electromagnetic effects in the dusty plasma of the rings.

First, it can be shown that variations of external conditions (such as temperature and density of the ambient plasma, intensity of the ultra-violet radiation, etc.) in an azimuthally bounded area can bring forth radial transport of material (Goertz and Morfill, 1983; Havnes, 1993). Indeed, dust grains in a ring drift relative the corotating plasma at the velocity

$$v_d \simeq R(\Omega_K - \Omega_p). \tag{27}$$

If external conditions were invariable, the charged grains would produce a d.c. ring current. Now let the plasma parameters be different over some portion of the ring. Accordingly, the grain charge will be different by some $\Delta Q$, bringing forth a variation $\Delta I$ of the linear density in the current passing through the entire depth of the ring,

$$\Delta I = \Delta Q n_{0,d} H_{eff} v_d, \tag{28}$$

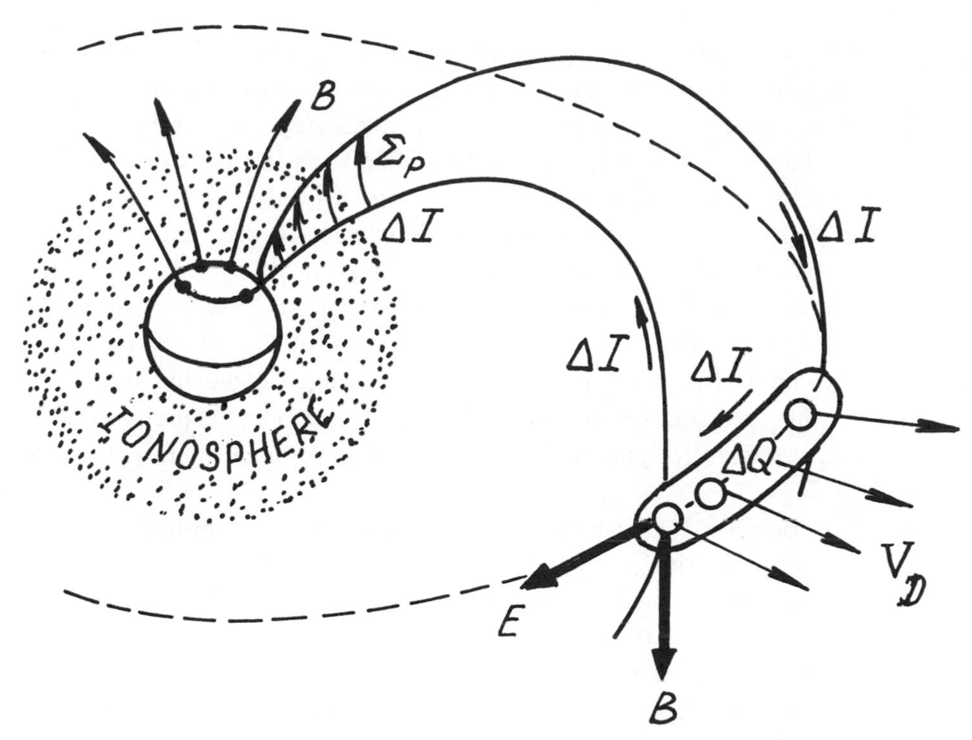


Figure 7: The Radial Drift of a Plasma Volume Owing to Variations in the Azimuthal Current over a Limited Part of the Ring.

where $H_{eff}$ is the effective ring thickness. This extra current existing over an azimuthally limited area will have to close through the ionosphere where it penetrates along the magnetic lines of force (Figure 7). The electric field acting in this circuit is oriented in the azimuthal direction. Its strength is

$$\Delta E_\phi = \Delta I/\Sigma_p, \tag{29}$$

where $\Sigma_p$ is the total Pedersen conductivity of the ionospheric part of the circuit. As a result, the charged particles of the ring are subjected to the action of crossed fields **B** and **E** bringing about a radial drift of the dusty plasma volume at the velocity

$$V_r = c\Delta E_\phi/B. \tag{30}$$

The direction of $\mathbf{V}_r$ is determined by the sign of $\Delta Q$ and orientation of the drift velocity $\mathbf{V}_d$, the latter depending in its turn on the orbit radius. The mass transport ultimately responsible for the radial structure in the ring may also result from peculiar trajectories of individual grains (*cf.* Chapter 2). The laws of gravitoelectrodynamics are not active until the grain leaves the ring plane, acquiring an extra charge and moving at a certain initial transverse velocity $V_{\perp}$ and orbital velocity $V_0(R)$. In Subsection 2.3.5 we discussed oscillating trajectories (see Figure 11 of Chapter 2) along which a particle that has left the ring plane at a radial distance $R$ can come back to an orbit of radius $R+\Delta R$. At this distance the dust grain will be reabsorbed by the ring, thus bringing there mass and the angular momentum corresponding to the take-off point. The change in the momentum of the target will result in its displacement along the radial, hence in a change of the ring structure.

Still another mechanism of the radial mass transfer, equally associated with gravitoelectrodynamic effects, is the gyrophase drift (see Subsection 2.3).

The gravitoelectrodynamic treatment of the motion of individual dust grains provides a description of only elementary acts of mass transfer. In order to analyze such integrated characteristics as the radial dependence of optical depth in the ring (or grain concentration), it is necessary to introduce the grain distribution function and formulate and solve the appropriate kinetic equation. A full-scale investigation of the kind was made by Shan and Goertz (1991) who suggested an explanation for the complex radial structure of Saturn's $B$-ring. They assumed that dust grains might leave the ring as a result of bombardment by meteoroids, and analyzed numerically the evolution of the initially uniform density distribution in the ring. The calculated optical depth profile $\tau(R)$ (see Figure 6) agreed well with the observations if the mass transfer was taken to last $(4 \div 8) \times 10^6$ years under meteoroid bombardment of constant intensity. This suggests that the currently observable system of Saturn rings might be much younger than the solar system ($\sim 4.5 \times 10^9$ years of age).

## 5.3 Electrostatic Waves in Planetary Rings

The low frequency electrostatic waves in the dusty plasma of planetary rings show some characteristic distinctions. First, there is the specific geometry. Strictly speaking, wave disturbances should be considered either for filamentary elliptic configurations representing narrow rings or for thin plasma layers in the case of broad rings. However, the standard simple models are rectilinear unbounded particle streams. Another effect to be taken into account throughout, except the synchronous orbit, is the drift of dust grains relative the plasma corotating with the planet. This can give rise to the plasma-beam instability if the drift velocity exceeds a threshold value determined by the Landau damping (Rosenberg, 1993; Melandsø, *et al.*, 1993).

We will analyze the effects controlled by the geometry of narrow and broad rings (*cf.* Subsection 4.1.2), and nonlinear waves in Saturn's broad ring, with allowance for the differential motion of dust grains (Bliokh and Yaroshenko, 1985). The effects of self-gravitation will be discussed for two narrow rings of Uranus.

### 5.3.1 A Multistream Model for Saturn Rings

The photographs transmitted by the space vehicle *Voyager-1* in November 1980 showed a distinct pattern of a few hundred (up to a thousand) concentric ringlets of which the system of Saturn's rings actually consists. Each ringlet represents a stream of electrically charged dust grains passing through the plasma connected to the planetary magnetosphere. The narrow ringlets are almost separated from each other by similarly narrow gaps. The entire structure is a current system along which wave disturbances can propagate. These can be associated with the azimuthal inhomogeneities (spokes) characteristic of the ring $B$. The life time $\tau_s$ of spokes varies between tens of minutes and a few hours. We will therefore concentrate on wave disturbances of frequencies $\omega \sim 10^{-3}\,\mathrm{s}^{-1}$ for which the set of thin ringlets (ring currents) may be considered as a stationary background. The many high-resolution images of the ring that are currently available suggest stability of the ringlets, clearly detectable in the region of spoke formation. This allows suggesting that the wavelike disturbances mainly involve longitudinal (i.e. azimuthal) particle displacements, while radial deviations of the moving grains never exceed the transverse size of the elemental rings. Besides, the scale size $\sim 10^6$ to $10^7\,\mathrm{m}$ of the disturbance is much smaller than the radius $R_{co} \simeq 1.1 \times 10^8\,\mathrm{m}$ of the synchronous orbit where spokes are mainly observed. Therefore, the curvature of the plasma beam can be neglected over a limited range of azimuthal angles. The frequencies of interest are quite high compared with the orbital frequency of the planet, $\Omega_p = 1.6 \times 10^{-4}\,\mathrm{s}^{-1}$ and collision frequencies $\nu \sim 10^{-5}\,\mathrm{s}^{-1}$ of the finer grains (Goertz and Morfill, 1983). Accordingly, the effects of the Coriolis force and collisions on the particle motion may be neglected. This brings us to the ring model as an ensemble of narrow rectilinear streams of diameter $d$, lying within the same plane $(xOy)$ at some space $l$ from one another (Figure 8). The solution obtainable within this model permits of two limiting transitions. If the separation $l$ between individual streams (ringlets) is large, then

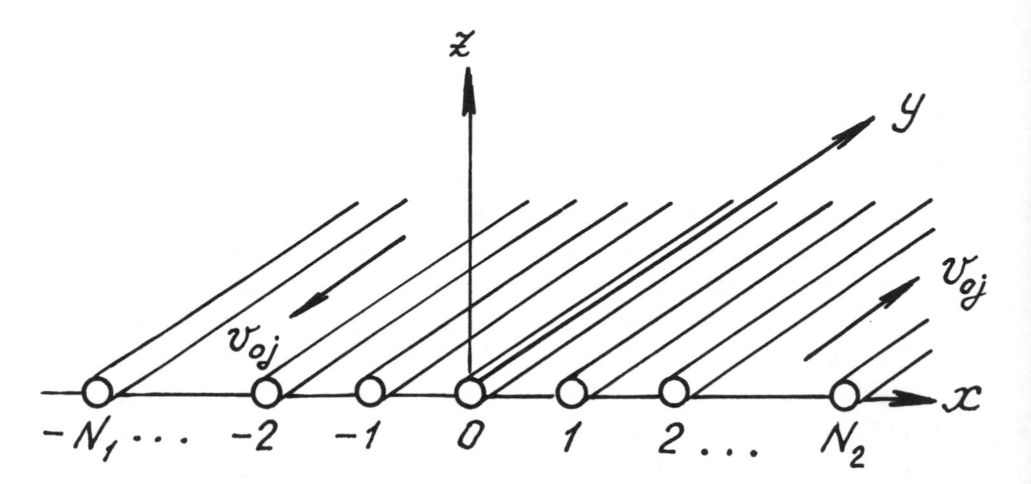


Figure 8: Model of a Broad Ring Consisting of Many Thin Ringlets.

the solution represents waves in an isolated thin ringlet. Contrary to this, if the ringlets are closely adjacent, then the structure represents a continuous dusty plasma layer.

The complete set of equations to describe the wave disturbances we are analyzing consists of linearized equations of motion and continuity for the particles of different species in the individual streams, plus Poisson's equation for the electric potential $\psi_E$,

$$\frac{\partial \mathbf{v}_{\alpha,j}}{\partial t} + (\mathbf{V}_{0,\alpha,j}\nabla)\mathbf{v}_{\alpha,j} = -\frac{Q_\alpha}{m_\alpha}\left[\nabla\psi_E + \frac{1}{c}\mathbf{v}_{\alpha,j} \times \mathbf{B}_{0,j}\right] - \frac{v_{T,\alpha}^2 \nabla n_{\alpha,j}}{n_{0,\alpha,j}}$$
$$\frac{\partial n_{\alpha,j}}{\partial t} + \nabla(n_{0,\alpha,j}\mathbf{v}_{\alpha,j} + n_{\alpha,j}\mathbf{V}_{0,\alpha,j}) = 0$$
$$\Delta\psi_E = -4\pi \sum_j \sum_\alpha Q_\alpha n_{\alpha,j}. \tag{31}$$

The subscript $\alpha$ indicates, as before, the particle species ($Q_\alpha$ denoting the charge and $m_\alpha$ the mass); $j$ is the ringlet number (with $j = 0$ corresponding to the synchronous orbit, $j < 0$ relating to the streams that are closer to the planet and $j > 0$ to those that are farther away). The subscript zero relates to unperturbed values of the velocity ($\mathbf{V}_{0,\alpha,j}$ is parallel to the $y$-axis), particle concentration and the magnetic field ($\mathbf{B}_0$ is along the axis $z$). Consider the wave of such polarization that $v_{\alpha,y} \neq 0$, assuming all the values to vary as $\exp[-i(\omega t - ky)]$. Then the two first Equations (31) yield

$$n_{\alpha,j} = k^2 Q_\alpha n_{0,\alpha,j} \psi_E m_\alpha^{-1} \left[ (\omega - kV_{0,\alpha,j})^2 - k^2 v_{T,\alpha}^2 - \omega_{B,\alpha,j}^2 \right]^{-1} . \qquad (32)$$

Re-writing the Poisson equation as

$$\Delta \psi_E = -4\pi \sum_\alpha Q_\alpha n_\alpha(\mathbf{r})$$

we can seek a solution of the form

$$\psi_E(\mathbf{r}) = \sum_\alpha Q_\alpha \int \frac{n_\alpha(\mathbf{r}')\, d\mathbf{r}'}{|\mathbf{r} - \mathbf{r}'|} .$$

If the elemental beams are assumed as infinitesimally thin (i.e. $r \gg d$), then the particle concentration becomes

$$n_\alpha(\mathbf{r}) = \sum_j \gamma_{\alpha,j} \delta(x - x_j) \delta(z) f_j(y),$$

where $\gamma_{\alpha,j}$ is the linear number density of particles of species $\alpha$ in the $j$-th beam, $\gamma_{\alpha,j} = n_{\alpha,j} \pi d^2/4$, and $f_j(y)$ a dimensionless function. The potential within the plane $z = 0$ is

$$\psi_E|_{z=0} = \sum_j \sum_\alpha Q_\alpha \gamma_{\alpha,j} \int_{-\infty}^{\infty} \frac{f_j(y')\, dy'}{\sqrt{(x - x_j)^2 + (y - y')^2}} .$$

For harmonically varying disturbances with $f_j(y) = e^{iky}$ this becomes

$$\psi_E|_{z=0} = 2\pi \sum_j \sum_\alpha Q_\alpha \gamma_{\alpha,j} K_0(k(x - x_j)) \Bigg|_{k|x - x_j| \gg 1} \longrightarrow$$

$$\longrightarrow \frac{\pi^2 d^2}{2} \sqrt{\frac{\pi}{2k}} \sum_j \sum_\alpha Q_\alpha n_{\alpha,j} |x - x_j|^{-1/2} \exp(-k|x - x_j|),$$

where $K_0$ is Macdonald's function of order zero. However, this equation does not permit calculating the potential at $x = x_j$. To remove the singularity resulting from the vanishingly small beam thickness, let us consider the field produced by a single beam of finite diameter $d$. Still neglecting the variations of $n_{\alpha,j}$ along $y$ inside the beam, the potential can be written as

$$\psi_{E_j}(0 \le x \le d/2) \simeq 2\pi \sum_{\alpha} Q_{\alpha} \int_0^r r n_{\alpha,j}\, dr = \pi \sum_{\alpha} Q_{\alpha} n_{\alpha,j} r^2.$$

This value should be averaged over the beam cross-section for a filamentary (one-dimensional) beam,

$$\overline{\psi_{E_j}}(0 \le x \le d/2) \simeq \frac{\pi d^2}{8} \sum_{\alpha} Q_{\alpha} n_{\alpha,j},$$

whence the total electric potential in the vicinity of the $j$-th beam is

$$\begin{aligned} \psi_E \;\simeq\; & \frac{\pi d^2}{8} \sum_{\alpha} Q_{\alpha} n_{\alpha,j} + \frac{\pi^2 d^2}{4} \sqrt{2\pi} \sum_{\alpha} \sum_{i \ne j} Q_{\alpha} n_{\alpha,i} \times \\ & \times \left(k|x_j - x_i|\right)^{-1/2} \exp\left(-k|x_j - x_i|\right). \end{aligned} \tag{33}$$

It can be conveniently expressed as

$$\psi_{E_j} = \sum_{\alpha} \sum_{i} S_{\alpha,ji} n_{\alpha,i}$$

where the matrix $S_{\alpha,ji}$ represents the contribution of $\alpha$-species particles from beam $i$ to the potential near beam $j$. The matrix elements drop off exponentially with the beam separation; therefore it seems allowable to assume, for simplifying the derivations,

$$S_{\alpha,ji} = \frac{\pi d^2}{8} Q_{\alpha} \begin{cases} 1 & \text{for } |j - i| \le N^* \\ 0 & \text{for } |j - i| > N^*, \end{cases} \tag{34}$$

where $N^* \sim (|k|l)^{-1}$ is the number of beams effectively involved in the interaction. As the disturbance wavelength $\lambda \sim k^{-1}$ is increased, the number of electrically interacting beams increases too. Note however that the estimate of $N^*$ has been obtained without allowance for the Debye screening of the beam field. Physically, this implies that all the plasma electrons have been absorbed by dust grains and the linear currents are separated by vacuum gaps. Otherwise the equation for $N^*$ should involve the Debye length $\lambda_D$ in lieu of the wavelength $\lambda$.

Substituting the $\psi_{E_j}$ of Equation (33) into Equation (32) we arrive at a set of homogeneous linear equations for $n_{\alpha,j}$,

$$n_{\alpha,j} - \frac{k^2 Q_{\alpha} n_{0,\alpha,j} \sum_i \sum_{\beta} S_{\beta,ji} n_{\beta,i}}{m_{\alpha}\left[(\omega - kV_{0,\alpha,j})^2 - k^2 v_{T,\alpha}^2 - \omega_{B,\alpha,j}^2\right]} = 0. \tag{35}$$

Equating the determinant of Equation (35) to zero yields the modal equation of azimuthal waves in the model multibeam ring.

The number of equations in the set is very high; therefore a straightforward analysis is hardly possible. Besides, such parameters as the number of beams in the ring, interbeam separation or number of different particle species (sizes) are quite arbitrary and should not participate in the final formulas. The ways of overcoming these difficulties will be discussed below, while now some estimates are to be made, concerning the coefficients in Equation (35). It is convenient to work in the reference system fixed to the planet magnetosphere. Then the unperturbed velocities of all the dust grains at $j = 0$ are zeros, while those for $j \neq 0$ are given by the equation

$$V_{0,\alpha,j} \simeq -\frac{1}{2} lj \left( \Omega_p + \frac{3}{2} \omega_{B,\alpha,j} \right) \simeq -\frac{1}{2} lj \Omega_p. \tag{36}$$

The second equality in Equation (36) corresponds to the estimate adopted in Subsection (2.3.1), namely $Q_\alpha / m_\alpha \sim 10\,\mathrm{C/kg}$ in the case of dust grains from the region of spokes. Accordingly, $\omega_{B,\alpha,j} \sim 10^{-5}\,\mathrm{s}^{-1}$ and $\Omega_p = 1.6 \times 10^{-4}\,\mathrm{s}^{-1}$, such that the inequality $\Omega_p \gg \omega_{B,\alpha,j}$ can be assumed to hold. (Apparently, this simplifying assumption is inapplicable to those cases where account of the scatter in grain velocities within a single narrow ring is essential, *cf.* Subsection 4.1.4).

With regard to ions, it is quite realistic to assume them corotating with the planetary magnetosphere, i.e. $V_{0,i,j} = 0$. The ion gyrofrequency is high ($\omega_{B,i,j} \sim 10^2\,\mathrm{s}^{-1}$) and the inequality it obeys is of opposite sense compared with the dust grains, $\Omega_p \ll \omega_{B,i,j}$. As for thermal velocities, $v_{T,\alpha}$ can be set equal to zero for dust grains and estimated as $(T/m_i)$ for ions with $T \simeq 10\,\mathrm{eV}$ (i.e. $v_{T,i} \sim 3 \times 10^4\,\mathrm{m/s}$).

### 5.3.2 Electrostatic Waves in Narrow Rings

First, let us assume the separation between individual streams so great as to allow neglecting their interaction. Then $S_{\alpha,ji} = 0$ for $j \neq i$ and the determinant of Equation (35) takes a diagonal form. The dispersion relation becomes a set of $N_1 + N_2$ independent equations, each representing waves in just one of the beams ($N_1$ is the total number of beams with $j < 0$ and $N_2$ with $j > 0$). All the equations are of the same form,

$$1 = \frac{k^2 d^2}{32} \sum_\alpha \omega_{p,\alpha,j}^2 \left[ (\omega - k V_{0,\alpha,j})^2 - k^2 v_{T,\alpha}^2 - \omega_{B,\alpha,j}^2 \right]^{-1}, \tag{37}$$

where $\omega_{p,\alpha,j} = (4\pi Q_\alpha^2 n_{0,\alpha,j}/m_\alpha)^{1/2}$ is the plasma frequency of $\alpha$-species particles in stream $j$. The factor $k^2d^2/32$ is characteristic of narrow beams ($kd \ll 1$) where the plasma frequency depends on the linear number density of particles, $\gamma_{0,\alpha} = \pi d^2 n_{0,\alpha}/4$, rather than their volume density (*cf.* Subsection 4.1.2).

For the low frequency disturbances we are discussing here ($\omega \sim 10^{-3}\,\mathrm{s}^{-1}$) the contribution of ions in the sum is small (formally because of the large value $\omega_{B,ij}^2 \sim 10^4\,\mathrm{s}^{-2}$ in the square brackets). Accordingly, the "ionic" terms can be neglected, while on the other hand there is no sense retaining $\omega_{B,d}^2 \sim 10^{-10}\,\mathrm{s}^{-2}$ in the brackets. Thus, the simpler version of Equation (37) is

$$1 = \frac{k^2d^2}{32}\sum_\alpha \frac{\omega_{p,\alpha,j}^2}{(\omega - kV_{0,\alpha,j})^2}, \tag{38}$$

which coincides with Equation (16) of Chapter 4, apart from the now specified numerical factor.

It might seem that with the simplifications we have assumed Equation (38) loses any dependence on the magnetic field. In fact this is not so, as the particle velocities $V_{0,\alpha,j}$ are field-dependent through the charge-to-mass ratio, which dependence may give rise to an instability of the narrow ring.

As was shown in Subsection 4.1.3, the structure can become unstable against excitation of electrostatic waves if the size distribution of dust grains is a discrete spectrum (see Equation (20) of Chapter 4). Some numerical estimates for Saturn's ring $F$ were also given in the Subsection.

In the case of a continuous power-law spectrum of sizes with $\beta = 2$, $a_{min} \leq a \leq a_{max}$, the instability does not develop and the dispersion relation Equation (38) involving many sorts of particles reduces to the form well known for a single-component beam with a "thermal" scatter of velocities, $\Delta V_j = (1/2)[V_{0,j}(a_{min}) - V_{0,j}(a_{max})]$, and a drift velocity $\bar{V}_{0,j} = (1/2)[V_{0,j}(a_{min}) + V_{0,j}(a_{max})]$,

$$1 = \frac{k^2d^2}{32} \cdot \frac{\omega_{p,j}^2(a_{min})}{(\omega + k\bar{V}_{0,j})^2 - k^2\Delta V_j^2} \qquad (a_{min} \ll a_{max}). \tag{39}$$

### 5.3.3 Wavelike Disturbances in Gravitation-Coupled Narrow - Rings

As was shown in Chapter 2, self-gravitational effects are of little importance for the cooperative phenomena involving micron and submicron-size particles with the electric charge of $10^2\,\mathrm{e} \div 10^4\,\mathrm{e}$. However, this conclusion resulted from comparison of the gravitational and electrostatic forces obeying Coulomb's laws of interaction with the same range dependence in both cases. The estimates are different if the space containing the probe particles is filled with a plasma. The electric fields then are greatly weakened owing to the Debye screening, while the forces of gravitation remain unchanged. As a result, self-gravitation can prove essential at distances $r \gg \lambda_D$, even if electric forces prevail at short distances.

The same is true for the coupled waves propagating along different orbits in a planetary ring. The estimate of Equation (34) of the number of effectively interacting particle streams remains valid in the presence of an ambient plasma if the coupling is related to gravitation disturbances. The corresponding dispersion relation can be obtained, following the general pattern of Section 4.2, provided the dispersion relation for electrostatic disturbances is known. By way of example, we will consider a simple system consisting of only two rings whose radii differ by $\Delta R < \lambda$. Further, let all the grains be of the same size. Then the coupling matrix has only four nonzero elements $S_{ij} = \pi d^2 Q/8$ $(i, j = 1, 2)$ and the equation set (35) reduces to

$$\begin{aligned} n_1 - \frac{\pi k^2 d^2 Q^2 n_{0,1}}{8m(\omega - kV_{0,1})^2}(n_1 + n_2) &= 0 \\ n_2 - \frac{\pi k^2 d^2 Q^2 n_{0,2}}{8m(\omega - kV_{0,2})^2}(n_1 + n_2) &= 0. \end{aligned} \tag{40}$$

A nonzero solution exists if the condition is met

$$1 = \frac{k^2 d^2}{32}\left[\frac{\omega_{p,1}^2}{(\omega - kV_{0,1})^2} + \frac{\omega_{p,2}^2}{(\omega - kV_{0,2})^2}\right]. \tag{41}$$

This can be recognized as the dispersion relation for waves in two *mutually penetrating* streams, despite the physical separation $\Delta R$ between the beams. If $\Delta R$ obeys the double inequality $\lambda > \Delta R > \lambda_D$, then the electrostatic coupling vanishes, while the gravitational remains. The modal equation for gravitational disturbances can be obtained from Equation (41) by substituting the Jeans frequencies $\omega_{G,j}^2$ instead of the plasma frequencies $\omega_{p,j}$ and changing the sign, *viz.*

$$1 = -\frac{k^2 d^2}{32}\left[\frac{\omega_{G,1}^2}{(\omega - kV_{0,1})^2} + \frac{\omega_{G,2}^2}{(\omega - kV_{0,2})^2}\right]. \tag{42}$$

An example of narrow rings largely separated from one another is given by the ring system of Uranus (Beaty *et al.*, 1981; Elliot and Nicholson, 1984). Of the nine rings about the planet, most are of the width $(2 \div 12) \times 10^3$ m, with the gaps between them $(5 \div 10) \times 10^5$ m. Let us try to apply Equation (42) to two adjacent rings, assuming $\omega_{G,1} = \omega_{G,2} = \omega_G$. In the frame of reference where $V_{0,1} = -V_{0,2} = V_0$, Equation (42) becomes a biquadratic equation with two pairs of complex-conjugate solutions. The approximate values can be written for $V_0 \gg \omega_G d/4\sqrt{2}$, which condition seems to be well satisfied (with $\omega_G$ corresponding to the surface density $\sim 25\,\mathrm{g/cm^2}$ and $V_0 \sim 10^2\,\mathrm{m/s}$, $d \sim 2 \times 10^3\,\mathrm{m}$). The roots with $\mathrm{Re}\,\omega > 0$ are

$$\omega_{1,2} \simeq kV_0 \pm i\frac{k\omega_G d}{4\sqrt{2}}. \tag{43}$$

They suggest a Jeans instability with the growth rate $\sim \omega_G$, drifting at the velocity $V_0$. To obtain a numerical estimate of the growth rate $\nu = \mathrm{Im}\,\omega$, we need to specify $k$. According to the above formulated conditions, the wavelength $\lambda = 2\pi/k$ should obey the double inequality $R \gg \lambda > \Delta R$. With $R \sim 4.5 \times 10^7\,\mathrm{m}$ and $\Delta R \sim 5 \times 10^5\,\mathrm{m}$, this implies $\lambda \sim 10^6\,\mathrm{m}$ and $k \sim 10^{-6}\,\mathrm{m}^{-1}$, or $\nu \sim 10^{-6}\,\mathrm{s}^{-1}$. Thus, the rings of Uranus may be characterized by small-scale inhomogeneities of mass density with the scale time of development about a month. We dare not make more positive conclusions in view of the limited applicability of the coarse model to the real structure. First, isolation of just two rings of the system of nine is rather conditional. To some extent, it can be justified for rings $\alpha$ and $\beta$ separated by $\Delta R \sim 10^3\,\mathrm{km}$ from one another but wider gaps of $2 \times 10^3\,\mathrm{km}$ and $3 \times 10^3\,\mathrm{km}$ from their nearest neighbors, however this difference of separations is not very significant. A more rigorous approach would be to consider the system of nine rings as a whole. That would become a necessity for analyzing the rings of Saturn with the great number of elemental ringlets.

### 5.3.4 Waves in a Broad Ring: Model of a Planar Plasma Layer

Analysis of the complex multicomponent system Equation (35) can be simplified by considering density perturbations averaged along the radial (i.e. performing summation of Equation (35) over $j$),

$$\sum_j n_{\alpha,j} = \frac{k^2 Q_\alpha}{m_\alpha} \sum_j \frac{n_{0,\alpha,j}}{(\omega - kV_{0,\alpha,j})^2} \sum_i \sum_\beta S_{\beta,ji} n_{\beta,i}. \tag{44}$$

(Once again, the terms with $v_{T,\alpha}$ and $\omega_{B,\alpha}$ in the denominators have been omitted and ions excluded from the sums over particle species). Making use of Equation (34), summation over $j$ and $i$ can be replaced by such over $j$ and $p = i - j$,

$$\sum_{i=-N_1}^{N_2} S_{\beta,ji} n_{\beta,i} \simeq \sum_{p=-N^*}^{N^*} S_{\beta,jp} n_{\beta,p+j} = \frac{\pi d^2 Q_\beta}{8} \sum_{p=-N^*}^{N^*} n_{\beta,p+j}. \tag{45}$$

Strictly speaking, the summation limits are not symmetrical for all the beams. The beams close to the ring edge may require special consideration, since $j - N_1$ or $N_2 - j$ can be less than $N^*$. The edge effects could be neglected if the number of effectively interacting beams were much less than their total number. Meanwhile, $N^*$is a large number, $N^* \simeq (|k|l)^{-1} \gg 1$, therefore the condition would be $N_1 + N_2 \gg N^* \gg 1$. In addition, it might be plausible to regard the ring uniform on the average, in which case the result of averaging $n_{\alpha,j}$ over a large number of elemental ringlets would be almost independent of the location where the averaging is done,

$$\bar{n}_\alpha = \frac{1}{N_1 + N_2} \sum_i n_{\alpha,i} \simeq \frac{1}{2N^*} \sum_{p=-N^*}^{N^*} n_{\alpha,j+p}. \tag{46}$$

By combining Equations (44) and (46) we arrive at the dispersion relation

$$1 = \frac{k^2 d^2}{16} \cdot \frac{N^*}{N_1 + N_2} \sum_\alpha \sum_j \frac{\omega_{p,\alpha,j}^2}{(\omega - kV_{0,\alpha,j})^2} \tag{47}$$

differing from Equation (38) only in the respect that the single sum has been replaced by a double. As a result, the probability of instability development increases (owing to the instability of slipping streams). The corresponding growth rates could be estimated if the distribution of particles of different sizes over individual beams were known. A rough estimate of the growth rate can be obtained by breaking the ring up into two segments, $j < 0$ and $j > 0$, with equal grain concentrations and certain mean velocities, $V_+ = -V_- = \bar{V}$. Then, apparently, $\nu_E \simeq k\bar{V}$, which is $\nu_E \sim 10^{-3}\,\mathrm{s}^{-1}$ for $k \sim \lambda^{-1} \sim 10^{-6}\,\mathrm{m}^{-1}$. The growth rate of a gravitational instability in a similar set of mutually slipping streams of electrically neutral particles is markedly different, namely $\nu_G \sim \omega_G/\sqrt{2} \sim 10^{-5}\,\mathrm{s}^{-1}$ (taking the $\omega_G$ for the average surface density of particles in ring $B$, $\sigma_0 \sim 10^2\,\mathrm{g \cdot cm^{-2}}$). In contrast to $\nu_G$, the electrostatic growth rate $\nu_E$ is a value of the same order as the reciprocal life time of "spokes", which implies some relation between electrostatic disturbances and spoke-like inhomogeneities of Saturn's ring.

The dispersion relation Equation (47) can be modified for a large number of closely lying beams and a continuous spectrum of particle sizes. The sums over $\alpha$ and $j$ are replaced by integrals over $a$ and $x = l_j$, such that

$$1 \simeq \frac{k^2 d^2 N^*}{16l(N_1 + N_2)} \int_{a_{min}}^{a_{max}} \int_{x_1}^{x_2} \frac{\tilde{\omega}_p^2(a, x)\, da\, dx}{[\omega - kV_0(a, x)]^2}, \tag{48}$$

with $x_1 = -N_1 l$ and $x_2 = N_2 l$. We have seen earlier that integration over grain sizes brings forth an effect similar to thermal scatter in velocities (*cf.* Equation (39)) that was discussed in detail in Chapter 4. Now we will concentrate on the integral over $x$ assuming, for the sake of simplicity, all the particles to be of the same size independent of $x$. Denoting $\overline{\omega_p^2} = \int_{a_{min}}^{a_{max}} \tilde{\omega}_p^2(a)\, da$ we bring Equation (48) to the integrable form

$$1 \simeq \frac{k^2 d^2 \overline{\omega_p^2} N^*}{16l(N_1 + N_2)} \int_{V_1}^{V_2} \frac{dx/dV_0}{(\omega - kV_0)^2} dV_0, \tag{49}$$

where $V_1 \simeq \Omega_p N_1 l/2$, $V_2 = -\Omega_p N_2 l/2$ and $dx/dV_0 \simeq -2/\Omega_p$ (see Equation (36)). With $N^* \simeq 1/(|k|l)$, the dispersion relation becomes

$$1 \simeq \frac{|k| d^2 \overline{\omega_p^2}}{16l[(\omega - k\bar{V})^2 - k^2 \Delta V^2]}, \tag{50}$$

$\bar{V} = (V_1 + V_2)/2$ and $\Delta V = (V_1 - V_2)/2$.

The equations analyzed heretofore have contained terms such as $-k^2\Delta V^2$ in the denominator on two occasions. First, they were the terms resulting from thermal motion of the plasma particles, i.e. $-k^2 v_T^2$. That was a well known situation in which dusty plasmas were not different from the conventional plasma media. Another case was when we discussed the velocity scatter $\Delta V$ produced by the different effect of the Lorentz force on particles with different charge-to-mass ratios. That scatter was peculiar to dusty plasmas with a continuous spectrum of grain sizes. Finally, the $\Delta V$ of Equation (50) is the difference of Kepler velocities at two edges of a broad planetary ring of width $\Delta R$. While effectively coupled through a wave disturbance might be just a fraction of the ringlets ($N^* \ll N_1 + N_2$), the scatter in grain velocities is still determined by the total width $\Delta R$ (or total number $N_1 + N_2$ of elemental ringlets). Obviously, the term $-k^2\Delta V^2$ in Equation (50) accounts for differential rotation in the broad ring.

A rigorous analysis should be based on a modal equation allowing for all three effects, i.e. thermal motion with the appropriate scatter in velocities; continuous spectrum of sizes, and differential rotation. However, the magnitudes of $\Delta V$ arising for the different reasons may prove greatly different, in which case the dispersion relation should include the highest $k\Delta V$. The region of spokes in Saturn's ring $B$ is characterized by $v_T \simeq 10^2$ cm/s (with $T = 10$ eV; $a \simeq 1\,\mu$ and $\rho = 1\,\mathrm{g}\cdot\mathrm{cm}^{-3}$); $\Delta V \simeq 10^3$ cm/s owing to the continuous spectrum of sizes (this is a maximum estimate for $a_{min} = 0.1\,\mu$ and $a_{max} \simeq 10\,\mu$); and $\Delta V \simeq 10^4$ cm/s owing to differential rotation in the range from $R_1 \simeq 9 \times 10^7$ m to $R_2 \simeq 1.2 \times 10^7$ m.

Representing Equation (50) as

$$1 \simeq \frac{|k| d\omega_p^{*2}}{16[(\omega - k\bar{V})^2 - k^2\Delta V]^2} \tag{51}$$

we can see its structure to coincide with Equation (15) of Chapter 4 that was derived for a plane plasma layer of thickness $d$ and particle concentration $n_0^* = n_0 d/l$. By introducing an effective surface density of the plasma particles, $\sigma_0 = d^2 n_0/8l$, the dispersion relation can be brought to the concise form

$$1 \simeq \frac{\omega_{p,\sigma}^2}{(\omega - k\bar{V})^2 - k^2\Delta V^2}, \tag{52}$$

with $\omega_{p,\sigma}^2 = 2\pi|k|\sigma_0 Q^2/m$, which is often quoted in the literature (e.g., Fetter, 1973). This equation can be derived in the model of a vanishingly thin plasma sheet with the surface density of particles $\sigma_0$. Let the particles move along the axis $y$ at an unperturbed velocity $\bar{V}$ and the thermal scatter in their velocities be $\Delta V$. The set of linearized basic equations includes

the equation of motion within the plane $z = 0$,

$$\frac{\partial v}{\partial t} + \bar{V}\frac{\partial v}{\partial y} + \frac{Q}{m}\cdot\frac{\partial \psi_E}{\partial y} + \frac{\Delta V^2}{\sigma_0}\cdot\frac{\partial \sigma}{\partial y} = 0, \tag{53}$$

and the continuity equation valid in the same plane,

$$\frac{\partial \sigma}{\partial t} + \frac{\partial}{\partial y}(\sigma_0 v + \sigma\bar{V}) = 0. \tag{54}$$

The Laplace equation off the plane $z = 0$ is

$$\frac{\partial^2 \psi_E}{\partial y^2} + \frac{\partial^2 \psi_E}{\partial z^2} = 0 \tag{55}$$

(the potential $\psi_E$ is independent of $x$ for plane waves traveling along the $y$-axis). We will seek the solutions of the form $\exp[-i(\omega t - k_y y) - k_z|z|]$, i.e. surface waves dying away from $z = 0$. Then Equation (55) yields $k_z = k_y$. The boundary condition at $z = 0$ is

$$\left.\frac{\partial \psi_E}{\partial z}\right|_{z=-0} - \left.\frac{\partial \psi_E}{\partial z}\right|_{z=+0} = 4\pi Q\sigma. \tag{56}$$

Combining this with Equations (53) and (54) we arrive at the dispersion relation of Equation (52).

### 5.3.5 Nonlinear Waves in a Plane Plasma Layer. Soliton Solutions for Envelopes

Analysis of the azimuthal disturbances of wavelength $\lambda \gg l$ in a thin broad ring can be greatly simplified by adopting the plasma layer model. We will employ it to investigate nonlinear waves of a certain kind that may have relation to spokes in Saturn's $B$-ring. Let us restore the nonlinear terms neglected in the equations of motion during linearization, namely $v\partial v/\partial y$ in Equation (53) and $\partial(\sigma v)/\partial y$ in Equation (54). Assuming $v$ and $\sigma$ to depend solely on $\zeta = \omega t - ky$, we arrive at the set of ordinary equations

$$(\omega - k\bar{V})v' - \frac{Q}{m}k\psi_E' - \frac{\Delta V^2}{\sigma_0}k\sigma' = kvv' \tag{57}$$

$$(\omega - k\bar{V})\sigma' - k\sigma_0 v' = k(\sigma v)', \tag{58}$$

where the prime denotes $d/d\zeta$. Note the potential $\psi_E$ to be a function of two variables, $\zeta$ and $z$. If the right-hand sides of Equations (57) and (58) are in some sense small, then $\psi_E(\zeta, z)$ can be sought as an expansion in $\cos n\zeta$ with the coefficients dependent on $z$. They are determined from the boundary condition Equation (56) and the demand that $\psi_E$ should remain finite at $|z| \to \infty$:

$$\begin{aligned}\psi_E(\zeta, t) &= \psi_0 \exp(-k|z|)\cos\zeta + \mu_1\psi_0^2\exp(-2k|z|)\cos 2\zeta + \\ &\quad \mu_2\psi_0^3\exp(-3k|z|)\cos 3\zeta + \cdots\end{aligned}$$

The constant factors $\mu_1$, $\mu_2 \ldots$ and the functions $v(\zeta)$ and $\sigma(\zeta)$ are found in the perturbation technique.

Thus we arrive at a nonlinear dispersion relation in which the $\omega = f(k)$ dependence is controlled by the amplitude $b$ of density perturbations,

$b = k\psi_0/2\pi Q$ (the first-order (linear) perturbation in the plasma density is $\sigma^{(1)} = k\psi_0 \cos\zeta/2\pi Q \equiv b\cos\zeta$). The nonlinear dispersion relation $f(\omega, k, b) = 0$ is (Bliokh *et al.*, 1986)

$$\omega = k\bar{V} + (\omega_{p,\sigma}^2 + k^2\Delta V^2)^{1/2}\left[1 + \frac{b^2}{8\sigma_0^2}\left(7 + 9\frac{k^2\Delta V^2}{\omega_{p,\sigma}^2}\right)\right]. \tag{59}$$

It can be simplified through estimating its terms for the plasma parameters characteristic of the region of spoke formation, namely $\bar{V} \sim 10^2$ m/s and $\Delta V \sim 10^4$ cm/s. Let the wavenumber $k$ be close to $k = 2\pi/\Lambda$, $\Lambda \sim 10^7$ m being the radial scale length of spokes. To estimate $\omega_{p,\sigma}^2$, let us assume a parabolic spectrum of grain sizes, with the parameters corresponding to literature data on the average surface density of material in Saturn's rings (i.e. 60 g/cm$^2$ — Smith *et al.*, 1982). Further, with the electric potential of dust grains corresponding to a few volts, or $|\psi| \sim 10^{-2}$ CGS, and $d^2/l \sim 10^2$ m we arrive at an order of magnitude estimate of $\omega_{p,\sigma}^2 \sim 10^{-5} \div 10^{-6}\,\mathrm{s}^{-2}$. This infers $\omega_{p.\sigma}^2 \gg k^2\bar{V}^2$, $k^2\Delta V^2$ and

$$\omega \simeq \omega_{p,\sigma}\left(1 + 7b^2/8\sigma_0^2\right), \tag{60}$$

and $b^2/\sigma_0^2$ becomes the principal small parameter.

Approximate solutions to the nonlinear Equations (57) and (58) may be sought in the form of a modulated plane wave with a slowly varying complex amplitude $B(y,t)$,

$$\sigma = \mathrm{Re}\,\{B(y,t)\exp\left[-i(\omega t - ky)\right]\}.$$

The amplitude satisfies a Schroedinger type equation that can be derived with the use of the nonlinear dispersion relation (see, e.g. Kadomtsev, 1976),

$$i\left(\frac{\partial B}{\partial t} + U_0\frac{\partial B}{\partial y}\right) + \frac{1}{2}U_0'\frac{\partial^2 B}{\partial y^2} - \kappa b^2 B = 0, \tag{61}$$

$$U_0 = \left.\frac{\partial\omega}{\partial k}\right|_{\substack{k=k_0\\b=0}} \simeq \frac{\omega_0}{2k_0}; \qquad U_0' = \left.\frac{\partial^2\omega}{\partial k^2}\right|_{\substack{k=k_0\\b=0}} \simeq -\frac{\omega_0}{4k_0^2}; \qquad \kappa = \left.\frac{\partial\omega}{\partial b^2}\right|_{b=0} \simeq \frac{7\omega_0}{8\sigma_0^2}.$$

$k_0$ and $\omega_0$ are the wavenumber and frequency, respectively, satisfying the linear dispersion relation Equation (52).

As can be seen, Equation (61) allows for a solution with the initial amplitude $\sqrt{2}\,b_0$ taking the form of a localized wave packet (soliton of the wave

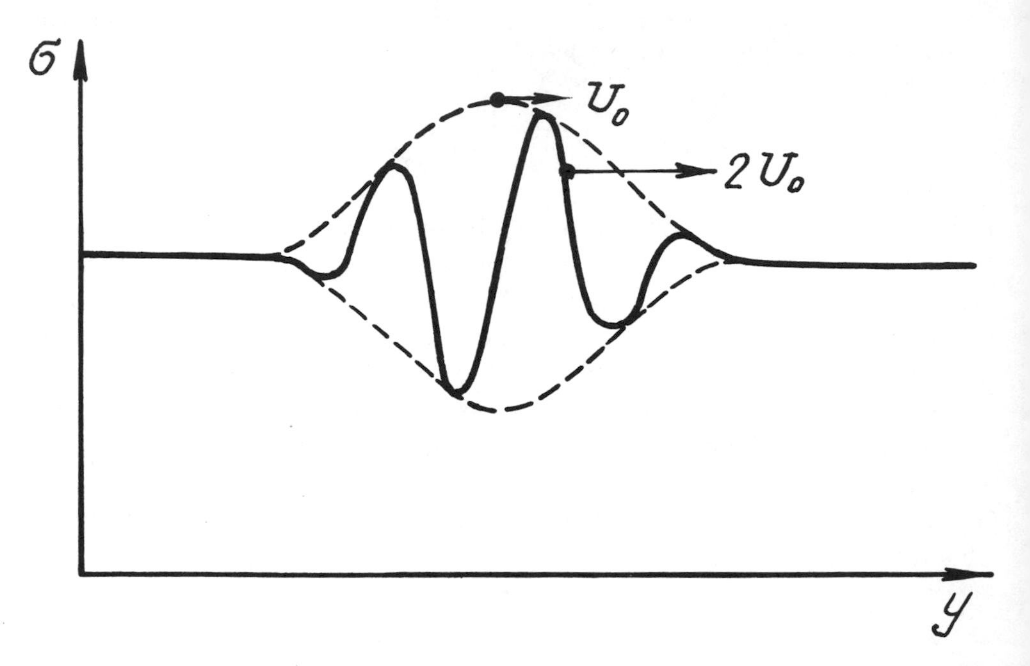


Figure 9: Variations of the Surface Density of Dust Grains along the Ring Resulting from Soliton Propagation.

envelope), $B = \exp(-i\kappa b_0^2 t)W(y - U_0 t)$, where

$$W(y - U_0 t) = \sqrt{2}\, b_0 \mathrm{sech}\left[\sqrt{-2\kappa b_0^2/U_0'}\,(y - U_0 t)\right]. \tag{62}$$

This packet travels at the group velocity $U_0$ giving rise to the density wave

$$\sigma = \cos\left[k_0 y - t(\omega_0 + \kappa b_0^2)\right] W(y - U_0 t). \tag{63}$$

Azimuthal variations of $\sigma$ are shown diagrammatically in Figure 9 for a fixed time moment $t$. The packet envelope propagates at the velocity $U_0$, while the internal filling at the velocity

$$\frac{\omega_0 + \kappa b_0^2}{k_0} \simeq \frac{\omega_0}{k_0} = 2U_0. \tag{64}$$

## 5.4 "Spokes" in Saturn's Ring

The quiet stream of new data on planetary rings that had been flowing for over 300 years turned into a violent torrent in late 1970s — early 1980s.

Radio messages from the space probes *Pioneer-11*, *Voyager-1* and *Voyager-2* brought much information that was quite unexpected, even on such objects as Saturn rings that seemed to have been studied in detail. We have already mentioned the disintegration of the old "classical" rings $A$, $B$ and $C$ into sets of narrow ringlets. But probably the most striking new result was the reported appearance and decay of some radially extended structures in the broad ring $B$. In photographs, they resembled spokes in a wheel and were so named (Collins *et al.*, 1980). Many of the newly discovered features of Saturn's rings have found a satisfactory interpretation within the classical celestial mechanics where gravitation is the only field of force. The spokes are an exception. The essential role of electromagnetic effects, along with gravitation, was soon understood.

### 5.4.1 Properties of Spokes. Electrostatic and Magnetostatic Hypotheses

The spokes are short-lived structures whose onset and disappearance can be followed by studying a series of photographs. The typical life time is tens of minutes to a few hours, which is much shorter than the planet's rotation period.

Spokes have been observed at radial distances ranging from 100,000 km to 177,500 km from the planet center. They appear particularly frequently at $(104 \div 110) \times 10^3$ km, where ring $B$ is characterized by its highest optical density. Geometrically, they are wedge-like, with the apex lying near the synchronous orbit $R_{co} = 112.5 \times 10^3$ km (Figure 10), and extend along the radial for 3,000 km to 12,000 km. The spokes are oriented almost radially, sometimes deviating toward the side where they rotate in a Keplerian way. In some spokes the Kepler velocity pertains to the leading edge of the wedge only, while the other edge remains radial as it corotates with the magnetosphere. The spoke activity is subject to periodic variations (its level is measured by the number of spokes observed through a given range of azimuthal angles), with the period coinciding with Saturn's rotation period of 10 hr 39.4 min to within the measurement accuracy (i.e. $\pm 22$ min). In fact, this coincidence is a sufficient indication of the relation existing between spokes and the planetary magnetic field rotating at the same angular velocity. Another evidence for such a relation is the remarkable correlation of the spoke activity with a certain magnetic longitude. Indeed, in the Saturn Longitude System (Kaiser, Desch and Lecacheux, 1981) the maximum activity is observed in the magnetic field quadrant centered around longitude

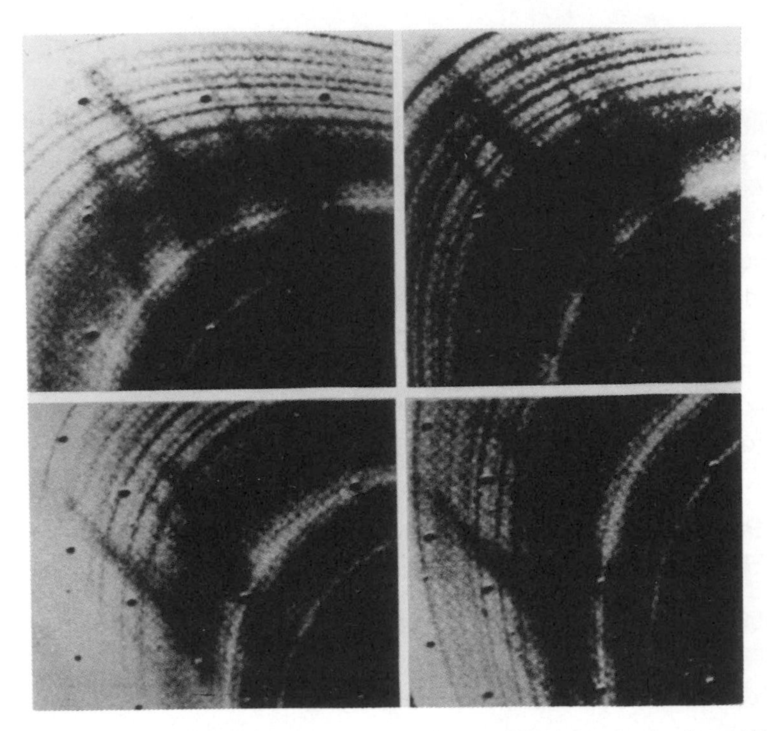

Figure 10: Spoke Evolution with Time (Smith *et al.*, (1982)).

115°. At the moments of this meridian passing the morning ansa, the spoke activity increases. Some other electromagnetic events are also associated with this magnetic quadrant, e.g. kilometer wavelength radiation (Kaiser *et al.*, 1981) or intense ultraviolet auroral radiation (Sandel *et al.*, 1981).

Spokes are observed throughout the light side of the ring, being especially distinct at the morning ansa. They are also seen quite clearly on that side of the ring which is illuminated by reflections from the planet rather than the Sun.

An important characteristic feature of the spokes is the variation of their contrast against the background dependent on the angle of illumination. They look $\sim 10\,\%$ darker than the background if observed in backscattered light but 10 to 15 % brighter in forward scattered light. A difference as considerable as this suggests that light is scattered by extremely small particles of micron and submicron size. Observations within the ring plane have shown spokes to occasionally go off the plane, rising up as high as 80 km above the level (Grün *et al.*, 1984). However, the small light-scattering particles may well be inside the ring (Cuzzi *et al.*, 1984). A detailed analysis of the observable properties of spokes and their evolution was given by Grün *et al.* (1983; 1992).

The challenge presented by spokes has immediately stimulated the appearance of various hypotheses to explain their origin and properties. Remarkably, all the hypotheses are based on the interaction of small dust grains with electromagnetic fields. One group of theories could be named static.

They relate the appearance of spokes with the preferred orientation of extended dust grains in a static electric or magnetic field. Polarized by a radial electric field $E_r$, dust grains may line up along the radial, giving rise to characteristic optical effects in the forward and backscattered light. The effects of dust grain polarization and radial orientation were discussed in Section 2.4. As was shown, grains of extended geometry placed in the unipolar induction field in the ring plane were not only oriented but could coalesce, making up chains of a few tens of micron size particles. However such chains can hardly be related to spokes. First, the unipolar induction field is very weak near the synchronous orbit where spokes arise, much below levels assumed in the estimates of the Section. Meanwhile, there may be radial electric fields of a different origin. An electric field may be generated by the currents in the ionosphere that are due to the eastward zonal winds of $\sim 100\,\mathrm{m/s}$ at latitudes between $\sim 46°$ N and $\sim 32°$ S. The circuit is closed along magnetic lines of force in the magnetosphere, producing a radial field of a few volts per kilometer in $B$-ring (Carbary *et al.*, 1982).

Still, the hypothesis of spoke origin from radial orientation of dust grains must be rejected. Had the structures really been observable only for a certain position angle of the Sun, owing to the high anisotropy of scatterers, the spoke activity should have demonstrated two maxima separated by one half of the ring circumference. Two maxima have not been observed. Besides, the wind-produced radial field $E_r$ seems too low for orientating micron size ice grains of rotational kinetic energy $\sim T \sim 10^{-2}\,\mathrm{eV}$ (Weinheimer and Few, 1982).

The magnetostatic hypothesis (Davydov, 1982) is in a sense contrary to the electrostatic. It assumes dust grains in the ring to be oriented in a roughly orderly way everywhere except the spokes where they are arranged randomly. The difference in scattering properties of the regular and random structures results in optical contrasts observable as the pattern of spokes. To be susceptible to magnetic orientation, dust grains should possess certain magnetic properties. Ferromagnetic and paramagnetic grains would be oriented along the magnetic lines of force, i.e. perpendicular to the ring plane, while diamagnetic (if such were present) would remain within the plane.

The grains should behave so almost everywhere, except areas of magnetic anomaly in the ring. The uniform orientation pattern would be violated there, producing the spoke effect. Within this theory, the "inclination" of spokes toward the synchronous orbit receives interpretation. The grains moving at the Kepler velocity are subjected for a longer time to the magnetic anomaly. The connection with a certain magnetic longitude is also understandable, as well as the preferable stretch from the synchronous orbit toward the planet and the higher magnetic field. However the questions that remain unanswered are crucial for the theory. They are, how can a magnetic anomaly now arise, now disappear and what should the dust grains consist of to effectively interact with the magnetic field. Therefore, the magnetostatic hypothesis is as unsatisfactory as the electrostatic.

### 5.4.2 Dynamic Theories of Spokes

The theories of this group explain the formation and evolution of spokes in terms of dynamics of charged dust grains in the electric and magnetic fields existing in and near $B$-ring. An important assumption, based on observational results, is that spokes consist of grains elevated above the ring plane. Of course, the dust component is present throughout the bulk of the ring, and special conditions are required for the steady-state density profile (see Subsections 5.2.1 and 5.2.2) to possess enhancements at $z \gg z_m$. These are the conditions providing for electrostatic levitation that were discussed in Subsection 1.6.3. To apply the analysis to $B$-ring, we shall need estimates of the grain charge for dust particles within the ring. It is quite understandable that micron and submicron-size dust grains cannot coexist in the same orbit with large fragments for a long time. The larger fragments move around the planet at the Kepler angular velocity $\Omega_K$, whereas the dust grains at $\Omega \neq \Omega_K$. As a result of collisions, smaller particles are sedimented on the surface of the larger body making up a regolith layer. The grain charge $Q$ is given in Subsection 1.3.3, i.e.

$$Q \simeq a^2 \psi / 4a_b, \tag{65}$$

where $a$ is the grain radius and $a_b$ that of the larger body characterized by the electric potential $\psi$. The average separation between particles in the ring is less than the Debye length $\lambda_D$, which allows considering the ring as a uniformly charged disk carrying the potential $\psi_R \sim -6\,V$ (Goertz, 1989). The potential $\Phi$ of the ambient plasma varies from $\psi_R$ at the ring to zero at $z \gg \lambda_D$. The electric field perpendicular to the ring plane reaches its maximum value near the ring surface (see Figure 10 of Chapter 1) and can be estimated as $E_z \simeq \psi_R/\lambda_D$. This field will be able to tear a negatively charged dust grain off the ring if $E_z > mQ/g_\perp$, where $g_\perp$ is the vertical component of gravity.

By formally applying Equation (65) to estimate the charge of a submicron grain we obtain $Q < e$, which implies that the probability of acquiring at least one excessive electron is very low for such grains. The few grains that do acquire the minimum excess charge $Q_{min} = e$ start moving upwards under the action of $E_z$. They are certainly not guaranteed to pass through the Debye layer without colliding with a positive ion whose number density in the region is higher than that of electrons. The probability of passing through the layer depends on the relation between two scale times, namely the re-charging (neutralization) time $t_c \sim (4\pi a^2 n_0 v_{T,i})^{-1}$ and the transit time $t_D \simeq \lambda_D(m/e\psi_R)^{1/2}$. The larger is the grain, the higher the ratio $t_D/t_c \sim a^{1/2}$, accordingly the lower its probability to leave the ring. On the other hand, the probability of acquiring an excess charge increases with the particle radius. As a result of competition of the two tendencies, the probability of levitation first grows with the grain radius and then decreases. The "optimum" size $a_0$ is given by the equation

$$a_0^{-7/2} = \frac{4\pi}{7} n_0 v_{T,i} \lambda_D (\pi\rho/3\psi_R)^{1/2}, \tag{66}$$

where $\rho$ is the mass density of the grain material (Goertz, 1984). The take-off probability $P(a)$ shows a sharp maximum at $a = a_0$. The optical thickness $\tau$ of the elevated dust layer is related to the optical depth $\tau_R = \int N(a)\pi a^2 \, da$ of dust in the ring plane itself ($N(a)$ is the concentration function of dust grains integrated along the ring depth) as

$$\tau_D \simeq \tau_R P(a_0) t_0/t_D, \tag{67}$$

where $t_0$ is the duration of the dust grains' stay above the ring. The order of magnitude estimate of $t_0$ is one half of the Kepler period, $t_0 \sim T_K/2 \simeq 5\,\text{hr}$. The optical thickness $\tau_D$ must be sufficiently large for the levitating dust to become visible. According to Equation (66), $\tau_D$ is dependent on the ambient plasma concentration $n_0$. If the plasma near the ring were of ionospheric origin, then its number density would be rather low, $n_0 \simeq 10^{-2}\,\text{cm}^{-3}$, and the corresponding estimates of other values would be $a_0 \simeq 10^{-4}\,\text{cm}$; $P(a_0) \simeq 3 \times 10^{-6}$; $t_D \simeq 100\,\text{s}$ and $\tau_D \sim \tau_R(5 \times 10^{-4})$. The elevated dust layer of an optical thickness as low as this would not be seen (Morfill and Goertz, 1983).

Should the concentration increase to $n_0 \sim 300\,\mathrm{cm}^{-3}$, all the parameters would alter significantly, *viz.* $a_0 \simeq 5 \times 10^{-5}\,\mathrm{cm}$; $P(a_0) \simeq 10^{-4}$; $t_D \simeq 0.1\,\mathrm{s}$ and $\tau_D \simeq \tau_R \cdot 10^{-3} t_0$. The scale time $t_0$ then should not be estimated in terms of the Kepler half-period $T_K/2$ but rather in terms of the life time $\Delta t$ of an anomalously dense plasma layer above some portion of the ring (of course, if $\Delta t < T_K/2$). To estimate $t_0 \sim \Delta t$, we are in need of an assumption on the reason for the increased $n_0$. A meteorite striking the ring could produce an extremely dense plasma cloud, $n_0 \sim 10^{15}\,\mathrm{cm}^{-3}$, expanding quickly in volume (Goertz and Morfill, 1983). The neutral gas released by the impact also turns to plasma owing to photoionization. The plasma particles first move at the velocities they gained in the collision and then get involved in corotation with the planetary magnetic field. By that time, the plasma concentration drops down to $n_0 \sim 10^2\,\mathrm{cm}^{-3}$. The joint action of the azimuthal current and the electric field $E_\phi$ brings forth radial drift of the particles (cf. Subsection 5.2.4 and Figure 7).

The plasma cloud moving above the ring pulls dust grains out of the bulk by the increased electric field $E_z$ beneath. Thus, the plasma column leaves behind a plume of dust that can be observed as a spoke. The whole structure continues moving around the planet. Since $\Omega_K > \Omega_p$ inside the synchronous orbit, the leading edge of the spoke moves at the Kepler velocity. The smallest grains almost corotate with the magnetic field, falling behind the Kepler motion. Accordingly, the rear edge of the wedge remains nearly radial.

Estimated values of the drift velocity $V_R$ for the radial distances where spokes are observed make a few tens of kilometers per second (Goertz, 1989). Taking the size $\Delta R$ of the plasma cloud to be roughly equal to the observed spoke width, $\sim 1,000\,\mathrm{km}$, we obtain for the cloud flight time over a given point on the ring $\Delta t \sim \Delta R/V_R \sim 100\,\mathrm{s}$. Then the optical thickness of the levitating dust is $\tau_D \sim 0.1\tau_R$. Assuming further that the smaller dust particles in the ring completely covered the underlying larger grains, we arrive at the estimate $\tau_R \sim 1$. The resultant estimate $\tau_D \sim 0.1$ is in agreement with the observational data. The estimated "optimum" size of levitating particles, $a_0 \sim 0.5\,\mu$, provides an equally good explanation to the optical contrasts shown by spokes and the background in the forward and backscattered light. The number of spokes observable at any given moment can be calculated from a known rate of meteorite bombardments. Taking reasonable figures for the latter, Morfill *et al.* (1983) obtained the number of spokes close to 10, which is also in agreement with the observations.

The meteorite hypothesis of spoke formation has been presented here rather sketchily. In fact, it provides reasonable explanations to yet other features shown by spokes (Goertz and Morfill, 1983; Morfill and Goertz, 1983), in particular radial limits to the active region. The minimal radius $R_{min} \simeq 100,000$ km practically coincides with the critical value $R_c \simeq 1.63R_p \simeq 98,000$ km that was defined in Subsection 2.3.5. At $R < R_c$, Saturn's gravity tends to push the plasma cloud away from the ring, so that spokes cannot arise. The outer boundary of the region of spoke activity is at the edge of ring $B$, or near the Cassini division. The content of dust grains is very low here; the plasma no longer interacts with the ring and stops its radial motion away from the synchronous orbit. The meteorite theory faces some difficulties in explaining the longitudinal nonuniformity of the spoke activity. A possible reason for the preference of some magnetic longitudes might be the tilt of the planet's magnetic dipole toward $\sim 300°$ in the Saturn longitude system. However, the tilt angle is very small, 0.7 to 0.8°.

The plasma columns of enhanced density that might provoke formation of spokes shall not necessarily result from meteorite bombardment. They could be produced as well by lightning discharges. Besides, the high electric field $E_z$ can also have an alternative source, namely streams of energetic electrons arriving to the ring from the ionosphere along magnetic lines of force (Hill and Mendis, 1981; 1982).

Let us consider the latter hypothesis in some detail. It suggests the existence of electrostatic double layers in the upper ionosphere of Saturn. The double layers (responsible for electron acceleration) concentrate about certain longitudes and cover the range of latitudes where the magnetic field lines come that have crossed the region of spokes (Mendis and Rosenberg, 1994). Similar double layers are known in the terrestrial magnetosphere.

As soon as a dust grain has left the ring, it follows the law of motion given by Equation (29) of Chapter 2. It is written here without external forces $\mathbf{F}(\mathbf{r})$, however with account of variations in the grain charge with time,

$$\ddot{\mathbf{r}} = \frac{Q(t)}{mc}\left[-(\boldsymbol{\Omega}_p \times \mathbf{r}) + \dot{\mathbf{r}}\right] \times \mathbf{B}(\mathbf{r}) - \frac{GM_p}{r^3}\mathbf{r}. \tag{68}$$

The magnitude of $Q(t)$ is the highest at the "take-off" moment when the grain leaves the ring, reducing afterwards to the level conditioned by the ambient plasma. For small-sized grains, the scale time of charge variation is quite long (about 1 hr for $a \sim 1\,\mu$).

Numerical results concerning the motion of dust grains in Saturn's magnetosphere are shown in Figure 11. Equation (68) was solved numerically in a Cartesian coordinate system with the $z$-axis parallel to $\boldsymbol{\Omega}_p$. The figure

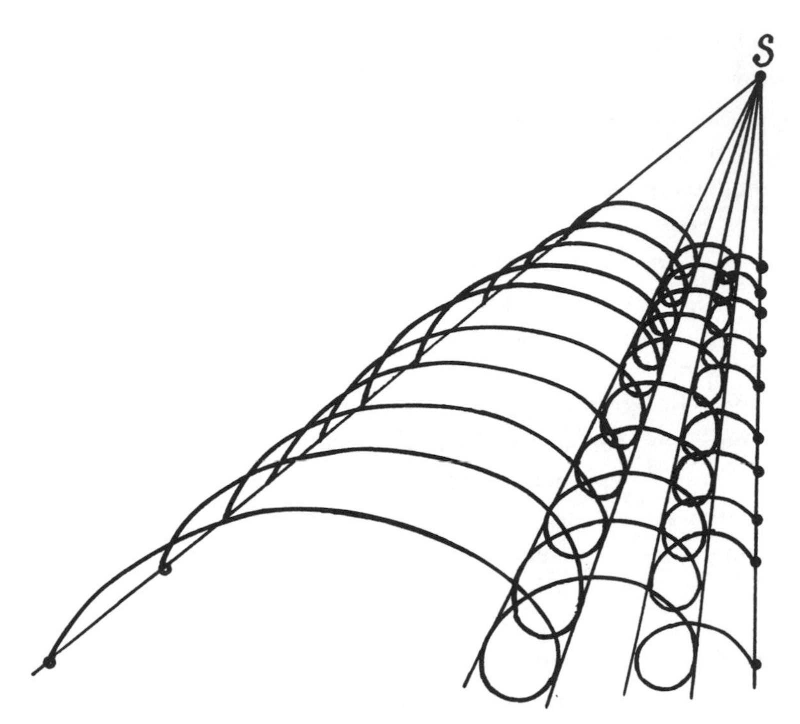


Figure 11: Schematic of Ten Different Trajectories of Dust Grains Departing from Different Points along the Radial. The Vertical Radius Represents the Traveling (Corotational) Spoke Edge and Point $S$ its Intersection with the Keplerian Edge. The Other Lines Represent "Envelopes" of the Synchronous Curves.

represents trajectories of dust particles of the same size, leaving the ring at the initial velocity $V_z \simeq 100\,\mathrm{m/s}$ from points of the same azimuth but spaced along the radius (Hill and Mendis, 1982). The vertical line in the Figure corresponds to synchronous rotation of the grains and the oblique left edge is close to the Kepler motion. The characteristic wedge-like geometry can be clearly seen. At the same time, the spoke structure resembles a lady's fan with a few "ribs". This analogy can be developed further. The spoke evolves in time in such a way that the first to appear is the radial, short-lived edge — the fan is closed. As the tilted Keplerian edge becomes gradually shaped, the fan opens. Figure 11 indicates an important property of spokes (equally efficient within the meteorite hypothesis), namely that the speed of spoke transport cannot be identified with the velocity of individual dust grains. It is rather plausible to speak of the *phase velocity* of some wave.

Each rib of the fan structure corresponds to the wave front of an ensemble of moving particles.

### 5.4.3 Wave Theóries

Apart from the enormous difference of sizes, the Saturn ring and spiral galaxies have very much in common. They are kept in equilibrium by the balance of gravitation and centrifugal forces and consist of a great many discrete objects moving along mostly circular trajectories. The important distinction is the nature of the gravitational forces involved. In the case of galaxies, they result from self-gravitation of all the objects, whereas in the Saturn ring self-gravitation equals a small fraction of the planetary gravitation field.

The formation and structure of spiral galaxies are discussed in a vast astrophysical literature. The major ideas of these works have found their reflection in theories of planetary rings. Wavelike disturbances in the large ensemble of particles such as the planetary ring owe to long-range interactions which are gravitation, electrostatic repulsion or attraction of charged particles and the magnetic forces between currents. Representing the ring in theory as a thin plane disk we consider the wave disturbances as spiral surface waves whose Fourier components can be written as

$$\Phi \exp\left[i(k_x x + k_y y - \omega t) - k_z|z|\right]. \tag{69}$$

Here $x$, $y$ and $z$ are the local Cartesian coordinates introduced at some point $R$, $\theta$, $z$ of the cylindrical system in the following way. The coincident $z$-axes are along the axis of the disk rotation; $x$ is along the radius $R$ and $y$ along the azimuthal direction $\vartheta$. In the cylindrical system, the wavenumber $k_y$ becomes $m/R$ (with an integer $m$). In the case of electrostatic waves, $\Phi$ is the electric potential, while for magnetohydrodynamic (MHD) waves it is the magnetic potential. The Laplace equation in a vacuum yields $k_z^2 = k_x^2 + k_y^2 \equiv k^2$. Since $\exp(-k_z|z|)$ in Equation (69) should decrease at larger $|z|$, $k_z$ is $k_z > 0$.

We will first consider the electrostatic waves propagating along the ring that are characterized by periodic variations in the number density of charged dust grains. Since $k_x = 0$ and $k_y = k$, the results of Subsection 5.3.4 are directly applicable. Shown in Figure 12 is the dispersion curve $\omega = \omega(k)$ plotted according to Equation (52) for the parameters of ring $B$. The curve passes through the area of the $(\omega, k)$ plane that can be associated with spokes if these are treated as a wave with $\omega \sim \tau_s^{-1} \sim 10^{-3}\,\mathrm{s}$ and $k \sim \delta_s^{-1} \sim 10^{-6}\,\mathrm{m}^{-1}$ (where $\tau_s$ is the spoke life time and $\delta_s$ the spoke width). This could be a

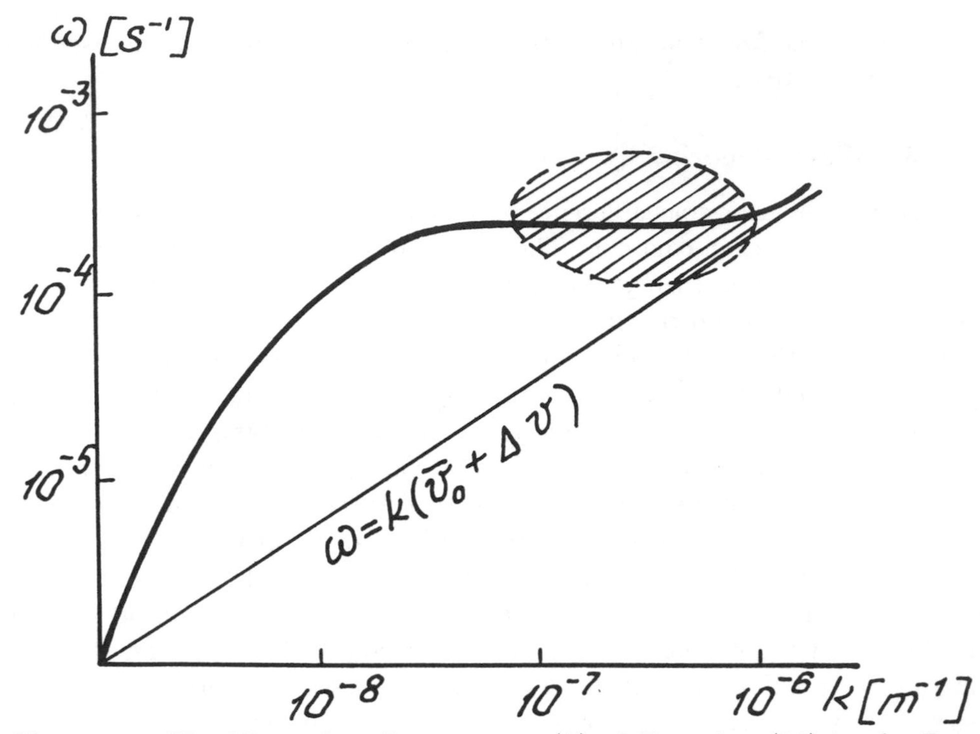


Figure 12: The Dispersion Curve $\omega = \omega(k)$ of Equation (52) in the Positive $\omega, k$ Domain. The Hatched Area Corresponds to Values of Plasma Parameters Characteristic of Spokes.

mere coincidence as the estimates of $\omega$ and $k$ are not sufficiently grounded. Yet, we can try discussing the arguments for and against the wave hypothesis of spoke origin. First, it is free of the assumption of grain levitation above the ring, since azimuthal condensations of dust within the ring would show exactly the same optical contrasts, both in the forward and backscattered light. Second, small grains can coexist with larger fragments for any long period only in the vicinity of the synchronous orbit, where $\Omega_K \simeq \Omega_p$, which explains the corresponding tendency shown by spokes. Meanwhile, Equation (52) was derived through averaging the disturbances along the radius (cf. the integral over $x$ in Equation (48)), which rules out the possibility of explaining the wedge-like geometry of individual spokes. And probably the greatest difficulty of the hypothesis is the uncertainty about quasiperiodicity in the azimuthal arrangement of spokes. A positive conclusion could be made after a statistical treatment of photographic images, aimed at establishing the spatial (azimuthal) spectrum $S(k)$ of a great set of spokes. A maximum in the spectrum near some $k = k_m \neq 0$ would be a strong argument in favor of the wave theory, at the same time allowing an objective estimate of the probable wavenumber $k$. Certainly, many questions would remain even after a positive outcome of the statistical test. One concerns the specific mechanism of density wave excitation. In principle, a stream of charged dust grains moving at the Kepler velocity can become unstable as the velocity of corotating electrons and ions is different (cf. Subsection 4.1.5). However, taking the numerical parameters of ring $B$, we have to conclude that the drift velocity is below the threshold magnitude, and hence the dust grain beam should remain stable (Goertz, 1989). Quite possibly the density wave is a manifestation of the slipping stream instability due to the radial gradient in the azimuthal velocity of dust grains, but this hypothesis has not been analyzed yet. The crude model in which $V_\vartheta(R)$ is discontinuous at the synchronous orbit does predict an instability if the discontinuity is of sufficient magnitude, however it is hardly applicable to the actual gradual variations of $V_\vartheta$. Anyway, the modal Equation (52) has only stable solutions. The part of a triggering mechanism for the excitation of density waves can be played by the abrupt variations of the charge density at the light-shadow boundary (see Section 4.3). This seems an attractive hypothesis as it relates quite naturally the spoke activity with the morning ansa.

A visual analysis of the spokes photographed in Figure 10 suggests treating the structure as a wave packet rather than an infinitely extending wave. Once again, more positive conclusions should be based on the results of a statistical analysis of many images. Yet, if the hypothesis were accepted, then the grouping of spokes could be interpreted within the nonlinear theory where the wave packet is regarded as a soliton of the wave envelope (Bliokh *et al.*, 1986; 1994; cf. Subsection 5.3.5). This idea seems quite credible when the photo of Figure 10 is confronted with the graph of Figure 9 and the wave crests of the internal soliton filling, where the dust concentration is the highest, are associated with spokes. Spokes travel along the ring at a velocity of the order of $v_{ph} = \omega/k \sim 10^3\,\mathrm{m/s}$. The value is much lower than the Kepler velocity near the synchronous orbit ($V_K \sim 2 \times 10^4\,\mathrm{m/s}$), therefore the wave drift relative the dust beam is hardly noticeable and the spokes appear "frozen" in ring $B$.

Now consider yet another wave theory where spokes are regarded as dust grain condensations produced by a magnetosonic wave (Tagger *et al.*, 1991). In contrast to galaxies, Saturn rings are characterized by a high electric conductivity, specifically of such components as the light-weight charged dust grains and the plasma proper. If the dust component is strongly coupled to the plasma, then the disturbances arising in such a medium can be analyzed within a single-liquid magnetohydrodynamic model. The MHD forces between electric currents are as long-range as the electric or gravitational, and in some cases can prevail in a structure like the conducting disk (planetary ring). MHD-type disturbances may be produced by inhomogeneities in the planetary magnetic field. The azimuthal phase velocity of such $B$-driven waves is close to $\Omega_p$, which suggests a connection with spokes. While in a "cold" conventional plasma the dispersion law of magnetosonic waves is

$$\omega^2 = k^2 v_{Ao}^2, \tag{70}$$

with $v_{Ao} = B_0/(4\pi\rho)^{1/2}$ being the Alfvén velocity, in a differentially rotating disk it becomes (Tagger *et al.*, 1991)

$$[\omega - m\Omega(R)]^2 - \kappa^2 = \frac{k^2}{|k|} B_0^2/2\pi\Sigma, \tag{71}$$

where $\Sigma$ is the surface mass density of the plasma and $m$ an integer number. Equation (71) allows for the Doppler frequency shift $m\Omega(R)$ owing to the motion along the circular orbit, and the epicyclic rotation at the frequency $\kappa$ within the ring plane. The term $[\omega - m\Omega(R)]^2$ is very small near the synchronous orbit, which implies the existence of a "stop band" $\Delta R$ where $k^2 < 0$, and hence the wave cannot propagate. The boundaries of this forbidden zone are determined by the so-called horizontal, or Lindblad resonances,

$$\omega - m\Omega(R_L) = \pm\kappa(R_L).$$

The solutions $R_{L,1} < R_{L,2}$ of this equation lie on opposite sides of the synchronous orbit.

Suppose a disturbance has appeared at some orbit $R_0 < R_{co}$. Waves of the type of Equation (71) will propagate from that orbit both toward the smaller and greater $R$. Starting from $R = R_{L,1}$, the wave traveling toward the synchronous radius is partially reflected but partially it penetrates into the forbidden zone where it is exponentially damped (see Figure 13). Along with the spiral wave, the equation has a different solution representing a radial bar-like feature. It shows a power-law behavior in the forbidden zone and can be tentatively identified with spokes. However the corresponding disturbances in the dust concentration are very low, and anyway they cannot account directly for the observed optical contrasts of the spoke structure.

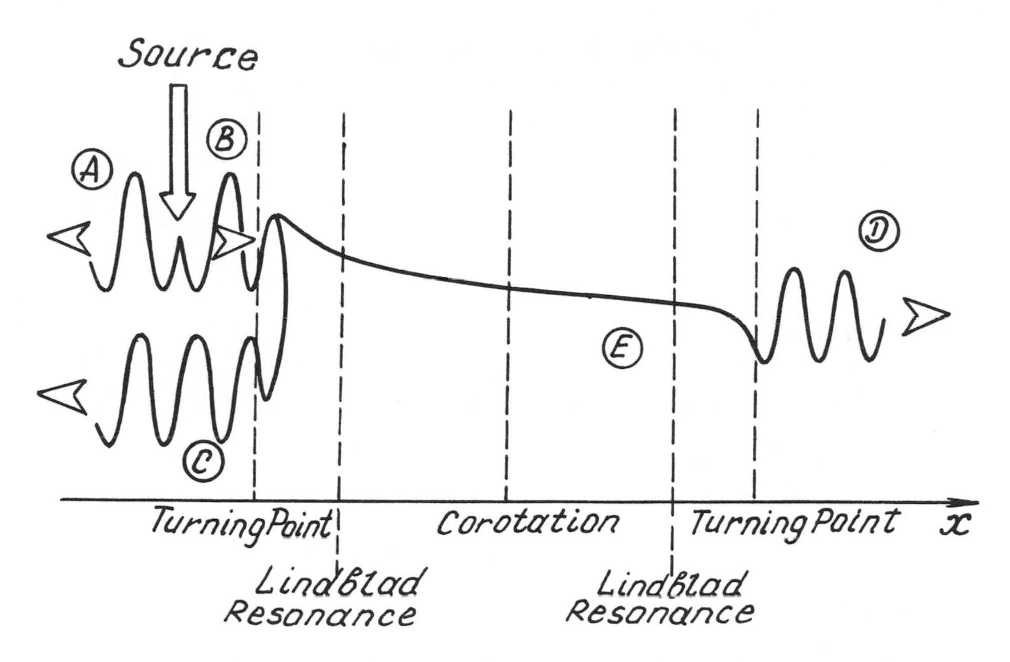


Figure 13: Wave Propagation Pattern with the Source in the Propagation Region: (A) the Emitted Wave Traveling away from the Lindblad Resonance Radius $R_{L1}$; (B) the Emitted Wave Moving toward $R_{L1}$; (C) the Wave Reflected from $R_{L1}$; (D) the Transmitted Wave and (E) the Bar-Like Feature Extending through the Forbidden Zone (after Tagger *et al.*, 1991).

The indirect mechanism to make the disturbances observable is a kind of chain reaction increasing the number density of dust grains. Indeed, velocity perturbations are sufficiently great to enable the accelerated dust grain tear a considerable number of small particles off the regolith, as a result of a collision with a larger fragment. The newly released grains will act in the same way. The initial source of perturbation can be the same as discussed above within other hypotheses, namely a stream of energetic electrons incident on the ring or impact of a meteorite. The assumption of dust levitation is not critically important for this theory, and in fact the observational evidence available sets only an upper limit of 80 km but no lower limit on the thickness of the layer of small grains above the ring. Interestingly enough, the theory permits of two directions for the developing spoke structure, both toward the synchronous orbit and away from it. Meanwhile, the Voyager photographs cannot completely rule out other models, with propagation only away from the synchronous radius.

As can be seen from this Section, there is no shortage of hypotheses to explain the origin of spokes in Saturn rings. They are rather too numerous to consider the theory of spokes completed (six of them have been discussed here!). We are witnessing a live process of developing such a theory that would synthesize rational elements from all the hypotheses compatible with observations.

## References

Aslaksen, T.K. and Havnes, O. 1992, *J. Geophys. Res.* **97**, 19175.

Beaty, J.K., O'Leary, B. and Chaikin, A. (Eds.) 1981. The New Solar System (Sky Publ. Corp.: Cambridge, MA).

Bliokh, P.V. and Yaroshenko, V.V. 1985, *Sov. Astron.* **29**, 330 (*in Russian*).

Bliokh, P.V., Khankina, S.I. and Yaroshenko, V.V. 1986, *Sov. Radiophysics and Quantum Electron.* **29**, 854 (*in Russian*).

Bliokh, P.V., Khankina, S.I. and Yaroshenko, V.V. 1994. In H. Kikuchi (Ed.) *Dusty Plasma, Noise and Chaos in Space and in the Laboratary*, (Plenum Publ. Corp.: New York), p. 29.

Burns, T.A., Showalter, M.R. and Morfill, G.E. 1984. In R. Greenberg and A. Brahic (Eds.) *Planetary Rings*, (Univ. of Ariz. Press: Tucson, AZ), p. 200.

Carbary, J.F., Bythrew, P.F. and Mitchell, D.G. 1982, *Geophys. Res. Lett.* **9**, 420.

Collins, S.A., Cook, A.F., Cuzzi, J.N. *et al.* 1980, *Nature* **288**, 439.

Cuzzi, J.N., Esposito, L.W., Holberg, J.B. *et al.* 1984. In R. Greenberg and A. Brahic (Eds.) *Planetary Rings*, (Univ. of Ariz. Press: Tucson, AZ), p. 73.

Davydov, V.D. 1982, *Kosmich. Issledovaniya* **20**, 460 (*in Russian*).

Elliot, J.L. and Nicholson, P.D. 1984. In R. Greenberg and A. Brahic (Eds.) *Planetary Rings*, (Univ. of Ariz. Press: Tucson, AZ), p. 25.

Fetter, A.L. 1973, *Ann. Phys.* **81**, 367.

Gail, H.P. and Sedlmayr, E. 1980, *Astron. Astrophys.* **86**, 380.

Goertz, C.K. 1984, *Adv. Space Res.* **4**, 137.

Goertz, C.K. 1989, *Rev. Geophys.* **27**, 271.

Goertz, C.K. and Morfill, G.E. 1983, *Icarus* **53**, 219.

Greenberg, R. and Brahic, A. (Eds.), 1984. Planetary Rings (Univ. of Ariz. Press: Tucson, AZ).

Grün, E., Goertz, C.K., Morfill, G.E., and Havnes O. 1992, *Icarus* **99**, 191.

Grün, E., Gregor, E.M., Richard, J.T. *et al.* 1983, *Icarus* **54**, 227.

Grün, E., Morfill, G.E. and Mendis, D.A. 1984. In R. Greenberg and A. Brahic (Eds.) *Planetary Rings*, (Univ. of Ariz. Press: Tucson, AZ), p. 275.

Hartquist, T.W., Havnes, O. and Morfill, G.E. 1992, *Fundamentals of Cosmic Physics* **15**, 107.

Havnes, O. and Morfill, G.E. 1984, *Adv. Space Res.* **4**, 85.

Havnes, O. 1986, *Astrophys. Space Sci.* **122**, 97.

Havnes, O. 1993, *Adv. Space Res.* **13**, 10153.

Havnes, O., Aslaksen, T.K., Melandsø, F. and Nitter, T. 1992, *Phys. Scripta* **45**, 491.

Hill, J.R. and Mendis, D.A. 1981, *Moon and Planets* **24**, 431.

Hill, J.R. and Mendis, D.A. 1982, *J. Geophys. Res.* **87**, 7413.

Kadomtsev, B. 1976. Cooperative Processes in Plasmas (Nauka: Moscow) (*in Russian*).

Kaiser, M.L., Desch, M.D. and Lecacheux, A. 1981, *Nature* **292**, 731.

Melandsø, E. and Havnes, O. 1991, *J. Geophys. Res.* **96**, 5837.

Melandsø, E., Aslaksen, T. and Havnes, O. 1993, *J. Geophys. Res.* **98**, 13315.

Mendis, D.A. and Rosenberg, M. 1994, *Ann. Rev. Astron. Astrophys.* **32**, (*in press*).

Morfill, G.E. and Goertz, C.K. 1983, *Icarus* **55**, 111.

Morfill, G.E., Fechtig, H., Grün, E. and Goertz, C.K. 1983, *Icarus* **55**, 439.

Northrop, T.G. and Hill, J.R. 1983, *J. Geophys. Res.* **88**, 6102.

Rosenberg, M. 1993, *Planet. Space Sci.* **41**, 329.

Sandel, B.R., Shemansky, D.E., Broadfoot, A.L. *et al.* 1982, *Science* **215**, 548.

Shan Lin-Hua and Goertz, C.K. 1991, *Astrophys. J.* **367**, 350.

Smith, B.A., Soderblom, L., Boyce, J. *et al.* 1982, *Science.* **215**, 504.

Tagger, M., Henriksen, R.N. and Pellet, R. 1991, *Icarus* **91**, 297.

Weinheimer, A.J. and Few, A.A., Jr. 1982, *Geophys. Res. Lett.* **9**, 1139.